Floods in a Changing Climate

Risk Management

Climate change and global warming of the atmosphere are very likely to lead to an increase in flooding, and there is now an urgent need for appropriate tools to tackle the complexity of flood risk management problems and environmental impacts. This book presents the flood risk management process as a framework for identifying, assessing, and prioritizing climate-related risks, and developing appropriate adaptation responses. It integrates economic, social, and environmental flood concerns, providing support for interdisciplinary activities involved in the management of flood disasters. Rigorous assessment is employed to determine the most suitable plans and designs for complex, often large-scale, systems, and a full explanation is given of the available probabilistic and fuzzy set-based analytic tools, when each is appropriate, and how to apply them to practical problems. Additional software and data, enabling readers to practice using the fuzzy and probabilistic tools, are accessible online at www.cambridge.org/simonovic.

This is an important resource for academic researchers in the fields of hydrology, climate change, environmental science and policy, and risk assessment, and will also be invaluable to professionals and policy-makers working in hazard mitigation, water resources engineering, and environmental economics.

This volume is the fourth in a collection of four books within the International Hydrology Series on flood disaster management theory and practice within the context of anthropogenic climate change. The other books are:

1 – Floods in a Changing Climate: Extreme Precipitation *by Ramesh Teegavarapu*
2 – Floods in a Changing Climate: Hydrologic Modeling *by P. P. Mujumdar and D. Nagesh Kumar*
3 – Floods in a Changing Climate: Inundation Modelling *by Giuliano Di Baldassarre*

SLOBODAN SIMONOVIĆ has over thirty years of research, teaching and consulting experience in water resources engineering, and has received a number of awards for excellence in teaching, research and outreach. Most of his research is being conducted through the Facility for Intelligent Decision Support (FIDS) at the University of Western Ontario, where he is a Professor of Civil and Environmental Engineering and the Director of Engineering Studies with the Institute for Catastrophic Loss Reduction. His primary research focus is on the application of systems approach to, and development of the decision support tools for, management of complex water and environmental systems and the integration of risk, reliability, uncertainty, simulation and optimization in hydrology and water resources management. Dr Simonovic teaches courses in civil engineering and water resources systems, plays an active role in national and international professional organizations, and has been invited to present special courses for practicing water resources engineers in many countries. He is Associate Editor of the *Journal of Flood Risk Management*, and *Water Resources Management*, and has published over 350 articles and two major textbooks.

INTERNATIONAL HYDROLOGY SERIES

The **International Hydrological Programme** (IHP) was established by the United Nations Educational, Scientific and Cultural Organization (UNESCO) in 1975 as the successor to the International Hydrological Decade. The long-term goal of the IHP is to advance our understanding of processes occurring in the water cycle and to integrate this knowledge into water resources management. The IHP is the only UN science and educational programme in the field of water resources, and one of its outputs has been a steady stream of technical and information documents aimed at water specialists and decision-makers.

The **International Hydrology Series** has been developed by the IHP in collaboration with Cambridge University Press as a major collection of research monographs, synthesis volumes, and graduate texts on the subject of water. Authoritative and international in scope, the various books within the series all contribute to the aims of the IHP in improving scientific and technical knowledge of fresh-water processes, in providing research know-how and in stimulating the responsible management of water resources.

Floods in a Changing Climate

Risk Management

Slobodan P. Simonović

University of Western Ontario

CAMBRIDGE
UNIVERSITY PRESS

University Printing House, Cambridge CB2 8BS, United Kingdom

One Liberty Plaza, 20th Floor, New York, NY 10006, USA

477 Williamstown Road, Port Melbourne, VIC 3207, Australia

4843/24, 2nd Floor, Ansari Road, Daryaganj, Delhi - 110002, India

79 Anson Road, #06-04/06, Singapore 079906

Cambridge University Press is part of the University of Cambridge.

It furthers the University's mission by disseminating knowledge in the pursuit of education, learning and research at the highest international levels of excellence.

www.cambridge.org
Information on this title: www.cambridge.org/9781108447058

First published 2012
First paperback edition 2017

A catalogue record for this publication is available from the British Library

Library of Congress Cataloging in Publication data
Simonovic, Slobodan P.
Floods in a changing climate : risk management / Slobodan P. Simonovic.
    pages   cm. – (International hydrology series)
Includes bibliographical references and index.
ISBN 978-1-107-01874-7
1. Flood control.   I. Title.
TC530.S565   2012
363.34´932 – dc23   2012010631

ISBN 978-1-107-01874-7 Hardback
ISBN 978-1-108-44705-8 Paperback

To Tanja, Dijana, and Damjan

What is the appropriate behavior for a man or a woman in the midst of this world, where each person is clinging to his piece of debris? What's the proper salutation between people as they pass each other in this flood?

Buddha (c. 563–483 BC)

There is a tide in the affairs of men, which, taken at the flood, leads on to fortune. . . . we must take the current when it serves, or lose our ventures.

William Shakespeare (1564–1616)

There can be no vulnerability without risk; there can be no community without vulnerability; there can be no peace, and ultimately no life, without community.

M. Scott Peck (1936–2005)

Decision is a risk rooted in the courage of being free.

Paul Tillich (1886–1965)

Contents

Color plates appear between pages 16 and 17.

Forewords

Almost every day, many people are affected by flooding. In 2011, cyclones and heavy monsoon rains triggered unusually severe seasonal flooding across Southeast Asia, affecting many nations including Thailand. The major floods in Bangkok had, by mid-December, killed at least 675 people and caused major economic impacts. In the fall 2011, there were also floods in Colombia, Australia, Kenya, and other places. Hydrologic disasters, dominantly floods but also including wet mass movements (mud slides), are responsible for just more than half of all the disasters in the period 2000–2010. During this period, these events killed, on average, more than 5,000 people per year with the total affected being about 100 million people. The total affected is the sum of injured, homeless, and people requiring immediate assistance during a period of emergency. Annual damage costs are about US$20 billion.

M. Wahlström, the United Nations Assistant Secretary-General for Disaster Risk Reduction, stated, "Over the last two decades (1988–2007), 76% of all disaster events were hydrological, meteorological or climatological in nature; these accounted for 45% of the deaths and 79% of the economic losses caused by natural hazards." She concluded her statement with: "The real tragedy is that many of these deaths can be avoided." This book on flood risk management, by Professor Slobodan Simonović, is about actions that can be taken to anticipate and prevent or mitigate harms that may be avoidable and reduce the number of deaths and lower the socio-economic impacts.

Although the impacts of a flood are usually less than an earthquake, floods occur more often. Meteorological events, such as storms, the next most common, occur less than half as often. Both are part of what we can call the climate system, and this book is addressing flood risk management in the context of climate change. In 2009, world leaders at the United Nations Climate Change Conference agreed to the Copenhagen Accord, which states in the opening paragraph: "We underline that climate change is one of the greatest challenges of our time.... We recognize the critical impacts of climate change and the potential impacts of response measures on countries particularly vulnerable to its adverse effects and stress the need to establish a comprehensive adaptation programme including international support." Since climate change adaptation is "the adjustment in natural or human systems in response to actual or expected climatic stimuli or their effects, which moderates harm or exploits beneficial opportunities," this book deals with how to moderate harm or specifically reduce flooding risk. Chapter 6 specifically addresses future perspectives in a changing climate.

In November 2011, governments approved the Summary for Policy Makers of the Intergovernmental Panel on Climate Change Special Report on Managing the Risks of Extreme Events and Disasters to Advance Climate Change Adaptation. The Summary for Policy Makers includes the statement: "A changing climate leads to changes in the frequency, intensity, spatial extent, duration, and timing of extreme weather and climate events, and can result in unprecedented extreme weather and climate events." The Special Report specifically concluded: "It is likely that the frequency of heavy precipitation or the proportion of total rainfall from heavy falls will increase in the 21st century over many areas of the globe." The Special Report states that the changes in projected precipitation and temperature changes imply possible changes in floods, but notes that there is *low confidence* in projections of changes in fluvial floods due to the *limited evidence* and because the causes of regional changes are complex. This book is addressing those relationships so with further studies based on these principles the evidence should become clearer in the future.

This book provides methods and approaches to reduce the impacts of floods that have for millennia been affecting people around the world and usually most on those most vulnerable. Now through the actions of people collectively, and specifically mostly those in developing countries, the atmospheric greenhouse gas concentrations have increased and are changing the climate. With that climate change there will be more intense precipitation events and warmer temperatures which based on physical logic implies more flooding events. Hence, the impacts on the vulnerable will increase more and raise the need for actions. Among those actions needed is the reduction of risk of flooding and that is the topic addressed in this important and timely book.

Dr. Gordon McBean, C.M., O. Ont., FRSC
President ICSU
University of Western Ontario
London, Ontario, Canada
December 2011

Climate change is undoubtedly one of the most pressing issues facing society today and the potential impacts of climate change are currently a prime concern for water resource professionals. The possible impacts of climate change have the potential to dramatically alter the temporal and spatial distribution and availability of water on the Earth's surface with consequences that could be disastrous. Furthermore, climate change could, paradoxically, lead to both more frequent and severe drought conditions and flooding events that are of greater frequency and magnitude. In this book, Professor Slobodan Simonović addresses the potential impacts of increases in flood event magnitude and frequency through a comprehensive analysis of the interplay between climate change and flooding conditions with a particular focus on the role of the management of flood risk.

Professor Simonović has had a distinguished career as a water professional, consultant, educator, and researcher. I first met Slobodan more than 25 years ago when we were both faculty members at the University of Manitoba in Winnipeg, Manitoba, Canada. Through collaboration on research and consulting projects I came to appreciate the wealth of knowledge that he brings to his professional activities. Towards the end of our time in Winnipeg, we were both involved in different aspects of the Red River Flood of 1997, the so-called "Flood of the Century", in the Red River valley, a flood-prone area of Canada and the United States. Slobodan's involvement with this major flood event, and its aftermath, is but one of many examples of the practical expertise that he brings to the writing of this important and timely book. We frequently hear reports in the media of devastating flooding events, in various parts of the world, of seemingly unprecedented scope, geographic extent, and magnitude. There often follows natural speculation that the occurrence of such a flooding event, or events, must be further evidence of the impacts of climate change. This book helps to make sense of these events and provides the water professional with important tools to cope with the impacts of increases in the frequency and magnitude of flooding events and the associated societal consequences.

In this book, Professor Simonović considers not just what the impacts of climate change may be on water, and flooding in par-

ticular, but also looks at flood risk management, which can be usefully applied, as he suggests, as an effective form of climate change adaptation. Climate change adaptation through flood risk management is one of several themes that tie the parts of this book together. The book consists of four parts. The first part, entitled "Setting the Stage", deals with the central topic of flood risk management and introduces climate change and the interplay between climate change and flood risk management. An important and interesting section in this part is the very detailed case study of climate change impacts on municipal infrastructure within the City of London, Ontario, Canada. This extensive example application very nicely draws together the common intertwining threads of floods, risk, and climate change within a real-world application. The research described in this section of the book is one of the strengths of this publication. The second and third parts of the book deal with flood risk management from the perspective of a probabilistic and a fuzzy set approach, respectively. In both of these parts, flood risk management is introduced from a systems analysis context; systems analysis is another common and unifying theme for much of the material in this book. The final part of the book looks at "Future Perspectives" and again provides an essential link between the potential impacts of climate change and flood risk management with a particular focus on the importance of both of these issues to the public and also the overarching need to effectively communicate climate change impacts and flood risk management issues to the general public. These are again topics with which Professor Simonović has considerable experience.

This is an important book that will be of interest to water professionals, policy makers, researchers, and others concerned with the potential impacts of climate change on flooding events and on flood risk management.

Dr. Donald H. Burn
University of Waterloo
Waterloo, Ontario, Canada
January 2012

Preface

I have stated many times that I am one of the lucky few who have the opportunity to work all their professional lives in an area that they enjoy. The most enjoyable activity for me is to integrate knowledge from different fields into an approach for solving complex problems that include uncertainty. My work has brought me into contact with many people, responsible professionals, talented engineers, capable managers, and dedicated politicians. In my capacity as an academic I have also had an opportunity to work with young talented people – the future of our workforce. I learned a lot from all of them. I learned many things about the profession, I learned a lot about different cultures, and most importantly I learned about life. Thank you.

My interest in risk and flooding as a natural disaster grew from my main area of expertise – water resources systems management. From the early days of my professional career I was involved with floods and flood management, first from an engineering point of view and then later from a management point of view. Flood problems along the Morava, Sava, and Danube rivers in my country of origin – Serbia – were among the first professional challenges I had to deal with after graduation. In 1997, I was teaching at the University of Manitoba and living in Winnipeg. That was the year of the "Flood of the Century." The governments of Canada and the USA have agreed that steps must be taken to reduce the impact of future flooding on the Red River. In June 1997, they asked the International Joint Commission (IJC) to analyze the causes and effects of the Red River flood of that year. The IJC appointed the International Red River Basin Task Force to examine a range of alternatives to prevent or reduce future flood damage. I was appointed to the task force and the subsequent experience changed my life.

My work has taken me all over the world. I have had an opportunity to see flood problems in the developed and developing world, in small villages and large urban centers. Projects I have been involved with range in scale from the local to the international. I have discussed flooding issues with farmers of the Sihu area in China as well as the Minister for Irrigation and Water Resources of Egypt. I hope that my professional expertise continues to contribute to the solution of some of these problems. It definitely inspires me to continue to work with greater effort and more dedication.

For more than 35 years of personal research, consulting, teaching, involvement in policy, implementation of projects, and presentation of experiences through the pages of many professional journals, I have worked hard to raise awareness of the importance of uncertainty – objective and subjective – in the solution of complex problems. The main thrust of my work is the use of a systems approach in dealing with complexity. I have accumulated tremendous experience over the years. In that time I realized that there is an opportunity to contribute to the area of flood risk management by transferring some of the knowledge and experience from the implementation of systems thinking and systems tools to various steps of the flood risk management cycle. Writing this book offered me a moment of reflection, and it elaborates on lessons learned from the past to develop ideas for the future.

Acknowledgments

Publishing this book was made possible through the contributions of many people. I would like to start by acknowledging the publication support provided by the International Hydrologic Programme of UNESCO, and the Water Science Division team including Siegfried Demuth and Biljana Radojevic. Most of the knowledge contained in this book came from my numerous interactions with teachers, students, and colleagues throughout the world. They taught me all I know. I would like particularly to thank the students whose work is used in this text. In order of appearance in the text, they are Hyung-Ill Eum (Chapter 3), Dragan Sredojevic (Chapter 3), Lisa Bowering-Taylor (Chapter 3), Angela Peck (Chapter 3), Dejan Vucetic (Chapters 4 and 5), Ozren Despic (Chapter 5), Ibrahim El-Baroudi (Chapter 5), Taslima Akter (Chapter 5), and Mike Bender (Chapter 5). A special thank you goes to Veerakcudy Rajasekaram, who is the developer of the computer programs.

The support of my family, Dijana, Damjan, and Tanja, was of the utmost importance in the development of this book. They provide a very large part of my motivation, my goals, my energy, and my spirit. Without the endless encouragement, criticism, advice, and support of my wife Tanja this book would never have been completed.

Definitions

Uncertainty: lack of certainty; a state of having limited knowledge where it is impossible to exactly describe the existing state or future outcome; more than one possible outcome. Sometimes the implications of uncertainty involve risk – a significant potential unwelcome effect of system performance. For example, if you do not know whether it will rain tomorrow, then you have a state of uncertainty. If you apply probabilities to the possible outcomes using weather forecasts, you have quantified the uncertainty. Suppose you quantify your uncertainty as a 90% chance of sunshine. If you are planning a major, costly, outdoor event for tomorrow then you have risk, since there is a 10% chance of rain and rain would be undesirable. Furthermore, if this is a business event and you would lose $100,000 if it rains, then you have quantified the risk (a 10% chance of losing $100,000).

Vagueness or ambiguity is sometimes described as "second-order uncertainty," where there is uncertainty even about the definitions of uncertain states or outcomes. The difference here is that this uncertainty is about human definitions and concepts, not an objective fact of nature. It has been argued that ambiguity, however, is always avoidable while uncertainty (of the "first order") is not necessarily avoidable.

Uncertainty may be purely a consequence of a lack of knowledge of obtainable facts. That is, you may be uncertain about whether a new dyke design will work, but this uncertainty can be removed with further analysis and experimentation.

There are other taxonomies of uncertainties and decisions that include a broader sense of uncertainty and how it should be approached from an ethics perspective (Tannert *et al.*, 2007). Figure 1 shows the taxonomy of uncertainties and decisions according to Tannert *et al.*

The first form of uncertainty in this scheme is objective uncertainty, which can be further divided into epistemological uncertainty and ontological uncertainty. The former is caused by gaps in knowledge that can be closed by research. In this case, research becomes a moral duty that is required to avoid dangers or risks, to realize possible benefits, or to balance risks and benefits in a rational and responsible way. On the other hand, ontological uncertainty is caused by the stochastic features of a situation, which will usually involve complex technical, natural, and/or social systems. Such complex systems are often characterized by non-linear behavior, which makes it impossible to resolve uncertainties by deterministic reasoning and/or research.

The second main form of uncertainty in Tannert's taxonomy is subjective uncertainty, which is characterized by an inability to apply appropriate moral rules. These types of uncertainty can lead to societal anxiety or conflict. Again, we can distinguish between two sub-forms of subjective uncertainty. The first is uncertainty with respect to rule-guided decisions. This is caused by a lack of applicable moral rules and we call these situations "moral uncertainties." In this case, decision-makers have to fall back on more general moral rules and use them to deduce guidance for the special situation in question. The second sub-form is uncertainty with respect to intuition-guided decisions – that is, uncertainty in moral rules. In specific situations, we can make decisions only by relying on our intuition rather than knowledge, or explicit or implicit moral rules. This means that we act on the basis of fundamental pre-formed moral convictions in addition to experiential and internalized moral models. As with rule-guided decisions, a level of deduction is used here, but in a subconscious and intuitive way. We call the decisions that stem from internalized experiences and moral values "intuitional."

Risk and reliability: An attempt by risk analysis experts in the late 1970s to come up with a standardized definition of risk concluded that a common definition is perhaps unachievable, and that authors should continue to define risk in their own way. As a result, numerous definitions can be found in recent literature, ranging from the vague and conceptual to the rigid and quantitative. At a conceptual level, we define risk (i) as a significant potential unwelcome effect of system performance, or (ii) as the predicted or expected likelihood that a set of circumstances over some time frame will produce some harm that matters, or (iii) as future issues that can be avoided or mitigated, rather than present problems that must be immediately addressed. More pragmatic treatments view risk as one side of an equation, where risk is equated with the probability of failure or the probability of load exceeding resistance. Other symbolic expressions equate risk with the sum

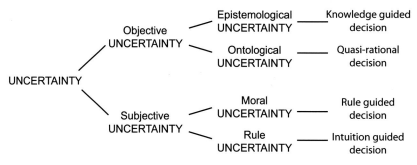

Figure 1 Taxonomy of uncertainties and decisions (after Tannert *et al.*, 2007).

of uncertainty and damage, or the quotient of hazards divided by safeguards (Simonovic, 2009).

Because there is a need to understand how a potential loss might affect and be perceived by the various stakeholders, it is insufficient, and indeed can be quite misleading, for the decision-maker to consider risk solely in terms of probability and consequence. Risk involves three key issues: (i) the frequency of the loss, that is, how often the loss may occur; (ii) the consequences of the loss, that is, how large might the loss be; and (iii) the perception of the loss, that is, how a potential risk is viewed by affected stakeholders in terms of its effect on their needs, issues, and concerns.

Perhaps the most expressive definition of risk is the one that conveys its multi-dimensional character by framing risk as the set of answers to three questions: What can happen? How likely is it to happen? If it does happen, what are the consequences? (Simonovic, 2009 after Kaplan and Garrick, 1981). The answers to these questions emphasize the notion that risk is a prediction or expectation that involves a hazard (the source of danger), uncertainty of occurrence and outcomes (the chance of occurrence), adverse consequences (the possible outcomes), a time frame for evaluation, and the perspectives of those affected about what is important to them. The answers to these questions also form the basis of conventional quantitative risk analysis methodologies.

Three cautions surrounding risk must be taken into consideration: risk cannot be represented objectively by a single number alone, risks cannot be ranked on strictly objective grounds, and risk should not be labeled as real. Regarding the caution of viewing risk as a single number, the multi-dimensional character of risk can only be aggregated into a single number by assigning implicit or explicit weighting factors to various numerical measures of risk. Since these weighting factors must rely on value judgments, the resulting single metric for risk cannot be objective. Since risk cannot objectively be expressed by a single number, it is not possible to rank risks on strictly objective grounds. Finally, since risk estimates are evidence-based, risks cannot be strictly labeled as real. Rather, they should be labeled inferred at best.

Reliability is directly related to risk. In general, reliability is the ability of a system to perform and maintain its functions in routine circumstances, as well as hostile or unexpected circumstances. In engineering, for example, reliability refers to the ability of a system or component to perform its required functions under stated conditions for a specified period of time.

Vulnerability: Generally vulnerability is the susceptibility to physical or emotional injury or attack. In relation to hazards and disasters, vulnerability is a concept that links the relationship that people have with their environment to social forces and institutions and the cultural values that sustain and contest them. The concept of vulnerability expresses the multi-dimensionality of disasters by focusing attention on the totality of relationships in a given social situation which constitute a condition that, in combination with environmental forces, produces a disaster.

Vulnerability is also the extent to which changes could harm a system. In other words, it is the extent to which a community can be affected by the impact of a hazard. In global warming, vulnerability is the degree to which a system is susceptible to, or unable to cope with, adverse effects of climate change, including climate variability and extremes.

Risk management: Activities undertaken by an individual, organization, or government all involve some degree of risk. All activities expose people or groups to a potential loss of something they value: their health, money, property, the environment, etc. Individuals, groups, or organizations who are able to affect, who are affected by, or believe they may be affected by, a decision or activity are called stakeholders. Because different stakeholders may place different values on things, they may also view the acceptability of risk differently. As such, attempts to manage risk may be unsuccessful if one fails to recognize its complex nature.

The objective of risk management is to ensure that significant risks are identified and that appropriate action is taken to minimize these risks as much as is reasonably achievable. Such actions are determined based on a balance of risk control strategies, their effectiveness and cost, and the needs, issues, and concerns of stakeholders. Communication among stakeholders throughout the process is a critical element of this risk management process. Decisions made with respect to risk issues must balance the technical aspects of risk with the social and moral considerations that often accompany such issues.

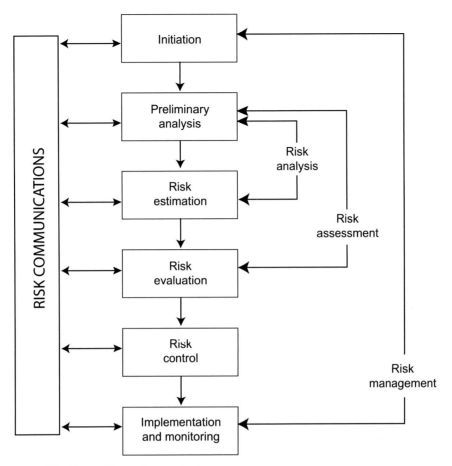

Figure 2 Steps in the process of decision-making under uncertainty.

Risk management can be defined as the systematic application of management policies, procedures, and practices to the tasks of analyzing, evaluating, controlling, and communicating about risk issues (Canadian Standards Association, 1997). Risk management for health and environmental risks uses scientific risk assessments to estimate the probable harm to persons and environments resulting from specific types of substances or activities. As such, even when decision-makers seek honestly to take into account varying perceptions of the risk in question, they are necessarily and properly constrained by the scope and limitations of their scientific assessment in recommending specific courses of action. This is an inescapable part of their duty to protect public interests and well-being to the best of their ability, taking into account the uncertainties that are always a factor in risk estimates. Leiss (2001) distinguishes between risk and risk issue management as two fundamentally different processes. The most important difference is that risk issues, as they play out in society at large, are not primarily driven by the state of scientific risk assessments. Rather, such assessments are just one of a series of contested domains within the issue. The phrase *risk issue* refers to: (i) stakeholder confrontation, or the existence of some dispute about the scope or existence of a risk and how it should be managed; (ii) intractable behavior, or the inability

of professionals to change the public's risk-taking behavior; and (iii) high uncertainty, or the public expressions of concern over risk factors that are poorly characterized from a scientific standpoint, or where uncertainties in risk assessments are quite large.

Decision-making under uncertainty: Statistician George Chacko (1991) defines decision-making as "the commitment of resources today for results tomorrow." Because decisions involve expectations about the future, they always involve uncertainty. Decision-making in general is concerned with identifying the values, uncertainties, and other issues relevant in a given decision, its rationality, and the resulting optimal decision. The practical application of decision theory (how people actually make decisions) is called decision analysis, and is aimed at finding tools, methodologies, and software to help people make better decisions. The most systematic and comprehensive software tools developed in this way are called decision support systems.

Decision-making under uncertainty will be described in this book as a systematic process that consists of six steps that follow a standardized management or systems analysis approach as summarized in Figure 2. Each step in the process is followed by a decision. It should be noted that the process is iterative.

Acronyms and abbreviations

ACO	Ant Colony Optimization	LP	Linear Programming
AMO	Atlantic Multi-decadal Oscillation	MCS	Monte Carlo Simulation
CCGP	Chance Constrained Goal Programming	NAM	Northern Annular Mode
CC_LB	Climate Change Lower Bound	NAO	North Atlantic Oscillation
CC_UB	Climate Change Upper Bound	NatCatSERVICE	Natural Catastrophe Service
CP	Circulation Pattern	NDBC	National Data Buoy Center
CSA	Canadian Standards Association	NEXRAD	Next Generation Radar
CRED	Centre for Research on the Epidemiology of Disasters	NN	Neural Network
		NWS	National Weather Service
CUP	Composition Under Pseudomeasures	NWIS	National Water Information System
DA	Dissemination Area	OF	Objective Function
DICE	Dynamic Integrated Climate Change	OWA	Ordered Weighted Averaging
DTM	Digital Terrain Model	PCA	Principal Component Analysis
EAD	Expected Annual Damage	PCP	Pollution Control Plant
EMS	Emergency Management Services	PDO	Pacific Decadal Oscillation
ENSO	El Niño–Southern Oscillation	PDF	Probability Density Function
EP	Evolutionary Programming	POT	Peak Over Threshold
FCP	Fuzzy Compromise Programming	P-CUP	Polynomial Composition Under Pseudomeasure
FLP	Fuzzy Linear Programming		
GA	Genetic Algorithm	RCM	Regional Climate Model
GCM	Global Climate Model	SAM	Southern Annular Mode
GDP	Gross Domestic Product	SFCP	Spatial Fuzzy Compromise Programming
GEV	Generalized Extreme Value Distribution	SPRC	Source–Pathway–Receptor–Consequence
GIS	Geographic Information System	SST	Sea Surface Temperature
GP	Genetic Programming	SSTA	Sea Surface Temperature Anomaly
HEC-HMS	Hydrologic Engineering Center's (US Army Corps of Engineers) Hydrologic Modeling System	TIN	Triangulated Irregular Network
		UNESCO	United Nations Educational, Scientific and Cultural Organization
HEC-RAS	Hydrologic Engineering Center's (US Army Corps of Engineers) River Analysis System	USACE	United States Army Corps of Engineers
		USGS	United States Geological Survey
IDF	Intensity–Duration–Frequency	UTRCA	Upper Thames River Conservation Authority
IFM	Integrated Flood Management	WCoG	Weighted Center of Gravity
IJC	International Joint Commission	WG	Weather Generator
IPCC	Intergovernmental Panel on Climate Change	WLD	West London Dyke
K-NN	K-Nearest Neighbor algorithm	WMO	World Meteorological Organization

Introduction

Flooding is a rising and overflowing of a body of water onto normally dry land. *This can occur (i) directly, when rainfall is of sufficient quantity that the land surface cannot absorb and redistribute water to prevent a surface accumulation, and/or (ii) indirectly, when a river overflows its banks.* River flooding is primarily caused by hydro-meteorological conditions. These can occur in the form of excess snowmelt runoff, rain, rain on snow, and ice jams or other natural dams. In this book, *flooding is considered as a general term that includes a variety of events and processes that lead to an accumulation of surface water on land that is usually dry.*

Floods are the most common natural disasters. Their frequency, magnitude, and cost are on the rise all over the world. Riverbanks and floodplains are very attractive for habitation as well as for agriculture, transportation, power generation, recreation, and disposing of wastes. Over thousands of years, cities and other infrastructure have increasingly encroached into the floodplains. As the encroachment increases, the actual damage and potential risks from flooding have increased.

Climate change and global warming of the atmosphere will increase the capacity of the atmosphere to hold water, and this will also accelerate many of the processes involved in the redistribution of moisture in the atmosphere. These changes alone suggest that floodgenerating processes linked to the atmosphere are likely to increase. The Intergovernmental Panel on Climate Change (IPCC) states that it is certain that floods will increase in future climates. However, there is considerable uncertainty in the exact nature of how this will evolve.

Flooding itself is simple and at the same time very complex. A flood is simply too much water; however, the main sources of uncertainty are in how and why there is too much water. Floods involve both atmospheric and hydrologic processes and various uncertainties associated with them. Changes in the climate will only increase the uncertainty in the wide variety of atmospheric and hydrologic processes with differing impacts on the magnitude and timing of floods. Flood severity and risk are to a large extent determined by human activities, both within the floodplain and outside. Changes in land use and drainage patterns may

greatly increase the risk of flooding. Population growth is likely to increase the number of people at risk.

Flood risk management is the attempt to anticipate and prevent or mitigate harm that may be avoidable. The main goal of this book is to introduce the systems approach to flood risk management as an alternative approach that can provide support for interdisciplinary activities involved in the management of flood disasters. The systems approach draws on the fields of operations research and economics to create skills in solving complex management problems. A primary emphasis of systems analysis in disaster management as I see it is on providing an improved basis for decision-making. A large number of analytical, computer-based tools, from simulation and optimization to multi-objective analysis, are available for formulating, analyzing, and solving disaster management problems.

Large and more frequent flood disasters in the last few decades have brought a remarkable transformation of attitude by the disaster management community toward (i) integration of economic, social, and environmental concerns related to disasters, (ii) serious consideration of both objective and subjective risks, and (iii) action to deal with the consequences of large flood disasters.

The early period of flood defense was characterized by taking knowledge from various fields of science and engineering that is applicable to flood hazards and using it in their management. The most significant contribution in the last 10 years is a fundamental shift in the character of how the citizens, communities, governments, and businesses conduct themselves in relation to the natural environment they occupy. Flood defense is being replaced with flood risk management that builds resiliency to flooding.

Flood risk management being divided among disciplinary boundaries has faced an uphill battle with the regulatory approaches that are used in many countries around the world. They have not been conducive to the integrative character of the systems approach that is inherent in simulation and optimization management models. Fortunately, recent trends in regulation include consideration of the entire region under flood threat, explicit consideration of all costs and benefits, elaboration of a large number of alternatives to reduce the damages, and the greater participation of all stakeholders in decision-making. Systems approaches based on simulation, optimization, and multi-objective analyses have great potential for providing appropriate support for effective disaster management in this emerging context.

Uncertainty, defined as lack of certainty, has important implications for what can be achieved by flood risk management. All flood risk management decisions should take uncertainty into account. Sometimes the implications of uncertainty involve *risk*, in the sense of significant potential unwelcome effects of flooding. Then managers need to understand the nature of the underlying threats in order to identify, assess, and manage the risk. Failure to do so is likely to result in adverse impacts and extensive damage. Sometimes the implications of uncertainty involve an opposite form of risk, significant potentially welcome effects. Then managers need to understand the nature of the underlying opportunities in order to identify and manage the associated decrease in risk. Failure to do so can result in a failure to capture good luck, which can increase the risk.

Systems can be defined as a collection of various structural and non-structural elements that are connected and organized in such a way as to achieve some specific objective through the control and distribution of material resources, energy, and information. The systems approach is a paradigm concerned with systems and interrelationships among their components. Today, more than ever, we face the need for appropriate tools that can assist in dealing with difficulties introduced by the increase in the complexity of flood risk management problems, consideration of environmental impacts, and the introduction of the principles of sustainability. The systems approach is one such tool. It uses rigorous methods to help determine the preferred plans and designs for complex, often large-scale systems. It combines knowledge of the available analytic tools, an understanding of when each is appropriate, and a skill in applying them to practical problems. It is both mathematical and intuitive, as is all the disaster management cycle of flood hazard mitigation, preparation, emergency/event/crisis management, and recovery.

The aim of this book is directly related to the current state of systems thinking as an approach within the social sciences and especially flood disaster management. Its purpose is to offer systems thinking as a coherent approach to inquiry and flood risk management. With this book I would like to contribute to the change of flood risk management practice and respond to a clear need to redefine the education of disaster management professionals and increase their abilities to (i) work in an interdisciplinary environment; (ii) develop a new framework for hazard mitigation, preparation, emergency/event/crisis management, and recovery that will take into consideration current complex socio-economic conditions; and (iii) provide the context for disaster management in conditions of uncertainty.

The innovative aspects of this book include: (i) parallel use of probabilistic and fuzzy set approaches to flood risk management;

(ii) the concept of adaptive risk management as a system that is able to learn, adapt, prevent, identify, and respond to new/unknown threats in critical time; and (iii) the notion of integrated flood risk management based on joint consideration of objective and subjective uncertainty and the use of risk communication to link together risk assessment, risk management, and the decision-making process. Simulation, optimization, and multi-objective analysis algorithms are presented for risk-based management and computational tools will be developed and included on the website for this publication.

THE ORGANIZATION OF THE BOOK

The material presented in the book is organized in six chapters and four parts. The introductory part of the book is focused on terminology and proper definitions of uncertainty, risk, reliability, vulnerability, risk management, and decision-making under uncertainty.

The first part of the book is divided into three chapters that set the stage for the rest of the material presented in the book. In the first chapter, the problem of flood risk management is clearly defined together with a presentation of what is flood risk. A comprehensive review of the current approaches to flood risk management is presented here. The following chapter deals with climate change and risk of flooding. The purpose of this section is to link climate change to flood risk. This will be done for floods caused by hydrometeorological events (snowmelt-runoff floods, storm-rainfall floods, rain-on-snow floods, and ice-jam floods) and natural dams (outburst floods and landslide floods). Climate change is reviewed to elaborate all impacts that affect flooding. Mitigation and adaptation are presented as the two main approaches for dealing with climate change impacts. One of the main contributions of the book is the definition of climate change adaptation as a risk management procedure. This idea is elaborated in the context of flood management in the third and final chapter of this part of the book. In the same chapter, both probabilistic and fuzzy approaches to flood risk management are introduced.

The second part of the book is technical in nature and is used to introduce methodological aspects of probabilistic flood risk management in Chapter 4. It reflects the systems view of risk management under uncertainty using a probabilistic approach. The presentation includes three main types of system tools: simulation, optimization, and multi-objective analysis. These tools are presented in probabilistic form. Simulation is represented using a Monte Carlo procedure. Evolutionary optimization is presented as an optimization technique in probabilistic form. A probabilistic multi-objective goal programming technique is selected as an example of multi-objective analysis techniques. LINPRO and EVOLPRO computer programs are introduced and selected examples demonstrate their utility.

The third part of the book presents a fuzzy set approach to flood risk management in Chapter 5. Basic explanations are provided on fuzzy sets, development of fuzzy membership functions, and operations on fuzzy numbers. A new definition of fuzzy risk is provided and fuzzy simulation, optimization, and multi-objective analyses tools are presented. The FUZZYCOMPRO computer program is introduced for the implementation of fuzzy multi-objective analysis in the discrete form.

The second and third parts of the book contain two chapters that present the theory and implementation of flood risk management tools. Implementation of each methodological approach is illustrated with a real example. The set of flood risk management problems in these two chapters includes flood-plain zoning, flood forecasting, flood emergency management, design of municipal water infrastructure, operations of a flood management reservoir, and ranking of flood risk management alternatives.

The final part, Chapter 6, provides an insight into the future of flood risk management under changing climate. The very slow process of dealing with emissions of greenhouse gases leaves us with the need to seriously consider adaptation as almost the sole approach to minimizing the negative impacts of climate change on flooding. In Chapter 6, flood risk management under uncertainty is once more revisited as the climate change adaptation approach. Practical guidelines are provided for the implementation of the tools introduced in the book. Risk communication is also discussed as a supporting tool for effective implementation of the tools presented in the book and some important

characteristics of the risk communication process are elaborated in detail.

USE OF THE BOOK

Computer programs presented in the text are included on the website (www.cambridge.org/simonovic). The text and the accompanying website have four main purposes. (i) They provide material for an undergraduate course in flood risk management under climate change. A course might be based on Chapters 1 through 5. (ii) They also provide support for a graduate course in flood risk management under climate change, with an emphasis on analytical aspects of application of systems tools to management of flood disasters. Such a course might draw on details from Chapters 4 or 5. Both undergraduate and graduate courses could use the computer programs provided on the website. (ii) Flood disaster management practitioners should find the focus on the application of the methodologies presented to be particularly helpful, and could use the programs for the solution of real flood disaster management problems. There is discussion of a number of specific applications in Chapters 4 and 5 that may be of assistance. (iv) Specific parts of the book can be used as a tool for specialized short courses for practitioners. For example, material from Chapters 3 and 4 could support a short course on "Probabilistic approach to flood risk management under changing climate." A course on "Fuzzy set approach to flood risk management under changing climate" could be based on Chapters 3 and 5.

Part I
Setting the stage

1 Flood risk management

A flood is a very simple natural phenomenon that occurs when a body of water rises to overflow land that is not normally submerged (Ward, 1978). At the same time, a flood is a very complex phenomenon that connects the natural environment, people, and the social systems of their organization. Flooding causes loss of human life. It damages infrastructure such as roads, bridges, and buildings, and hurts agricultural productivity because of lost crops and soil erosion. Flood disaster relief often requires enormous funding. Connectivity increases risks. As more links are present among the elements of natural, social, and technological systems, these systems develop unexpected patterns of connections that make breakdown more likely.

We are witnessing many catastrophic flood disasters. European floods in 2002 caused more than €7 billion damage. Hurricane Katrina caused flooding in 2005 that was the costliest natural disaster, as well as one of the five deadliest, in the history of the USA. At least 1,836 people lost their lives in the actual hurricane and in the subsequent floods; total property damage was estimated at US$81 billion. In June of 2006, northeastern Bangladesh disappeared under monsoon floods as rains drenched the region. The floods stretched across hundreds of kilometers of what had been dry land a month earlier and inundated two thirds of the territory of the country. Typhoon Morakot of 2009 was the deadliest typhoon to impact Taiwan in recorded history. It created catastrophic damage in Taiwan, leaving 461 people dead and 192 others missing, and roughly US$3.3 billion in damage. The storm produced huge amounts of rainfall, peaking at 2,777 mm (109.3 in). The extreme amount of rain triggered enormous mudslides and severe flooding throughout southern Taiwan. One mudslide buried the entire town of Xiaolin, killing an estimated 500 people. In the wake of the flood, Taiwan's President Ma Ying-jeou faced extreme criticism for the slow response to the disaster, having initially deployed only roughly 2,100 soldiers to the affected regions. Later additions of troops increased the number of soldiers to 46,000. Days later, the president publicly resigned due to the government's slow response to the disaster. The 2010 China floods began in early May. The total death toll as of August 5 was 2,507. More than 305 million people in 28 provinces, municipalities, and regions were affected, while at least 12 million people

had been evacuated because of the risk of flooding and landslides by early August. As I am writing these words, Pakistan's deadliest floods in decades have killed more than 1,500 people and overwhelmed government efforts to provide aid. The flood's death toll may rise to 3,000. Approximately 20 million people had been affected by floods by early August. Regions downstream in the Indus River valley, where most of Pakistan's 162 million people live, are bracing for floods that may damage crops.

1.1 THE GLOBAL FLOOD PROBLEM

Assessing the global flood problem is not an easy task due to gaps and numerous deficiencies in statistics, the highly variable quality of the available data, and the problems of comparing flood impacts across the wide socio-economic development spectrum. Most of the information to be presented here is from the Dartmouth Flood Data Observatory (2010) in Germany, the Emergency Events Database EM-DAT of the Centre for Research on the Epidemiology of Disasters (CRED, 2009) in Belgium and the Munich Re NatCatSERVICE online database (Munich Re, 2011).

The longer time period records (traced back to 1900, although more reliable after 1950) show a relentless upward movement in the number of natural disasters (Figure 1.1) and their human and economic impact (Figure 1.2). Black indicates the number and impacts of flood disasters. It is troubling that disaster risk and impacts have been increasing during a period of global economic growth. On the good side, a greater proportion of economic surplus could be better distributed to alleviate the growing risk of disaster. On the bad side, it is possible that development paths are themselves creating the problem: increasing hazards (for example through global climate change and environmental degradation), human vulnerability (through income poverty and political marginalization), or both.

The information on flood disasters presented in Figures 1.3 to 1.6 is taken from EM-DAT: The CRED International Disaster Database for the period 1950–2010. In order for a disaster to be entered into the database at least one of the following criteria has to be fulfilled: 10 or more people reported killed; 100 people reported

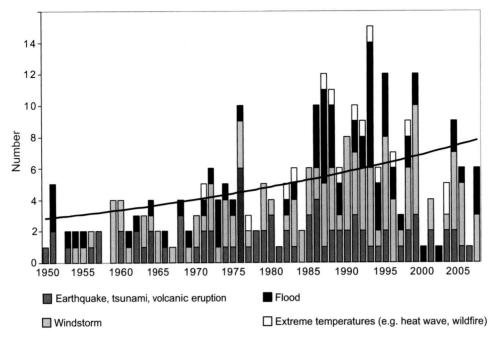

Figure 1.1 Great natural disasters 1950–2007: number of events (source: Munich Re, NatCatSERVICE).

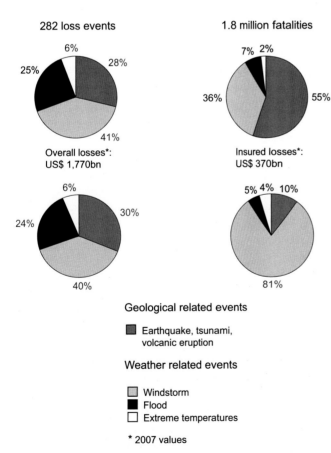

Figure 1.2 Great natural disasters 1950–2007: percentage distribution worldwide (source: Munich Re, NatCatSERVICE).

affected; a call for international assistance; and declaration of a state of emergency.

The damage related to floods is direct and indirect. Deaths of people and animals, damage to houses, properties, and standing crops, damage to physical infrastructure, etc. may be the direct consequence of floods. Other effects, such as a change in ecosystem or spread of diseases, may be indirect damage due to floods.

Throughout the world, floods are inflicting substantial damage year after year. According to the information presented in Figures 1.4 to 1.6 and the statistics of the International Red Cross organization, the average number of people who suffered from flood damage during the period from 1973 through to 1997 amounted to more than 66 million a year. This makes flooding the worst of all natural disasters (including earthquakes and drought). The average number of people affected by flooding for the five-year period from 1973 through to 1977 was 19 million and escalated sharply to 111 million for the period 1988 through 1992 and still further to 131 million for the 1993 to 1997 five-year period. The average death toll per year has been recorded as approximately 7,000 people for the last 25 years. In 1998 alone, this figure reportedly came close to 30,000.

A typology for floods is provided in Table 1.1 with estimates of the frequency of occurrence for the period between 1985 and 2009. More than 90% of floods are attributed to meteorological processes, with hydrologic types only contributing about 5% of the events.

Floods are the most common and widespread of all natural disasters, besides fire. They are also the number one killer. Flood disasters account for about a third of all natural disasters (by

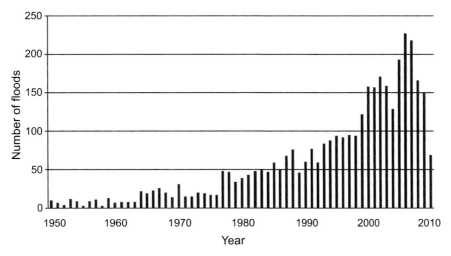

Figure 1.3 The number of flood disasters (source: EM-DAT CRED).

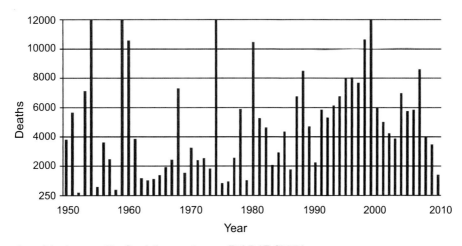

Figure 1.4 The total number of deaths caused by flood disasters (source: EM-DAT CRED).

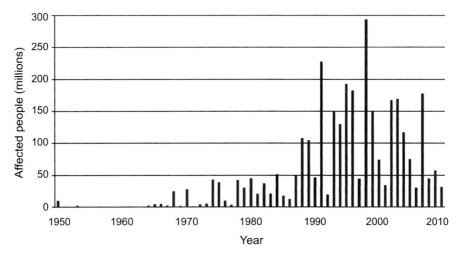

Figure 1.5 The total number of people affected by flood disasters (source: EM-DAT CRED).

Table 1.1 *Types of floods and their frequency of occurrence 1885–2009*

		Number	Frequency (%)	Total for class (%)
Human induced	Dam/levee break	47	1.34	1.34
Natural event	Avalanche-related	2	0.06	0.14
	Landslide	1	0.03	
	Outburst flood	2	0.06	
Hydrologic	Rain on snow	84	2.40	4.97
	Snowmelt	60	1.72	
	Ice jam/breakup	30	0.86	
Meteorological	Brief torrential rain	297	8.49	92.40
	Extra-tropical storm	19	0.54	
	Heavy rain	2235	63.89	
	Monsoon rain	280	8.00	
	Torrential rain	27	0.77	
	Tropical cyclone	348	9.95	
	Tropical storm	26	0.74	
Other	Tidal surge	4	0.11	1.14
	Not determined	36	1.03	
Total		**3498**		**100**

Source: the Dartmouth Flood Data Observatory

number and economic losses). They are responsible for over half of the deaths associated with all such disasters.

The estimates of damage by floods given here are estimates in the pure sense of the word. It is very difficult to calculate the damage from floods in terms of numerical values due to their wide regional coverage and also due to the fact that much of the damage (e.g., ecological damage, human pain, suffering, deaths, distress) cannot be directly expressed in terms of monetary values.

Floods form one of the most important parts of the world's natural disasters today, and there is an increasing trend in the resulting damage and deaths due to them. The flood disasters that were strongly evident throughout the second millennium have always been a part of the human experience. They are going to continue to be so in the third millennium. They continue to be destructive and they are more widespread and harmful now than in the past.

1.2 PROBLEM CONTEXT

The beginning of the third millennium is characterized by the major and widespread change known as the ***global change***. It encompasses the full range of global issues and interactions concerning natural and human-induced changes in the Earth's environment. In more detail, it is defined by the US Global Change Research Act of 1990 as "changes in the global environment – including alterations in climate, land productivity, oceans or other water resources, atmospheric chemistry, and ecological systems – that may alter the capacity of the Earth to sustain life." Global change issues include understanding and predicting the causes and impacts of, and potential responses to: long-term climate change and greenhouse warming; changes in atmospheric ozone and ultraviolet (UV) radiation; and natural climate fluctuations over seasonal to inter-annual time periods. Other related global issues include desertification, deforestation, land use management, and preservation of ecosystems and biodiversity. They are all directly related to flooding. Flood hazards and disasters have always been part of human history. They will continue to be so into the future. In the context of global change they are as much a

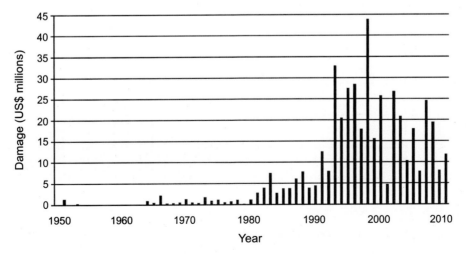

Figure 1.6 The total estimated damage (US$) caused by flood disasters (source: EM-DAT CRED).

part of the history of humankind as population growth, settlement, industrialization, computerization, and repeated cycles of recession and expansion. Flood hazards and disasters are an intrinsic component of accumulated cultural experience which manifests itself in complex social structures and practices. Flood hazard and risk are inherent in these structures and practices. The key issue for the global community today is the extent to which flood hazards and disasters can be contained and reduced.

Floods as physical features are most affected by *climate change* (a very detailed discussion of climate change and risk of flooding follows in Chapter 2). Many of the impacts of climate variations and climate change on society, the environment, and ecosystems are caused by (i) changes in the frequency or intensity of extreme weather and climate events, and (ii) sea level rise. The IPCC Fourth Assessment Report (IPCC, 2007) concluded that many changes in extremes had been observed since the 1970s as part of the warming of the climate system. These included: more frequent hot days, hot nights, and heat waves; fewer cold days, cold nights, and frosts; more frequent heavy precipitation events; more intense and longer droughts over wider areas; an increase in intense tropical cyclone activity in the North Atlantic; and sea level rise.

Recent climate research has found that rain is more intense in already-rainy areas as atmospheric water vapor content increases (The Copenhagen Diagnosis, 2009). Recent changes have occurred faster than predicted by some climate models, emphasizing that future changes could be more severe than predicted. In addition to the increases in heavy precipitation, there have also been observed increases in drought since the 1970s. This is consistent with the decrease in mean precipitation over land in some latitude bands. The intensification of the global hydrologic cycle with climate change is expected to lead to further increases in precipitation extremes, both increases in very heavy precipitation in wet areas and increases in drought in dry areas. While precise predictions cannot yet be given, current studies suggest that heavy precipitation rates may increase by 5–10% per °C of warming, similar to the rate of increase of atmospheric water vapor.

Population density in coastal regions and islands is about three times higher than the global average. Currently 160 million people live less than 1 meter above sea level. This allows even a small sea level rise to have disastrous consequences. Effects may be caused by coastal erosion, increased susceptibility to storm surges and resulting flooding, groundwater contamination by salt intrusion, loss of coastal wetlands, and other issues. Sea level rise is an inevitable consequence of global warming for two main reasons: ocean water expands as it heats up, and additional water flows into the oceans from the ice that melts on land. Since 1870, global sea level has risen by about 20 centimeters (IPCC, 2007). The average rate of rise for 1993–2008 as measured by satellite is 3.4 millimeters per year, while the IPCC projected a best estimate of 1.9 millimeters per year for the same period. Actual rise has thus been 80% faster than projected by models (Rahmstorf

et al., 2007). Future sea level rise is highly uncertain. The main reason for the uncertainty is in the response of the big ice sheets of Greenland and Antarctica. Sea level will continue to rise for many centuries after global temperature is stabilized, since it takes that much time for the oceans and ice sheets to fully respond to a warmer climate. The future estimates highlight the fact that unchecked global warming is likely to raise sea level by several meters in the coming centuries, leading to the loss of many major coastal cities and entire island states.

The Population Division of the Department of Economic and Social Affairs of the United Nations Secretariat (2009) reports that the world population, at 6.8 billion in 2009, is projected to reach 7 billion in late 2011 and 9 billion in 2050. Most of the additional 2.3 billion people expected by 2050 will be concentrated in developing countries, whose population is projected to rise from 5.6 billion in 2009 to 7.9 billion in 2050. The world is undergoing the largest wave of *population and urban growth* in history. In 2008, for the first time in history, more than half of the world's population was living in towns and cities. By 2030 this number will swell to almost 5 billion, with urban growth concentrated in Africa and Asia. While mega-cities have captured much public attention, most of the new growth will occur in smaller towns and cities, which have fewer resources to respond to the magnitude of the change. In principle, cities offer a more favorable setting for the resolution of social and environmental problems than rural areas. Cities generate jobs and income. With good governance, they can deliver education, health care, and other services more efficiently than less densely settled areas simply because of their advantages of scale and proximity. Cities also present opportunities for social mobilization and women's empowerment. Most of the world cities are along rivers, lakes, and ocean shores. The first attractions of such locations were as sources of food and drinking water. Later, the attractions also included water for irrigation and transportation. In the present, the closeness to water also provides power generation, commerce, and recreation. Riverbanks and flood plains are also attractive for agriculture, aesthetics, and as a way of disposing of wastes. Over several thousand years cities, settlements, and other infrastructure have increasingly encroached into the floodplains. Negative impacts of flood disasters are directly related to population trends and changes. A larger number of people translates directly into larger exposure and potentially higher risk from flooding.

Population and climate change are connected. Most climate change is attributed to anthropogenic impacts. Flood disasters, climate change, and population interactions will have one more dimension of complexity in the future – climate migrations. One of the observations of the IPCC (2007) is that the greatest single impact of climate change could be on human migration – with millions of people displaced by shoreline erosion, coastal flooding, agricultural disruption, etc. Since 2007 various analysts have tried to put numbers on future flows of climate migrants (sometimes called "climate refugees"). The most widely repeated prediction is 200 million by 2050 (IOM, 2008). But repetition does not make the number any more accurate. The scientific argument for climate change is increasingly confident. The consequences of climate change for human population distribution are unclear and unpredictable. The available science translates into a simple fact – the population of the world today is at higher level of risk. The disasters that will move people have two distinct drivers: (i) climate processes such as sea level rise, salinization of agricultural land, desertification, and growing water scarcity, and (ii) climate hazard events such as flooding, storms, and glacial-lake outburst floods. It is necessary to note that non-climate drivers, such as government policy, general population growth, and community-level resilience to natural disaster, are also important. All contribute to the degree of vulnerability people experience. The problem is one of time (the speed of change) and scale (the number of people it will affect).

Issues of *sustainable development* were drawn closer to water resources systems management after the publication of the Brundtland Commission's report "Our Common Future" (WCED, 1987), which introduced the concept of sustainable development as "The ability to meet the needs of the present, without compromising the needs of future generations." This vision of sustainable development may never be realized, but it is clearly a goal worthy of serious consideration. There is an increasing realization that exposure and vulnerability of population and environment to flooding are important dimensions of sustainable communities.

Applying principles of sustainability to management of flood risks requires major changes in the objectives on which decisions are based and an understanding of the complicated interrelationships between existing ecological, economic, and social factors. The broadest objectives for achieving sustainability are environmental integrity, economic efficiency, and equity (Simonovic, 2009). Another important aspect of sustainable flood risk management is the challenge of time (i.e., identifying and accounting for long-term consequences). We are failing to provide basic flood protection for a large portion of the world population, and therefore are not at the starting point in terms of dealing with the needs of future generations. For some developments, the prediction of long-term consequences is difficult. The third aspect of the sustainable flood risk management context is the change in procedural policies (implementation). Pursuing sustainable flood risk management through the implementation of structural projects and use of non-structural solutions will require major changes in both substantive and procedural policies. The diverse policy questions raised include: How should the flood risk management decision methods and processes be used? What should be the reliance on market as opposed to regulatory mechanisms? And what should be the role of public and interest groups in flood risk management decision-making?

Flood risk management involves complex interactions within and between the natural environment, human population (actions, reactions, and perceptions), and built environment (type and location). A different thinking is required to address the complexity of flood risk management. Mileti (1999, page 26) and Simonovic (2011, page 48) strongly suggest adaptation of a global *systems perspective*. Systems theory is based on the definition of a system – in the most general sense – as a collection of various structural and non-structural elements that are connected and organized in such a way as to achieve some specific objective through the control and distribution of material resources, energy, and information. The basic idea is that all complex entities (biological, social, ecological, or other) are composed of different elements linked by strong interactions, but a system is greater than the sum of its parts. This is a different view from the traditional analytical scientific models based on the law of additivity of elementary properties that view the whole as equal to the sum of its parts. Because complex systems do not follow the law of additivity, they must be studied differently. A systemic approach to problems focuses on interactions among the elements of a system and on the effects of these interactions. Systems theory recognizes multiple and interrelated causal factors, emphasizes the dynamic character of processes involved, and is particularly interested in how a system changes with time – be it a flood, floodplain, or disaster-affected community. The traditional view is typically linear and assumes only one, linear, cause-and-effect relationship at a particular time. A systems approach allows a wider variety of factors and interactions to be taken into account. Using a systems view, Mileti (1999) and Simonovic (2011) state that flood disaster losses are the result of interactions among three systems and their many subsystems: (i) the Earth's physical systems (the atmosphere, biosphere, cryosphere, hydrosphere, and lithosphere); (ii) the human systems (e.g., population, culture, technology, social class, economics, and politics); and (iii) the constructed systems (e.g., buildings, roads, bridges, public infrastructure, and housing).

All of the systems and subsystems are dynamic and involve constant interactions between and among subsystems and systems. All human and constructed systems and some physical ones affected by humans are becoming more complex with time. This *complexity* is what makes flood disaster problems difficult to solve. The increase in the size and complexity of the various systems is what causes increasing susceptibility to disaster losses. Changes in size and characteristics of the population and changes in the constructed environment interact with changing physical systems to generate future exposure and define future disaster losses. The world is becoming increasingly complex and interconnected, helping to make disaster losses greater (Homer-Dixon, 2006).

The first component of the complexity paradigm is that flood risk management problems in the future will be more complex. Domain complexity is increasing. Further population growth, climate change, and regulatory requirements are some of the factors that increase the complexity of flood risk management problems. Flood risk management strategies are often conceived as too short-sighted (design life of dams, levees, etc.). Short-term thinking must be rejected and replaced with flood risk management schemes that are planned over longer temporal scales in order to take into consideration the needs of future generations. Planning over longer time horizons extends the spatial scale. If resources for flood risk management are not sufficient within an affected region, transfer from neighboring regions should be considered. The extension of temporal and spatial scales leads to an increase in the complexity of the decision-making process. Large-scale flood risk management processes affect numerous stakeholders. The environmental and social impacts of complex flood risk management solutions must be given serious consideration.

The second component of the complexity paradigm is the rapid increase in the processing power of computers. Since the 1950s, the use of computers in water resources management has grown steadily. Computers have moved from data processing, through the user's office and into information and knowledge processing. Whether the resource takes the form of a laptop PC or a desktop multiprocessing workstation is not important any more. What is important is that the computer is used as a partner in more effective flood risk management (National Research Council, 1996; Global Disaster Information Network, 1997; Stallings, 2002). The main factor responsible for involving computers in flood risk decision-making processes is the treatment of information as the sixth economic resource (besides people, machines, money, materials, and management).

The third component of the complexity paradigm is the reduction in the complexity of contemporary systems tools. The most important advance made in the field of management in the last century was the introduction of systems analysis. Systems analysis is defined here as an approach for representing complex management problems using a set of mathematical planning and design techniques. Theoretical solutions to the problems can then be found using a computer. In the context of this book, systems analysis techniques, often called "operations research," "management science," and "cybernetics," include simulation and optimization techniques that can be used in the four-phase flood risk management cycle (discussed in detail in Section 1.3). Systems analysis is particularly promising when scarce resources must be used effectively. Resource allocation problems are very common in the field of flood risk management, and affect both developed and developing countries, which today face increasing pressure to make efficient use of their resources.

1.3 FLOOD RISK

The terms "floods," "flooding," "flood hazard," and "flood risk" cover a very broad range of phenomena. Among many definitions of floods that do not incorporate only notions of *inundation* and *flood damage*, for the purpose of this text I will stay with the definition provided by Ward (1978) that a flood is a body of water which rises to overflow land which is not normally submerged. This definition explicitly includes all types of surface inundation, but flood damage is addressed only implicitly in its final three words. Both inundation and damage occur on a great range of scales.

According to Smith and Ward (1998) the impact of floodwaters through deposition and erosion, or through social and economic loss, depends largely on the combination of water quality, depth, and velocity. Flood hazards result from the potential for extreme flooding to create an unexpected threat to human life and property. When severe floods occur in areas occupied by humans, they create natural disasters, which may involve the loss of human life and property together with disruption to existing activities of urban or rural communities. Flooding of a remote, unpopulated region is an extreme physical event – usually only of interest to hydrologists.

Terms such as "flood risk" and "flood losses" are essentially our interpretations of the negative economic and social consequences of natural events. Human judgment is subject to value systems that different groups of people may have and therefore these terms may be subject to different definitions. The flood risk, at various locations, may be increased by human activity – such as inappropriate land use practices. Also, the flood risk may be reduced by flood management structures and/or effective emergency planning. The real flood risk therefore, stems from the likelihood that a major hazardous event will occur unexpectedly and that it will impact negatively on people and their welfare (Smith and Ward, 1998). Flood hazards result from a combination of physical exposure and human vulnerability to flooding. Physical exposure reflects the type of flood event that can occur, and its statistical pattern, at a particular location. Human vulnerability reflects key socio-economic factors, such as the number of people at risk on the floodplain, the extent of flood defense works, and the ability of the population to anticipate and cope with flooding.

The philosophical discussion of risk definition is well documented in Kelman (2003). It ends with the statement that to understand risk, we must understand ourselves. For the purpose of this text the formal definition of **flood risk** is a combination of the chance of a particular event with the impact that the event would cause if it occurred. Flood risk therefore has two components – the chance (or probability) of an event occurring and the impact (or consequence) associated with that event. The consequence of an event may be either desirable or undesirable. In some, but not all, cases (Sayers *et al.*, 2002), therefore, a convenient single measure of the importance of a flood risk is given by:

$$Risk = Probability \times Consequence \qquad (1.1)$$

If either of the two elements in (1.1) increases or decreases, then risk increases or decreases respectively.

It is important to avoid the trap that risks with the same numerical value have equal significance, since this is often not the case. In some cases, the significance of a risk can be assessed by multiplying the probability by the consequences. In other cases it is important to understand the nature of the risk, distinguishing between rare, catastrophic events and more frequent, less severe events. For example, risk methods adopted to support the targeting and management of flood warning represent risk in terms of probability and consequence, but low probability/high consequence events are treated very differently than high probability/low consequence events. An additional factor to include is how society or individuals perceive a risk (a perception that is influenced by many factors including, for example, the availability of insurance, government assistance, or similar) and uncertainty in the assessment.

1.4 HOW DO WE MANAGE FLOOD RISK?

In many countries flood risk management is evolving from traditional approaches based on design standards to the development of risk-based decision-making, which involves taking account of a range of loads, defense system responses and impacts of flooding (Sayers *et al.*, 2002). The difference between a risk-based approach and other approaches to design or decision-making is that it deals with outcomes. Thus, in the context of flooding it enables intervention options to be compared on the basis of the impact that they are expected to have on the frequency and severity of flooding in a specified area. A risk-based approach therefore enables informed choices to be made based on comparison of the expected outcomes and costs of alternative courses of action. This is distinct from, for example, a standards-based approach that focuses on the severity of the load that a particular flood defense is expected to withstand.

The World Meteorological Organization (WMO, 2009) is promoting the principle of integrated flood management – IFM – that has been practiced at many places for decades. Integrated flood management integrates land and water resources development in a river basin and aims at maximizing the net benefits from the use of floodplains and minimizing loss of life from flooding. Globally, both land – particularly arable land – and water resources are scarce. Most productive arable land is located on floodplains. When implementing policies to maximize the efficient use of the resources of the river basin as a whole, efforts should be made to maintain or augment the productivity of floodplains. On the other hand, economic losses and the loss of human life due to flooding cannot be ignored. Integrated flood management recognizes the river basin as a dynamic system in which there are many interactions and flux between land and water bodies. In IFM the starting point is a vision of what the river basin should be, followed by the identification of opportunities to enhance the performance of the system as a whole. Integrated flood management takes a participatory, cross-sectoral and transparent approach to decision-making. The defining characteristic of IFM is integration, expressed simultaneously in different forms: an appropriate mix of strategies, carefully selected points of intervention, and appropriate types of intervention (structural or non-structural, short or long term). An IFM plan should address the following six key elements: (i) Manage the water cycle as a whole; (ii) integrate land and water management; (iii) manage risk and uncertainty; (iv) adopt a best mix of strategies; (v) ensure a participatory approach; and (vi) adopt integrated hazard management approaches.

Flood risk management, according to Equation (1.1), aims to reduce the likelihood and/or the impact of floods. Experience has shown that the most effective approach is through the development of flood risk management programs (Simonovic, 2011) incorporating the following elements:

- *Prevention*: preventing damage caused by floods by avoiding construction of houses and industries in present and future flood-prone areas; by adapting future developments to the risk of flooding; and by promoting appropriate land use, agricultural, and forestry practices.
- *Protection*: taking measures, both structural and non-structural, to reduce the likelihood of floods and/or the impact of floods in a specific location.
- *Preparedness*: informing the population about flood risks and what to do in the event of a flood.
- *Emergency response*: developing emergency response plans in the case of a flood.
- *Recovery*: returning to normal conditions as soon as possible and mitigating both the social and economic impacts on the affected population.

A change to proactive flood risk management requires identification of the risk, the development of strategies to reduce that risk, and the creation of policies and programs to put these strategies into effect.

1.5 SYSTEMS VIEW OF FLOOD RISK MANAGEMENT

Flood risk management is a part of all social and environmental processes aimed at minimizing loss of life, injury, and/or material damage. Mileti (1999) and Simonovic (2011) advocate a systems view of flood risk management processes in order to address their complexities, dynamic character, and interdisciplinary needs of management options. A primary emphasis of systems analysis in flood risk management is on providing an improved basis for effective decision-making. A large number of systems tools, from simulation and optimization to multi-objective analysis, are available for formulating, analyzing, and solving flood risk management problems.

The question I would like to answer in this section is: What are we trying to manage? We keep trying to manage environments (water, land, air, etc.). We keep trying to manage people within environments. It seems that every time we push at one point, it causes unexpected change elsewhere – the first fundamental systems rule. Perhaps it is time to sit back and rethink what we are trying to manage.

A model: In order to apply a continuous improvement approach to flood risk management it is essential to have a way of thinking – a model – of what is being managed. Without this it is not possible to see where energy or resources are being wasted, or might significantly alter outcomes. Up to now, no such general model has been proposed, let alone accepted, as a basis for predicting outcomes from different flood risk management interventions, and their combinations.

The system in our focus is a social system. It describes the way floods affect people. The purpose of describing the system is to help clarify the understanding and determine the best points of systems intervention.

Management systems principles: The flood risk management system comprises four linked subsystems: individuals, organizations, and society, nested within the environment. Individuals are the actors that drive organizations and society to behave in the way they do. They are decision-makers in their own right, with a direct role in mitigation, preparedness, response, and recovery from flooding. Organizations are the mechanism people use to produce outcomes that individuals cannot produce. Organizations are structured to achieve goals. Structure defines information and/or resource flows and determines the behavior of the organization. The concept of society is different from those of individuals and organizations, being more difficult to put boundaries around. In general, society itself is a system of which individuals and organizations are subsets and contains the relationships people have with one another, the norms of behavior, and the mechanisms that are used to regulate behavior. The environment includes concrete elements such as water and air, raw materials, natural systems, etc. It also encompasses the universe of ideas, including the concept of "future." This concept is important in considering flood risk management – it is the expectation of future damage and future impacts that drives concern for sustainable management of flood disasters.

Management principle 1: To achieve sustainable management of flood risk, interactions between the four subsystems: individual, organization, society and environment, must be appropriately integrated.

A second principle we can use in developing our framework is that we can order systems inputs and outputs into three categories: resources, information, and values. These connect individuals, organizations, society, and environment, linking the four subsystems. Only information and resources flow link people and organizations. Value systems are influenced by these two flows, but operate in a different way. Value systems are generated within the individual or organization but feed off information and resource flows.

Management principle 2: Two flows – resource flows and information flows – link the individual, organization, society, and environment subsystems. Value systems are the means through which different values are attached to information and resource flows.

All open systems require input of energy – resources – to produce output. The need to constantly access resources is a major mechanism for the operation of subsystems. Each subsystem relies on other subsystems and on the environment for its resources. In an ideal state, the goals of each subsystem, and performance relative to those goals, must represent a gain for other subsystems for all to continue to receive resources. The physical environment exerts passive pressure on the subsystems to ensure fit. In addition, the environment can limit the action by running out of a resource or by changing circumstances to make the resource more precious – for example changing climate.

Management principle 3: The ongoing need of subsystems for resources from one another sets the limits of their exploitation of one another and of the environment, and is a determinant of behavior within the system.

Information is used by each of the subsystems to make decisions required to ensure fit with other subsystems and the environment. Without flows of information from outside the system – or subsystem – the system must rely on its own internal information (knowledge) to make decisions. Such a circumstance increases the risk that the subsystem will drift out of fit with its context. Regardless, it constantly receives signals from the outside world, and is itself sending signals to other systems. Well-functioning systems have structures built into them to capture relevant information and use that information to maximize their chances of utilizing resources to achieve their systems goals.

Management principle 4: Information is used by subsystems to make decisions intended to ensure fit with the needs of other subsystems and the environment.

Data do not in themselves have meaning. A process of interpretation occurs between information and meaning, and this process is driven by existing values. Value systems determine what individuals, organizations, and societies find important: (i) the sorts of resources they will pursue; and (ii) the interpretation of information received and used. Value systems are embedded in the culture of society and organizations, and in the values held by individuals, and they determine how subsystems behave. Use of value systems may be triggered by information, and shaped by flow of resources.

Management principle 5: Values provide meaning to information flows that are then used to determine resource use by subsystems.

The reality of linking mechanisms indicates that it is the availability of resources that largely conditions choice. It is information about availability that signals to the decision-maker (individuals, organizations, or society) whether it is implementing appropriate management strategies. It is through the process of optimizing resource access that learning takes place and significant changes in culture and values are achieved. So, the most powerful management strategies will go directly to resource access, and will initiate signals that show which social or environmental performance will allow for access to resources on improved terms.

Management principle 6: The most effective management strategies for sustainable management of flood risk are those that condition access to resources.

Each subsystem utilizes different mechanisms for minimizing negative impacts of flood disasters. Within each subsystem there are many different interactions and many different options to optimize resource use.

***No "right" management strategy*:** Flood risk management is a process of managing behavior. There is no one strategy that will be optimal for any situation. Neither regulation, nor economic incentives, nor education, nor shifts in property rights, is the "right" management strategy. What will work will vary with the social system being managed, in response to three variables: the information and resource flows, and the value systems that are in place. The challenge for the flood risk manager is to manage these three elements, across individuals, organizations, and the society and within the environment, to achieve the most effective outcome that is possible.

Management principle 7: More intensive focus on the systems view of flood risk management will accelerate understanding of what management strategies work, and particularly why they might work.

For example, when one program deals with economic incentives, another deals with improving information flows, and a third is focused on regulatory enforcement, it is very easy to believe that they are focused on different aspects with fragile links. What is necessary is a systems model to make sense of the interactions and dynamics that are being managed. This will allow us to learn from what we are so "clumsily" doing, so that eventually we can do it better.

1.6 CONCLUSIONS

Flood risk management involves complex interactions within and between the natural environment, human population (actions, reactions, and perceptions) and built environment (type and location). A different thinking is required to address the complexity of flood risk management. Adaptation of a global systems perspective is strongly recommended. For the purpose of this text

the formal definition of flood risk is a combination of the chance of a particular event with the impact that the event would cause if it occurred. Flood risk management aims to reduce the likelihood and/or the impact of floods. Experience has shown that the most effective approach is through the development of flood risk management programs incorporating prevention, preparedness, emergency response, and recovery to normal conditions.

A change to proactive flood risk management requires identification of the risk, the development of strategies to reduce that risk, and the creation of policies and programs to put these strategies into effect. Flood risk management is a part of all social and environmental processes aimed at minimizing loss of life, injury, and/or material damage.

A systems view of flood risk management is recommended in order to address the complexity, dynamic character, and interdisciplinary needs of management options. A primary emphasis of systems analysis in flood risk management is on providing an improved basis for effective decision-making. A large number of systems tools, from simulation and optimization to multiobjective analysis, are available for formulating, analyzing, and solving flood risk management problems. The main objective of this book is to present a variety of systems tools for flood risk management.

1.7 EXERCISES

1.1. Describe the largest flood disaster experienced in your region.

  a. What were its physical characteristics?
  b. Who was involved in the management of the disaster?
  c. What is, in your opinion, the most important flood management problem in your region?
  d. What lessons can be learned from the past management of flood disasters in your region?
  e. What are the most important principles you would apply in future flood disaster management in your region?

1.2. Review the literature and find a definition of integrated flood management. Discuss the example from Exercise 1.1 in the context of this definition.

1.3. Discuss characteristics of the flood disaster from Exercise 1.1.

  a. What are the complexities of the problem in Exercise 1.1?
  b. Identify some uncertainties in the problem.
  c. Can you find some data to illustrate the natural variability of regional conditions?
  d. How difficult is it to find the data? Why?

Figure 3.13 The floodplain depth grid and bounding polygon for the CC_UB 250-year flood at one location along the North Thames.

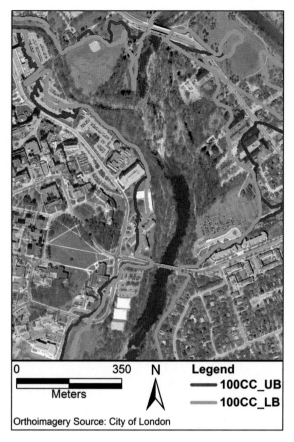

Figure 3.14 Comparison of floodplain boundaries for 100 CC_LB and 100 CC_UB at the location of the University of Ontario (North Thames).

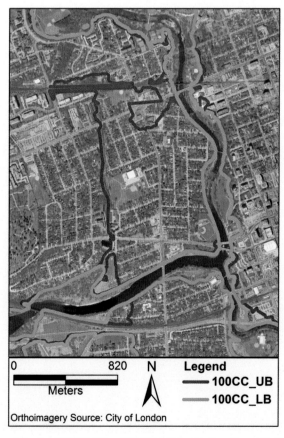

Figure 3.15 Comparison of floodplain boundaries for 100 CC_LB and 100 CC_UB at the location of the Forks (confluence of the North Thames and the South Thames).

Figure 3.16 Comparison of floodplain boundaries for 100 CC_LB and 100 CC_UB at the location of the Broughdale Dyke (North Thames).

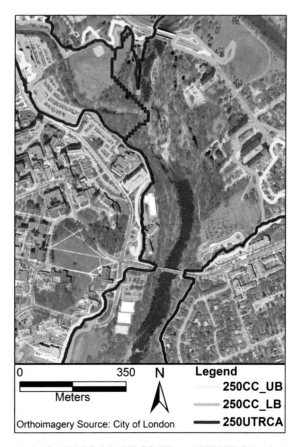

Figure 3.17 Comparison of floodplain boundaries for 250 CC_LB, 250 CC_UB, and 250 UTRCA at the location of the University of Ontario (North Thames).

Figure 3.18 Comparison of floodplain boundaries for 250 CC_LB, 250 CC_UB, and 250 UTRCA at the location of the Forks (confluence of the North Thames and the South Thames).

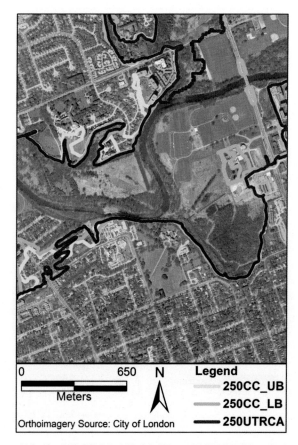

Figure 3.19 Comparison of floodplain boundaries for 250 CC_LB, 250 CC_UB, and 250 UTRCA at the location of the Broughdale Dyke (North Thames).

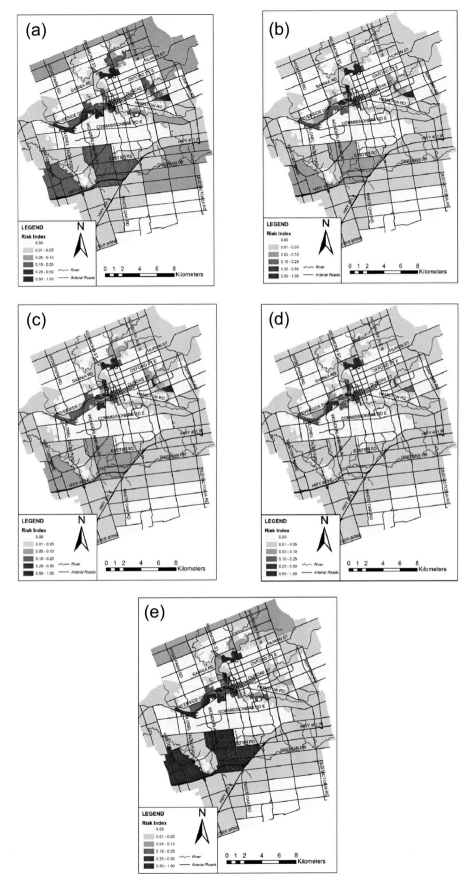

Figure 3.21 Risk to all infrastructure for (a) 100 CC_LB, (b) 100 CC_UB, (c) 250 CC_LB, (d) 250 CC_UB, and (e) 250 UTRCA.

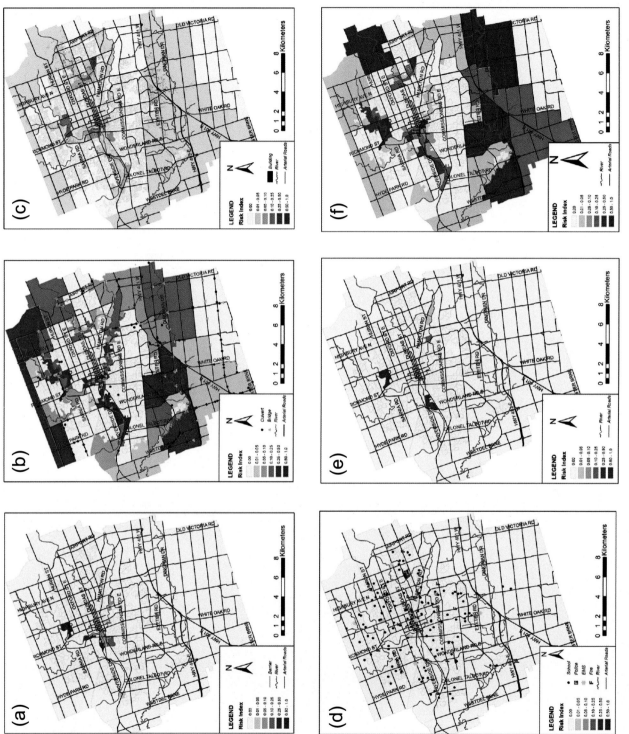

Figure 3.22 100 CC_UB infrastructure risk indices for (a) barriers, (b) bridges, (c) buildings, (d) critical infrastructure, (e) PCPs, and (f) roads.

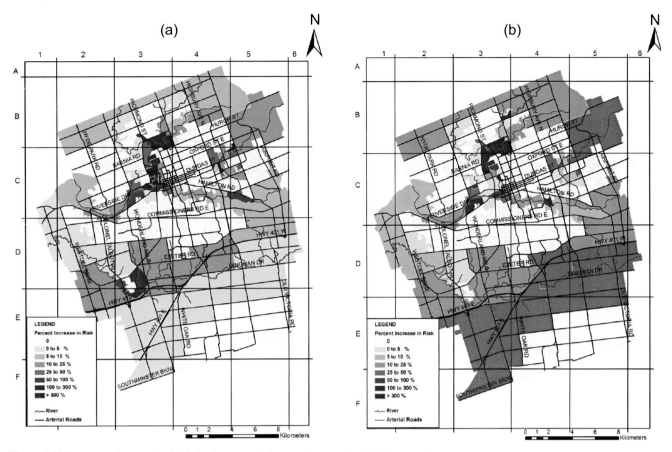

Figure 3.24 Percentage increase in risk index between the lower and upper bound, 100- and 250-year return events.

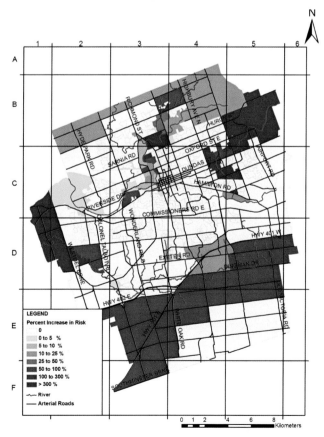

Figure 3.25 Percentage increase in risk between the 250 UTRCA and 250 CC_UB scenarios.

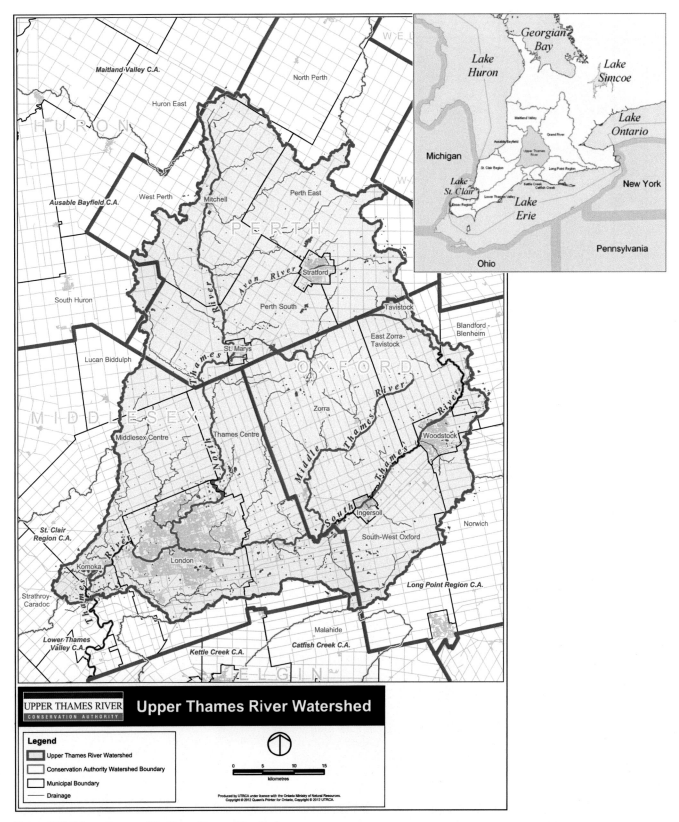

Figure 5.27 Schematic map of the Upper Thames River basin.

2 Climate change and risk of flooding

Climate change is a change in the statistical distribution of weather over long periods of time that range from decades to millions of years. Weather refers, generally, to day-to-day temperature and precipitation activity. Climate change may be limited to a specific region, or may occur across the whole Earth. In the context of environmental policy, climate change usually refers to changes in modern climate. It may be qualified as anthropogenic climate change, more generally known as *global warming*. Global warming is the increase in the average temperature of Earth's near-surface air and oceans since the mid twentieth century. According to the IPCC Fourth Assessment Report (IPCC, 2007), global surface temperature increased 0.74 ± 0.18 °C during the twentieth century.

The warming of the atmosphere increases the capacity of the atmosphere to hold water, and accelerates many of the processes involved in the redistribution of moisture in the atmosphere. These changes alone suggest that flood-generating processes linked to the atmosphere are on the increase. However, warming also alters many other aspects of the global water cycle: (i) increasing evaporation, (ii) changing precipitation patterns and intensity, (iii) affecting the processes involved in surface storage of water including snowpack generation, snowmelt, river ice breakup, and glacial melt. Most of these processes are active in flood generation. The general view of scientists is that flooding is increasing with changing climate (The Copenhagen Diagnosis, 2009).

In this chapter a brief review of the physical causes and characteristics of floods is provided first. It is followed by a discussion of climate change and its impacts. At the end a review of approaches for dealing with climate change is provided, including mitigation and adaptation.

2.1 FLOODS AND THEIR PHYSICAL CHARACTERISTICS

Floods are an integral part of the inherent variability of nature. Much as we may seek to control and eliminate floods, they will continue to occur. Most people are increasingly finding a common interest in accepting the inevitability of floods that are larger

Table 2.1 *Major factors initiating floods*

Flood type	Flood-initiating process
River floods	• Heavy rain • Rapid snowmelt • Rapid ice melt • Glacial lake breaches
Ice jam floods	Ice breakup
Debris-jam floods	Debris entrapment
Landslide blockages	Landslides

than expected, and in accepting that many human activities can magnify both the size and the impact of floods. While the most common type of flood is generated by the rapid input of excessive quantities of water, floods may also be created by blockages within the drainage network, and by high tides and onshore winds that can create both sea and estuarine floods (sea and estuarine floods will not be discussed further in this text). All inputs of water are modified by the topography and the hydraulic properties of the surface and/or subsurface materials they encounter. Table 2.1 provides a list of major factors that initiate floods. Flood volume and timing are affected by natural interactions with environment and humans. Table 2.2 lists the flood modifying factors.

The hydrograph of river flow synthesizes flood hydrology without necessarily producing more information on the main flood-producing processes. Hewlett (1982) stated: "it is not the peak discharge in the headwaters that produces the downstream flood, but rather the volume of stormwater released by the headwater areas."

2.1.1 Floods caused by rainfall

Floods in most river basins are caused almost entirely by excessive heavy and/or excessively prolonged rainfall or by periods of prolonged and/or intense snowmelt. In each case all processes result in a large volume of quick flow, which reaches the stream channel very rapidly during and immediately after a rainfall or melt event. Quick flow is the outcome from rain falling on a

Table 2.2 *Flood-modifying factors*

Basic interactions	Human effects
Basin morphology	Water supply engineering
Hillslope properties	• Dams
• Morphology	• River regulation
Material properties	• Inter-basin transfer
Channel properties	• Wastewater release
• Network morphology	• Water abstraction and irrigation
• Channel form and materials	Land use changes
	• Urbanization
	• Deforestation
	• Agriculture
	Channel modification
	• Land drainage
	• Channel straightening
	• Flood protection works
	Weather modification
	• Rainmaking

river basin and originates from the interaction of rainfall and catchment conditions (Smith and Ward, 1998). In the early stages of a storm, all rainfall (P) infiltrates the soil surface. Then, as a result of infiltration and flow in the soil profile (Q_t), the lower valley slopes become saturated as the shallow water table rises to the ground surface. In these saturated areas infiltration capacity is zero so that all precipitation falling on them, at whatever intensity, becomes overland flow (Q_o).

Although, at the onset of rainfall, variable source areas may be restricted to the stream channels themselves and to adjacent valley bottom areas, convergence of shallow subsurface flow paths may also result in surface saturation and saturation overland flow in slope concavities and areas of thinner soils throughout the catchment. Continuing throughflow from upslope unsaturated areas of the catchment will result in the spatial growth of source areas wherever they are located initially. Consequently, source areas that often cover less than 5% of the catchment at the beginning of rainfall may expand to cover 20–25% of the catchment as a storm continues, thereby resulting in a five-fold increase in the volume of quick flow generated by a given rate of rainfall.

Where rainfall is particularly intense, or where natural infiltration capacities have been reduced by anthropogenic effects such as soil compaction or overgrazing, overland flow may result from the process described by Horton (1933). During those parts of a storm when rain falls at a rate that is greater than the rate at which it can be absorbed by the ground surface there will occur an excess of precipitation, which will flow over the ground surface as overland flow:

$$Q_o = P_e = (i - f)t \qquad (2.1)$$

where i is rainfall intensity, f is infiltration capacity, t is time, P_e is precipitation excess, and Q_o is overland flow.

A combination of surface and soil profile effects normally causes a rapid reduction in infiltration capacity soon after the beginning of rain so that rain falling at moderate intensity, although incapable initially of generating overland flow, may do so once the early high infiltration rate has declined. Since infiltration capacity is likely to show a continued decrease through a sequence of closely spaced storms, it is commonly found that rain falling late in the storm sequence will generate more overland flow and therefore more severe flooding than the same amount of precipitation falling early in the storm sequence. When the storm area (the area on to which rain falls simultaneously) covers all or most of the catchment, and when the rainfall duration is prolonged, most of the catchment will eventually contribute quick flow simultaneously. Then, aside from initial infiltration or saturation conditions, the amount of quick flow generated for a given depth of precipitation will be about the same from a forested, agricultural, or urbanized area. This has important implications for the effectiveness of catchment management strategies aimed at reducing flood runoff.

2.1.2 Snowmelt and ice melt

Where snowmelt is a major component of flooding, as in high-latitude and high-altitude catchments, variable source areas and flood intensity may decline, rather than increase, with time. The main reasons for this are that, first, the overland flow of melt water at the base of the snowpack will be more efficient with a frozen than with an unfrozen ground surface. Second, the volume of the residual snowpack, which ultimately determines the maximum volume of melt water that can be produced, declines as melting proceeds. And third, since melting normally proceeds from lower to higher altitude, the leftovers of a melting snowpack tend to be located at an increasing distance from mainstream channels.

Ice melting normally takes place more slowly than snow melting and by itself is rarely responsible for severe flooding. However, glacial lake breach floods may occur when the melting of glacier ice suddenly releases large volumes of melt water that have ponded back within the glacier system. Similarly, floods may occur when the breakup of ice pack in a river results in an ice jam, which may hold back large volumes of water.

2.1.3 Volume and timing modification

EFFECTS OF BASIN AND CHANNEL
MORPHOLOGY
The uniformity of runoff times from different parts of the basin can be critical in creating a flood. The more uniform the response and travel times, the greater the likelihood of the river flow building

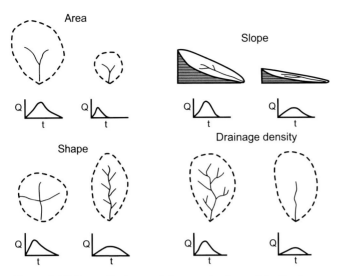

Figure 2.1 Effects of basin morphology on the shape of the flood hydrograph.

up into a high peak flow. Basins that are highly uniform in both process and transmission time may create flood flows out of runoff volumes that in other basins would pass as non-significant events.

Soils with low permeability generate floods more rapidly. The speed of runoff is as important as the actual creation of the runoff. Speed is critical in determining whether runoff arrives at a certain point in the channel in time to add to peak flow or to be regarded as a normal runoff. High slope gradients and high drainage densities contribute to speed of runoff, as well as a compact basin form and a compact channel network. Figure 2.1 illustrates the effects of the main morphologic variables. The importance of stream network density lies in the great contrast in speed of transmission between hill slope processes and channel flow. The shorter the length of hill slope that the water has to pass, by whatever route, the more quickly it will reach the channel. Typical open channel velocities during quick flow events are of the order of a few meters per second. This compares with average measurements of 300 mm/s for pipe flow, 50 mm/s for overland flow, 3 mm/s for macro-pore flow and so on (Parker, 2000).

In a feather-shaped elongated basin, water drains from different parts of the basin with different travel times and the peak discharges from the tributaries arrive at the basin outfall at different times, spreading the peak and reducing flood risk. Travel times are more equal when tributaries are of similar length, which is more commonly the case in more circular basins.

Of all the morphologic variables, basin area (A) is generally considered the most important in controlling the volume of discharge, and is the only one used in the United States Geological Survey (USGS) Index Flood formula for the mean annual flood (which has a return period of 2.33 years):

$$Q_{2.33} = C \times A^{0.7} \qquad (2.2)$$

where C is an empirically derived coefficient. The UK Flood Studies formulae (Smith and Ward, 1998) also consider stream frequency, the average slope of the main channel, and the effects of lake storage and urban drainage. Some more recent updates of this approach add special attention to soil properties, including the presence of impermeable layers and the thickness of groundwater layers.

EFFECTS OF HUMAN INTERVENTION

The surface of the Earth has been significantly modified by human activities. This has affected local and regional water balances and river regimes. These impacts are either direct, by manipulation of water resources, or indirect, through the effects of forest clearance and agriculture. Parker (2000) estimates that 20% of the land area of the globe has been drastically changed by human activities.

Some modifications have intensified the flood hazard while others have reduced it. Removal of the natural vegetation tends to reduce evapotranspiration losses and expose the soil surface to the full kinetic energy of the falling rain, causing breakup of the soil crumbs, clogging pores, reducing infiltration capacity, maybe even forming impermeable patches. Vegetation removal reduces water loss through transpiration and interception, and evaporation losses from the soil, because the reflectivity of the bare soil is higher. Higher surface wind speeds following removal of vegetation could counter this trend, but in the long term actual evaporation will tend to be reduced by lower soil moisture storage.

The range of activities that have reduced flood risk may be equally extensive. All activities that increase the transmission time or reduce the net water surplus within a basin will tend to reduce the flooding risk. This means that any activity that increases evapotranspiration losses or infiltration capacity, improves soil structure, exports water from the basin or consumes it will reduce river discharges. Denser vegetation, rougher surfaces, and barriers designed to retard runoff all delay overland flow, which is one of the prime sources of floodwaters. Activities designed for purposes such as rainwater harvesting, stormwater irrigation, and soil retention all delay and/or reduce the quick flow discharge into the mainstream.

Increasing the public water supply by inter-basin transfer of water resources can act in two ways. It can reduce the flood risk in the source basin and increase it in the receiving basin. Even if there is no net transfer of water in or out of a basin, the effects of water exploitation on levels of river flow can affect flood risk in both directions, depending on the timing and the rate of freshwater abstraction and wastewater return.

Forest modifications Deforestation may intensify river flooding by adversely affecting soil structure and volume, reducing infiltration rates, and reducing water storage. These influences are normally more significant during frequent low-magnitude storms. Their effect during increasingly severe flood-producing storms diminishes as prolonged heavy rainfall and/or melting fills available storage and creates widespread conditions of surface saturation and zero infiltration.

The initial modification of river flood behavior is brought about not by afforestation or deforestation but by associated, temporary, mechanical procedures. For example, ploughing and drainage prior to planting, and skid-road construction and more general compaction during felling may have dramatic, but short-lived effects on runoff volumes and suspended solids loads. More generally, in the longer term, afforestation of previously non-forested areas appears to cause an average decrease of flow equivalent to 200 mm of precipitation. Although average first-year increases of streamflow after forest removal are about 300 mm, the effect diminishes rapidly with time after cutting.

Forest clearance for agriculture has frequently been blamed for increased flooding downstream. The reduced vegetation cover allows more water to reach the soil more rapidly and returns less to the atmosphere. Surface runoff begins to occur more frequently and rates of erosion begin to rise. Erosion rapidly reduces the depth of the soil and thus its capacity to store water. Flooding is more serious because of the increased volume of water in the environment, because of the increased frequency and volume of surface runoff, and because of the rising levels of affected river beds.

As with other aspects of the relationship between flood hydrology and environmental change, there still remain gaps and uncertainties in our understanding.

Agriculture and land drainage The long-term effects of land clearance and agricultural development are difficult to assess. However, they are likely to include increases in the total volume of runoff and flood peaks. There are some estimates (Parker, 2000) that vegetation clearance reduces actual evapotranspiration by over 400 mm/year in the humid tropics and 200 mm/year in mid latitudes.

More recent flood problems are related to land drainage. Experimental work in Britain indicates that open drains in peat soils increase runoff peaks. The general tendency is that underdrainage increases runoff rates when the soil is saturated in extreme events. In North America and most of Europe much land drainage has been part of extensive reclamation of wetlands. It is now appreciated that the loss of wetland storage can have a crucial effect on flood hazard.

Urbanization Although urban areas occupy less than 3% of the Earth's land surface, the effect of urbanization on flood hydrology and flood hazard is disproportionately large. In part this reflects the large and rapidly growing urban population. In part the hydrologic importance of urbanization reflects the wide diversity of flow-producing processes. It has long been accepted, for example, that urban areas make their own climate and there is strong evidence that more rain falls over urban areas than elsewhere. It

Table 2.3 *Impermeability of typical urban surfaces*

Type of surface	Impermeability (%)
Roof surfaces	70–95
Asphalt paving	85–90
Stone, brick paving	
with tight joints	75–85
with open joints	50–70
Macadam roads and paths	25–60
Gravel roads and paths	15–30
Unpaved surfaces	10–30
Parks, gardens, lawns	5–25

After Smith and Ward, 1998

can also be shown that flood conditions within and downstream of urban areas are modified by changes in channel morphology, which are either engineered to improve channel efficiency and capacity or are brought about by downstream channel adaptation to modify outputs of water and sediment from urbanized areas. In addition, flood hydrology is influenced directly through the changed timing and volume of quick flow generated from the extensive impermeable surfaces of urban areas, by the import of clean water for domestic and industrial use, and by the generation and export of wastewater through stormwater and sewerage systems.

The extent to which flood characteristics are modified by urbanization depends very much on the nature of the modified urban surface, on the design of the urban hydrologic system, and on the climate. The main distinguishing characteristic of urban surfaces is that they are less permeable than most of the surfaces they replace. As a result they are effective source areas for quick flow and their flood hydrographs tend to have both higher and earlier peaks. Table 2.3 illustrates, however, that the permeability of urban surfaces varies considerably, so that careful urban design planning can do much to minimize the adverse hydrologic effects of urbanization.

The effect of urbanization on the flood hydrograph is also influenced by climate, especially rainfall, conditions (discussed in the following section). The threat of subsurface flooding has increased in major urban areas that have complex underground transport and service systems.

2.2 CLIMATE CHANGE AND VARIATION IMPACTS

Variations in climate and climate change are another main cause of significant trends in the occurrence of flooding over time. However, their explanations and interpretations have not yet produced convincing evidence on the character of these impacts. At a global scale, the warming of the atmosphere will increase the capacity of the atmosphere to hold water, and this warming will also accelerate many of the processes involved in the redistribution of moisture in the atmosphere. These changes alone suggest that flood-generating processes linked to the atmosphere are likely to increase. However, warming will also alter many other aspects of the global water cycle, increasing evaporation, changing precipitation patterns and intensity, and also affecting the processes involved in surface storage of water including snowpack generation, snowmelt, river ice breakup, and glacial melt. Many of these processes are active in flood generation.

However, these processes take place not at the global scale, but at relatively local scales, making generalizations about flooding in future climates difficult. Also, at the global scale the role of land use in flooding is generally unresolved, yet, at a watershed scale land use effects can be as great as or even more important than changes in the meteorological processes. While the general view of scientists is that flooding will increase with changing climate, making a precise pronouncement is unwise.

Research performed since the IPCC (2007) has reinforced the conclusion that for the period since 1950 it is *very likely* that there has been a decrease in the number of unusually cold days and nights, and an increase in the number of unusually warm days and nights on both a global and regional basis (where the respective extremes are defined with regard to the 1960–1990 base period). In addition, from a limited number of regional analyses and implicit from the documented mean changes in daily temperatures, it is *likely* that warm spells, including heat waves, have increased since the middle of the twentieth century. The few studies since the IPCC (2007) of annual maximum daily maximum and minimum temperatures suggest that human emission of greenhouse gases has *likely* had a detectable influence on extreme temperatures at the global and regional scales. Post IPCC (2007) studies of temperature extremes have utilized larger model ensembles and generally reinforce the projections of changes in temperature extremes as well as providing more regional detail (i.e., *virtually certain* warming trends in daily temperature extremes and *very likely* increases in heat waves over most land areas; the temperature extremes being defined with respect to the 1960–1990 base period).

Many studies conducted since the IPCC (2007) support its conclusion that increasing trends in precipitation extremes have *likely* occurred in many areas over the world. Overall, new studies have substantially strengthened the assessment that it is *more likely than not* that anthropogenic influence has contributed to a global trend towards increases in the frequency of heavy precipitation events over the second half of the twentieth century. The IPCC (2007) projected that it is *very likely* that the frequency of heavy precipitation will increase over most areas of the globe in the twenty-first century. In some regions, heavy daily precipitation events are projected to increase even if the annual total precipitation is projected to decrease. The analyses of climate model

simulations after the IPCC (2007) generally confirm the IPCC assessment.

Current understanding of climate change in the monsoon regions remains of considerable uncertainty with respect to circulation and precipitation. The IPCC (2007) projected that there "is a tendency for monsoonal circulations to result in increased precipitation mainly in the form of extremes due to enhanced moisture convergence, despite a tendency towards weakening of the monsoonal flows themselves. However, many aspects of tropical climatic responses remain uncertain." Work since the IPCC (2007) has not substantially changed these conclusions. At regional scales, there is little consensus in climate models regarding the sign of future change in the monsoons. Land use changes and aerosols from biomass burning have emerged as important forcings on the variability of monsoons, but are associated with large uncertainties.

Research since the IPCC (2007) has not shown clear and widespread evidence of observed changes in floods at the global level except for the earlier spring flow in snow-dominated regions. After a period of frequent occurrence at the end of the Little Ice Age and a more stable period during the twentieth century, glacial-lake outburst floods have increased in frequency in many regions. It is *more likely than not* that anthropogenic greenhouse gas emissions have affected floods because they have influenced components of the hydrologic cycle, but the magnitude and even the sign of this anthropogenic influence is uncertain. The causes of regional changes in floods are complex. It is *likely* that anthropogenic influence has resulted in earlier spring flood peaks in snowmelt rivers. A few recent studies for Europe (Dankers and Feyen, 2008, 2009) and one global study (Hirabayashi *et al.*, 2008; Hirabayashi and Kanae, 2009; Kundzewicz *et al.*, 2010) have projected changes in the frequency and/or magnitude of floods in the twenty-first century at a large scale. However, the sign of any projected trend varies regionally.

It is important to highlight that our confidence in past and future changes, including the direction and magnitude in extremes, depends on the type of extreme, as well as on the region and season, linked with the level of understanding of the underlying processes and the reliability of their simulation in models. The different levels of confidence need to be taken into consideration in management strategies for disaster risk reduction involving climate and weather extremes (The Copenhagen Diagnosis, 2009). The following three sections provide more details on the changes in extreme temperature, precipitation, and monsoons. The ending addresses the link between climate extremes and flooding.

2.2.1 Temperature

Temperature is associated with flood extremes. Temperature extremes often occur on weather time scales that require daily or higher time scale resolution data to accurately assess possible changes. However, paleoclimatic temperature reconstructions can offer further insight into long-term changes in the occurrence of temperature extremes and their impacts. Where instrumental data are used, it is important to distinguish between mean, maximum, and minimum temperatures, as well as between cold and warm extremes, due to their differing impacts on flood generation processes.

The latest IPCC report (2007) provides an extensive assessment of observed changes in temperature extremes. Global mean surface temperatures rose by 0.74 °C ± 0.18 °C over the 100-year period 1906–2005. The rate of warming over the 50-year period 1956–2005 is almost double that over the last 100 years (0.13 °C ± 0.03 °C vs. 0.07 °C ± 0.02 °C per decade). Trends over land are stronger than over the oceans. For the globe as a whole, surface air temperatures over land rose at about double the ocean rate after 1979 (more than 0.27 °C per decade vs. 0.13 °C per decade), with the greatest warming during winter (December to February) and spring (March to May) in the Northern Hemisphere.

Regarding changes in temperature extremes on a global scale, the IPCC (2007) reports an increase in the number of warm extremes and a reduction in the number of daily cold extremes in 70–75% of the land regions where data are available. The most marked changes are for cold nights, which have become rarer over the 1951–2003 period, whilst warm nights have become more frequent. From 1950 to 2004, the annual trends in minimum and maximum land surface air temperature averaged over regions with data were 0.20 °C per decade and 0.14 °C per decade, respectively. For 1979 to 2004, the corresponding linear trends for the land areas where data are available were 0.29 °C per decade for both maximum and minimum temperature.

Alexander *et al.* (2006) and Caesar *et al.* (2006) have brought many regional results together based on data for the period since 1951. According to their work, over 70% of the global land area sampled shows a significant decrease in the annual occurrence of cold nights; a significant increase in the annual occurrence of warm nights took place over 73% of the area. This implies a positive shift in the distribution of daily minimum temperature (T_{min}) throughout the globe. Changes in the occurrence of cold and warm days show warming as well, but generally less marked. This is consistent with T_{min} increasing more than maximum temperature (T_{max}). More recently, Meehl *et al.* (2009) found the ratio of the number of record daily maximum temperatures to record daily minimum temperatures averaged across the USA is now about 2 to 1, whereas in the 1960s the ratio was approximately 1 to 1.

In summary, regional and global analyses of temperature extremes nearly all show patterns consistent with a warming climate. Only a very few regions show changes in temperature extremes consistent with cooling, most notably the southeastern USA, which has a documented decrease in mean annual temperatures over the twentieth century. Research performed since the IPCC (2007) reinforces the conclusions that for the period since 1950 it is *very likely* that there has been a decrease in the number of both unusually cold days and nights, and an increase in the number of unusually warm days and nights on both a global and regional basis. Furthermore, based on a limited number of regional analyses and implicit from the documented changes in daily temperatures, it appears that warm spells, including heat waves defined in various ways, have *likely* increased in frequency since the middle of the twentieth century in many regions.

Compared with studies on mean temperature, studies of the attribution of extreme temperature changes are limited. Regarding possible human influences on these changes in temperature extremes, the IPCC (2007) concludes that surface temperature extremes have *likely* been affected by anthropogenic forcing. This assessment is based on multiple lines of evidence of temperature extremes at the global scale including an increase in the number of warm extremes, and a reduction in the number of cold extremes. There is also evidence that anthropogenic forcing may have significantly increased the likelihood of regional heat waves (Alexander *et al.*, 2006).

Detection studies of external influences on extreme temperature changes at the regional scale are also very limited. Regional trends in temperature extremes could be related to regional processes and forcings that have been a challenge for climate model simulations. Results from two global coupled climate models with separate anthropogenic and natural forcing runs indicate that the observed changes are simulated with anthropogenic forcings, but not with natural forcings (even though there are some differences in the details of the forcings).

Recent work done in Canada by Zwiers *et al.* (2011) compared observed annual temperature extremes, including annual maximum daily maximum and minimum temperatures and annual minimum daily maximum and minimum temperatures, with those simulated responses to anthropogenic (ANT) forcing or anthropogenic and natural external forcings combined (ALL) by multiple Global Climate Models (GCMs). They fitted generalized extreme value (GEV) distributions to the observed extreme temperatures with a time-evolving pattern of location parameters as obtained from the model simulation, and found that both ANT and ALL influence can be detected in all the extreme temperature variables at the global scale over the land, and also regionally over many large land areas. They concluded that the influence of anthropogenic forcing has had a detectable influence on extreme temperatures that have impacts on human society and natural systems at global and regional scales. External influence is estimated to have resulted in large changes in the likelihood of extreme annual maximum and minimum daily temperatures. Globally, waiting times for events that were expected to recur once every 20 years in the 1960s are now estimated to exceed 30 years for extreme annual minimum daily maximum temperature and 35 years for extreme annual minimum daily minimum temperature, and to have decreased to less than 10 or 15 years for annual maximum daily minimum and daily maximum temperatures respectively.

Regarding projections of extreme temperatures, the IPCC (2007) states that is *very likely* that heat waves will be more intense, more frequent, and longer lasting in a future warmer climate. Cold episodes are projected to decrease significantly in a future warmer climate. Almost everywhere, daily minimum temperatures are projected to increase faster than daily maximum temperatures, leading to a decrease in diurnal temperature range. Decreases in frost days are projected to occur almost everywhere in the middle and high latitudes. Temperature extremes were the type of extremes projected to change with most confidence in the IPCC (2007). If changes in temperature extremes scale with changes in mean temperature (i.e., simple shifts of the probability density function (PDF)), we can infer that it is *virtually certain* that hot (cold) extremes will increase (decrease) in the coming decades (if these extremes are defined with respect to the 1960–1990 climate).

2.2.2 Precipitation

Because climates are so diverse across different parts of the world, it is difficult to provide a single definition of "extreme precipitation." In general, three different methods have been used to define extreme precipitation: (i) relative thresholds, i.e., percentiles; (ii) absolute thresholds; or (iii) return values. As an example of the first case, a daily precipitation event with an amount greater than the 90th percentile of daily precipitation for all wet days within a 30-year period can be considered as extreme. Regarding the second type of definition, a precipitation amount that exceeds predetermined thresholds above which damage may occur can also be considered as an extreme. For example, 50 mm or 100 mm per day of rain in Canada is considered as extreme. A drawback of this definition is that such an event may not occur everywhere, and the damage for the same amount of rain in different regions may be quite different. Engineering practice often uses return values associated with a predetermined level of probability for exceedance as design values, estimated from annual maximum one-day or multi-day precipitation amounts over many years. Return values, similarly to relative thresholds, are defined for a given time period and region and may change over time. Climate models are important tools for understanding past changes in precipitation and projecting future changes. However, the direct use of output from climate models is often inadequate for studies of attribution changes of precipitation in general and of extreme precipitation in particular. The most important reason is related to the fact that precipitation is often localized, with very high variability across space, and extreme precipitation events are often of very short duration. The spatial and temporal resolutions of climate models are not fine enough to well represent processes and phenomena that are relevant to precipitation. Some of these modeling

shortcomings can be partly addressed with (dynamical and/or statistical) downscaling approaches (Teegavarapu, 2012; Mujumdar and Kumar, 2012). In addition, in some parts of the world, precipitation extremes are poorly monitored by very sparse network systems, resulting in high uncertainty in precipitation estimates, especially for extreme precipitation that is localized in space and of short duration, and thus limiting possibilities to thoroughly validate modeling and downscaling approaches.

Based on station data from Canada, the USA, and Mexico, Peterson *et al.* (2008) suggest that heavy precipitation was increasing over the period 1950–2004, as well as the average amount of precipitation falling on days with precipitation. The largest trends towards increased annual total precipitation, number of rainy days, and intense precipitation (e.g., fraction of precipitation derived from events in excess of the 90th percentile value) are focused on the central plains/northwestern Midwest.

Positive trends in extreme rainfall events are evident in southeastern South America, north central Argentina, northwest Peru and Ecuador (Marengo *et al.*, 2009). In the State of Sao Paulo, Brazil, there is evidence for an increase in magnitude and frequency of extreme precipitation events over 1950–1999 and 1933–2005. Increases in the annual frequencies in spatially coherent areas over the La Plata basin for both heavy and all (>0.1 mm) precipitation events during summer, autumn, and spring of 1950–2000 are reported too.

A number of recent regional studies have been completed for European countries. According to Moberg *et al.* (2006), averaged over 121 European stations north of 40° N, trends in 90th, 95th, and 98th percentiles of daily winter precipitation, as well as winter precipitation totals, have increased significantly over the 1901–2000 period. No overall long-term trend was observed in summer precipitation totals. In the United Kingdom (UK) there are widespread shifts towards greater contributions from heavier precipitation categories in winter, spring, and (to a lesser extent) autumn, and towards light and moderate categories in summer during 1961–2006. Extreme rainfall during 1961–2006 has increased up to 20% relative to the 1911–1960 period in the northwest of the UK and in parts of East Anglia. Similar seasonally dependent changes of precipitation extremes are reported over Germany during 1950–2004. Opposite trends occur in spring and the changes are spatially least coherent and insignificant in autumn. In contrast, in Emilia-Romagna, a region of northern Italy, the frequency of intense to extreme events decreases during winter, but increases during summer over the central mountains, while the number of rainy days decreased in summer during 1951–2004.

Several recent studies have focused on Africa and, in general, have not found significant trends in extreme precipitation. However, data coverage for large parts of central Africa is poor.

Observations at 143 weather stations in ten Asia-Pacific Network countries during 1955–2007 did not indicate systematic, regional trends in the frequency and duration of extreme precipitation events (Choi *et al.*, 2009). However, other studies have suggested significant trends in extreme precipitation at sub-regional scales in the Asia-Pacific region. Significant rising trends in extreme rainfall over the Indian region have been noted, especially during the monsoon seasons. Significant increases over the period 1951–2000 in extreme precipitation in western China, in the mid–lower reaches of the Yangtze River, and in parts of the southwest and south China coastal area are documented too, but a significant decrease in extremes is observed in north China and the Sichuan Basin. The precipitation indices show insignificant increases on average. A significant decreasing trend in the monsoon precipitation in the northwestern Himalaya during 1866–2006 has been observed. During the summers of 1978–2002, positive trends for heavy (25–50 mm per day) and extreme (>50 mm per day) precipitation near the east coasts of east Asia and southeast Asia were observed, while negative trends were seen over southwest Asia, central China, and northeast Asia. Summer extreme precipitation over south China has increased significantly since the early 1990s.

As atmospheric moisture content increases with increases in global mean temperature, extreme precipitation is expected to increase as well and at a rate faster than changes in mean precipitation content. In some regions, extreme precipitation is projected to increase, even if mean precipitation is projected to decrease. The observed change in heavy precipitation appears to be consistent with the expected response to anthropogenic forcing but a direct cause-and-effect relationship between changes in external forcing and extreme precipitation had not been established at the time of the IPCC (2007) Fourth Assessment. As a result, the IPCC only concludes that it is *more likely than not* that anthropogenic influence has contributed to a global trend. New studies since the Fourth Assessment Report have provided further evidence to support the assessment that it is *more likely than not* that anthropogenic influence has contributed to a trend towards increases in the frequency of heavy precipitation events over the second half of the twentieth century in many regions. However, there is still not enough evidence to make a more confident assessment regarding the causes of observed changes in extreme precipitation.

Regarding projected changes in extreme precipitation, the IPCC (2007) concluded that it is *very likely* that heavy precipitation events, i.e., the frequency (or proportion of total rainfall from heavy falls) of heavy precipitation, will increase over most areas of the globe in the twenty-first century. The tendency for an increase in heavy daily precipitation events in many regions was found to include some regions in which the mean precipitation is projected to decrease.

2.2.3 Monsoons

Changes in monsoon-related extreme precipitation due to climate change are still not well understood, but extremes such as floods may occur more or less frequently in the monsoon regions as a consequence of climate change. Generally, however, precipitation is the most important variable for inhabitants of monsoon regions, but it is also a variable associated with larger uncertainties in climate simulations. The IPCC (2007) concluded that the current understanding of climate change in the monsoon regions remains one of considerable uncertainty with respect to circulation and precipitation. With few exceptions in some monsoon regions, this has not changed since The Copenhagen Diagnosis (2009).

The observed negative trend in global land monsoon rainfall is better reproduced by atmospheric models forced by observed historical data than by coupled models without explicit forcing by observed ocean temperatures. The trend is strongly linked to the warming trend over the central eastern Pacific and the western tropical Indian Ocean. An important aspect for global monsoon patterns is the seasonal reversal of the prevailing winds. Changes in regional monsoons are strongly influenced by changes in the states of dominant patterns of climate variability such as the El Niño–Southern Oscillation (ENSO), the Pacific Decadal Oscillation (PDO), the Northern Annular Mode (NAM), the Atlantic Multi-decadal Oscillation (AMO), and the Southern Annular Mode (SAM) (Teegavarapu, 2012). However, it is not always clear how those modes may have changed in response to external forcing. Additionally, model-based evidence has suggested that land surface processes and land use changes could in some instances significantly impact regional monsoons.

The IPCC (2007) concluded that there "is a tendency for monsoonal circulations to result in increased precipitation due to enhanced moisture convergence, despite a tendency towards weakening of the monsoonal flows themselves. However, many aspects of tropical climatic responses remain uncertain." Work since the IPCC (2007) has not substantially changed these conclusions. As global warming is projected to lead to faster warming over land than over the oceans, the continental-scale land–sea thermal contrast, a major factor affecting monsoon circulations, may become stronger in summer and weaker in winter. Based on this hypothesis, a simple scenario is that the summer monsoon will be stronger and the winter monsoon will be weaker in the future than at present.

2.2.4 Flooding and climate change

The main causes of floods are intense and/or long-lasting precipitation, snow/ice melt, a combination of these types, dam break (e.g., glacial lakes), reduced conveyance due to ice jams or landslides, or a local intense storm (see Chapter 1). Climate-related floods depend on precipitation intensity, volume, duration, timing, phase (rain or snow), antecedent conditions of rivers and their drainage basins (e.g., presence of snow and ice, soil character and status, wetness, rate and timing of snow/ice melt, urbanization, existence of dykes, dams, and/or reservoirs). This section focuses on the spatial, temporal, and seasonal changes in high flows and peak discharge in rivers related to climate change.

The IPCC (2007) concluded that no gage-based evidence had been found for a climate-related trend in the magnitude and frequency of floods during the last decades. However, it was noted that flood damage was increasing (IPCC, 2007) and that an increase in heavy precipitation events was already *likely* in the late twentieth century trend. The IPCC Fourth Assessment Report concluded that abundant evidence was found for an earlier occurrence of spring peak river flows in snow-dominated regions. Research subsequent to the IPCC (2007) still does not show clear and widespread evidence of observed changes in flooding at the global level based on instrumental records, except for the earlier spring flow in snow-dominated regions.

Worldwide instrumental records of floods at gage stations are limited in spatial coverage and in time, and only a limited number of gage stations span more than 50 years, and even fewer over 100 years. Earlier flood data sources can be obtained from documentary records (archival reports, in Europe continuous over the last 500 years), and from proxy geological indicators known as paleofloods (sedimentary and biological records over centuries to millennia scales). Analysis of these long-past flood records have revealed:

- flood magnitude and frequency are very sensitive to small alterations in atmospheric circulation, with greater sensitivity for the largest "rare" floods (50-year flood and higher) than the smaller frequent floods (2-year floods);
- high inter-annual and inter-decadal variability is found in flood occurrences both in terms of frequency and magnitude, although in most cases cyclic or clusters of flood occurrence are observed in instrumental, historical, and proxy paleoflood records; and
- past flood records may contain analogues of unusual large floods, as large as those recorded recently, which are sometimes claimed to be the largest on record. However, the currently available pre-instrumental flood data are also limited.

OBSERVED CHANGES

Although flood trends might be seen in the north polar region and in northern regions where temperature change affects snowmelt or ice cover, widespread evidence of this (except for earlier spring flow) is not found. Cunderlik and Ouarda (2009) reported that snowmelt spring floods come significantly earlier in the southern part of Canada, and one fifth of all the analyzed stations show significant negative trends in the magnitude of snowmelt floods over the last three decades. On the other hand, there is no evidence of widespread common trends in the magnitude of extreme floods based on the daily river discharge of 139 Russian gage stations for

the past few to several decades, while a significant shift to earlier spring discharge is found as well (Shiklomanov *et al.*, 2007).

In Europe, significant upward trends in the magnitude and frequency of floods were detected in a considerable fraction of river basins in Germany for the period 1951–2002, particularly for winter floods (Petrow and Merz, 2009). Similar results are found for the Swiss Alps. In contrast, a slight decrease in winter floods and no change in summer maximum flow were reported in east and northeast Germany and in the Czech Republic (Elbe and Oder rivers). In France there is no evidence of a widespread trend in annual flow maxima over the past four decades, although there is evidence of a decreasing flood frequency trend in the Pyrenees, and increasing annual flow maxima in the northeast region. In Spain, southern Atlantic catchments showed a downward trend in flood magnitude and frequency, whereas in central and northern Atlantic basins no significant trend in frequency and magnitude of large floods was observed. Flood records from a network of catchments in the UK showed significant positive trends over the past four decades in high flow indicators primarily in maritime-influenced, upland catchments in the north and west of the UK (Hannaford and Marsh, 2008). In spite of the relatively abundant studies for rivers in Europe, a continental-scale assessment for Europe is difficult to obtain because geographically organized patterns are not seen.

The number of analyses for rivers in other parts of the world based on stream gage records is very limited. Some examples in Asia are as follows: annual flood maxima of the lower Yangtze region showed an upward trend over the last 40 years (Jiang *et al.*, 2008); an increasing likelihood of extreme floods during the last half of the century was found for the Mekong River (Delgado *et al.*, 2009); and both upward and downward trends were detected over the past four decades in four selected river basins of the northwestern Himalaya (Bhutiyani *et al.*, 2008). In the Amazon region in South America, the July 2009 flood set record highs in the 106 years of data for the Rio Negro at the Manaus gage site. However, such analyses cover only limited parts of the world. Evidence in the scientific literature from other parts of the world, and for other river basins, appears to be very limited.

In summary, except for the abundant evidence for an earlier occurrence of spring peak river flows in snow-dominated regions (*likely*), no clear and widespread observed evidence is found in the IPCC (2007) and subsequent research (The Copenhagen Diagnosis, 2009). In addition, instrumental records of floods at gage stations are limited in spatial coverage and in time, which limits the number of analyses. Pre-instrumental flood data can provide information for a longer period, but these data are also limited.

CAUSES OF CHANGES

Floods are affected by various characteristics of precipitation, such as intensity, duration, amount, timing, and phase (rain or snow). They are also affected by drainage basin conditions such as water levels in the rivers, presence of snow and ice, soil character and status (frozen or not, saturated or unsaturated), wetness (soil moisture), rate and timing of snow and ice melt, urbanization, and existence of dykes, dams, and reservoirs (see Chapter 1). A change in the climate physically changes many of these factors affecting floods and thus may consequently change the characteristics of floods. Engineering infrastructure such as dykes and reservoirs regulates flow, and land use may also affect floods. Therefore the assessment of causes of changes in floods is complex.

Many river systems are no longer in their natural state, making it difficult to separate changes in the streamflow data that are caused by changes in climate from those caused by regulation of the river systems. River engineering and land use alter flood probability. Many dams have a function to reduce the impacts of flooding. On the other hand, engineering infrastructure can contribute to flooding. For example, the largest and most pervasive contributors to increased flooding on the Mississippi River system over the past 100–150 years were wing dykes and navigation structures, followed by progressive levee construction (Pinter et al., 2008).

The possible causes for changes in floods were assessed in the IPCC Fourth Assessment Report (2007). A cause-and-effect relationship between external forcing and changes in floods has not been established. However, anthropogenic influence has been detected in the environments that affect floods, such as aspects of the hydrologic cycle including precipitation and atmospheric moisture. Anthropogenic influence is also clearly detected in streamflow regimes.

In climates where seasonal snow storage and melting plays a significant role in annual runoff, the hydrologic regime is directly affected by changes in temperature. In a warmer world, a smaller portion of precipitation will fall as snow (Hirabayashi et al., 2008) and the melting of winter snow will occur earlier in spring, resulting in a shift in peak river runoff to winter and early spring. This has been observed in the western USA and in Canada, along with an earlier breakup of river ice in Russian Arctic rivers. It is not clear yet if greenhouse gas emissions have affected the magnitude of the snowmelt flood peak, but projected warming may result in an increase in the spring river discharge where winter snow depth increases or a decrease in spring flood peak (Hirabayashi et al., 2008; Dankers and Feyen, 2009). There is still a lack of work identifying an influence of anthropogenic warming on peak streamflow for regions with little or no snowfall because of uncertainty in the observed streamflow data. However, evidence has emerged that anthropogenic forcing may have influenced the likelihood of a rainfall-dominated flood event in the UK. Additionally, it has been projected for many rain-dominated catchments that flow seasonality will increase, with higher flows in the peak flow season but little change in the timing of the peak or low flows. More recent hydrologic studies also show an increase in the probability of flooding due to a projected rainfall increase in rain-dominated catchments (e.g., humid Asia) where short-term extreme

precipitation and long-term precipitation are both projected to increase (see Section 2.2.2; Hirabayashi et al., 2008).

The IPCC (2007) in summary states that it is *more likely than not* that anthropogenic forcing leading to enhanced greenhouse gas concentrations has affected floods because they have detectably influenced components of the hydrologic cycle such as mean precipitation, heavy precipitation (see Section 2.2.2), and snowpack. Floods are also projected to change in the future due to anthropogenic warming (see Section 2.2.1), but the magnitude and even the sign of this anthropogenic influence have not yet been detected in scientific literature, and the exact causes for regional changes in floods cannot be clearly ascertained. It is *likely* that anthropogenic influence has resulted in earlier spring flood peaks in snow-melting rivers; the observed earlier spring runoff is consistent with expected change under anthropogenic forcing. It should be noted that these two assessments are based on expert judgment.

FUTURE CHANGES
The number of studies that showed the projection of flood changes in rivers, especially at a regional or a continental scale, was limited when the IPCC Fourth Assessment Report (2007) was published. The number of studies is still limited. Recently, a few studies for Europe (Dankers and Feyen, 2008, 2009) and a few studies for the globe (Hirabayashi et al., 2008, 2009) have demonstrated changes in the frequency and/or magnitude of floods in the twenty-first century at a large scale using daily river discharge calculated from Regional Climate Model or GCM outputs and hydrologic models at a regional or a continental scale. For Europe, the most notable changes are projected to occur in northern and northeastern areas in the late twenty-first century, but the results are varied. Three studies (Dankers and Feyen, 2008, 2009; Hirabayashi et al., 2008) show a decrease in the probability of extreme floods, which generally corresponds to lower flood peaks, in northern and northeastern Europe because of a shorter snow season, while some earlier work shows an increase in floods in the same region. Changes in floods in central and western Europe are less prominent and with not much consistency seen between the studies. For other parts of the world, Hirabayashi et al. (2008) show an increase in the risk of floods in most humid Asian monsoon regions, tropical Africa, and tropical South America, which were implied in earlier work that used annual mean runoff changes obtained from a coarse resolution GCM. This projected change was also implied in earlier studies by the changes in precipitation in monsoon seasons.

In summary, the number of projections on flood changes is still limited at a regional and continental scale, and those projections often show some degree of uncertainty. Projections at a catchment/river-basin scale are also not abundant in the peer-reviewed scientific literature. In particular, projections for catchments except for Europe and North America are very rare. In addition, considerable uncertainty has remained in the projections of flood changes, especially regarding their magnitude and frequency. The exception is the robust projection of the earlier shift of spring peak discharge in snow-dominated regions. Therefore, it is currently difficult to make a statement on the confidence of flood change projections due to anthropogenically induced climate change, except for the robustly projected earlier shift of spring floods (*likely*), because of insufficient reliability of climate models and downscaling methods.

2.3 APPROACHES FOR DEALING WITH CLIMATE CHANGE

Climate change as a global challenge is being addressed through two distinct approaches: mitigation and adaptation. Since the First Assessment Report of the IPCC in 1991, a very intensive discussion has been going on regarding: (i) the values and impacts of these two approaches; (ii) the priorities in their implementation; (iii) the links between them and the impacts of climate change; and (iv) the links between the two. The following discussion will briefly review the two approaches and focus on flood risk management as a component of the climate change adaptation portfolio.

2.3.1 Mitigation

Climate change mitigation is action to decrease the intensity of radiative forcing in order to reduce the potential effects of global warming. Most often, climate change mitigation scenarios involve reductions in the concentrations of greenhouse gases, either by reducing their sources or by increasing their sinks.

Scientific consensus on global warming, together with the precautionary principle and the fear of abrupt climate change is leading to increased effort to develop new technologies and sciences and carefully manage others in an attempt to mitigate global warming. Most means of mitigation appear effective only for preventing further warming, not at reversing existing warming. The Stern Review (2007) identifies several ways of mitigating climate change, such as: (i) reducing demand for emissions-intensive goods and services; (ii) increasing efficiency gains; (iii) increasing use and development of low-carbon technologies; and (iv) reducing non-fossil fuel emissions.

The energy policy of the European Union, later accepted by the larger international community, has set a target of limiting the global temperature rise to 2 °C (3.6 °F) compared to preindustrial levels, of which 0.8 °C has already taken place and another 0.5–0.7 °C is already committed. The 2 °C rise is typically associated in climate models with a carbon dioxide-equivalent (CO_2-eq) concentration of 400–500 ppm by volume; the current level of CO_2 alone is just above 390 ppm (NOAA, 2010) by volume, and rising at 2 ppm annually. Hence, to avoid a very likely breach of the 2 °C target, CO_2 levels would have to be stabilized very soon. This is generally regarded as unlikely, based on the current programs in place.

At the core of most mitigation proposals is the reduction of greenhouse gas emissions through reducing energy waste and switching to cleaner energy sources. Frequently discussed energy conservation methods include increasing the fuel efficiency of vehicles (often through hybrid, plug-in hybrid, and electric cars and improving conventional automobiles), individual-lifestyle changes, and changing business practices. Newly developed technologies and currently available technologies including renewable energy (such as solar power, tidal and ocean energy, geothermal power, and wind power) and more controversially nuclear power and the use of carbon sinks, carbon credits, and taxation are aimed more precisely at countering continued greenhouse gas emissions. More radical proposals that may be grouped with mitigation include biosequestration of atmospheric CO_2 and geoengineering techniques ranging from carbon sequestration projects, such as CO_2 air capture, to solar radiation management schemes, such as the creation of stratospheric sulfur aerosols. The ever-increasing global population and the planned growth of national economies based on current technologies are counter-productive to most of these proposals.

The Fourth Assessment Report of the IPCC (2007) assesses options for mitigating climate change. It introduces important "new" concepts such as cost–benefit analysis and regional integration. There is a clear and strong correlation between the CO_2-equivalent concentrations (or radiative forcing) of the published studies and the CO_2-only concentrations by 2100, because CO_2 is the most important contributor to radiative forcing. Essentially, any specific concentration or radiative forcing target, from the lowest to the highest, requires emissions to eventually fall to very low levels as the removal processes of the ocean and terrestrial systems saturate. To reach a given stabilization target, emissions must ultimately be reduced well below current levels and mitigation efforts over the next two or three decades will have a large impact on opportunities to achieve lower stabilization levels (*high agreement, much evidence*).

There are different measures for reporting costs of mitigation emission reductions, although most models report them in macroeconomic indicators, particularly gross domestic product (GDP) losses. The GDP is a measure of a country's overall economic output. It is the market value of all final goods and services made within the borders of a country in a year. It is often positively correlated with the standard of living. For stabilization at 4–5 W/m^2 (or ~590–710 ppmv CO_2-eq) macroeconomic costs range from −1 to 2% of GDP below baseline in 2050. For a more stringent target of 3.5–4.0 W/m^2 (~535–590 ppmv CO_2-eq) the costs range from slightly negative to 4% GDP loss (*high agreement, much evidence*). GDP losses in the lowest stabilization scenarios in the literature (445–535 ppmv CO_2-eq) are generally below 5.5% by 2050 (*high agreement, medium evidence*).

The risk of climate feedbacks is generally not included in the IPCC analysis. Feedbacks between the carbon cycle and climate change affect the required mitigation for a particular stabilization level of atmospheric CO_2 concentration. These feedbacks are expected to increase the fraction of anthropogenic emissions that remains in the atmosphere as the climate system warms. Therefore, the emission reductions to meet a particular stabilization level reported in the mitigation studies assessed here might be underestimated.

The economic potentials for mitigation at different costs have been reviewed for 2030 by the IPCC (2007). The review confirms that there are substantial opportunities for mitigation levels of about 6 Gt CO_2-eq involving net benefits (costs less than 0), with a large share being located in the buildings sector. Additional potentials are 7 Gt CO_2-eq at a unit cost (carbon price) of less than 20 US$/t CO_2-eq, with the total, low-cost, potential being in the range of 9 to 18 Gt CO_2-eq. The total range is estimated to be 13 to 26 Gt CO_2-eq, at a cost of less than 50 US$/t CO_2-eq and 16 to 31 Gt CO_2-eq at a cost of less than 100 US$/t CO_2-eq (370 US$/t C-eq) (*medium agreement, medium evidence*). No one sector or technology can address the entire mitigation challenge. This suggests that a diversified portfolio is required based on a variety of criteria.

2.3.2 Adaptation

Adaptation to climate change consists of initiatives and measures to reduce the vulnerability of natural and human systems to actual or expected climate change effects. The capacity and potential for human systems to adapt (called adaptive capacity) is unevenly distributed across different regions and populations (IPCC, 2007). Adaptive capacity is closely linked to social and economic development. The economic costs of adaptation are potentially large, but also largely unknown. Across the literature, there is wide agreement that adaptation will be more difficult for larger magnitudes and higher rates of climate change.

Because of the current and projected climate disruption precipitated by high levels of greenhouse gas emissions by the industrialized nations, adaptation is a necessary strategy at all scales to complement climate change mitigation efforts because we cannot be sure that all climate change can be mitigated. Indeed, the odds are quite high that in the long run more warming is inevitable, given the high level of greenhouse gases in the atmosphere, and the (several decade) delay between emissions and impact.

Adaptation has the potential to reduce adverse impacts of climate change and to enhance beneficial impacts, but will incur costs and will not prevent all damage. Extremes, variability, and rates of change are all key features in addressing vulnerability and adaptation to climate change, not simply changes in average climate conditions.

Human and natural systems will, to some degree, adapt autonomously to climate change. Planned adaptation can supplement autonomous adaptation, though there are more options and greater possibility for offering incentives in the case of adaptation of human systems than in the case of adaptation to protect natural systems.

SYSTEMS VIEW OF ADAPTATION

Adaptation can be defined as adjustments of a system to reduce vulnerability and to increase the resilience of a system to change, in this case in the climate system. Adaptation occurs at a range of interlinking scales, and can either occur in anticipation of change (anticipatory adaptation) or be a response to those changes (reactive adaptation). Most adaptation being implemented at present is responding to current climate trends and variability.

Adaptive capacity and vulnerability are important concepts for understanding adaptation. Vulnerability can be defined as the context in which adaptation takes place, and adaptive capacity is the ability or potential of a system to respond successfully to climate variability and change, in order to reduce adverse impacts and take advantage of new opportunities. High adaptive capacity does not necessarily translate into successful adaptation.

Adaptive capacity is driven by factors operating at many different scales, and it is important to understand the ways in which the different drivers of adaptive capacity interact. Physical constraints are important, but in most cases it is social processes that increase or decrease adaptive capacity. It is widely accepted that adaptive capacity is a social construct. The social drivers of adaptive capacity are varied but may include broad structures such as economic and political processes, as well as local structures such as access to decision-making and the structure of social networks and relationships within a community.

The social construction of adaptive capacity is very important when thinking about the risks and impacts of a changing climate. It is not just the change in climate that will affect vulnerability and livelihoods, but the way in which these changes propagate through complex social systems. Adaptation can be seen as a social and institutional process that involves reflecting on and responding to current trends and projected changes in climate.

Adaptation processes change with temporal and spatial scales. Much adaptation takes place in relation to short-term climate variability; however, this may cause maladaptation to longer-term climatic trends. It is clear from history that people have always adapted to a changing climate and that coping strategies already exist in many communities.

METHODS OF ADAPTATION

The following criteria may be of value to policy-makers in assessing adaptation strategies for dealing with global warming:

- *Economic efficiency*: Will the initiative yield benefits substantially greater than if the resources were applied elsewhere?

- *Flexibility*: Is the strategy reasonable for the entire range of possible changes in temperature, precipitation, and sea level?
- *Urgency*: Would the strategy be successful if implementation were delayed 10 or 20 years?
- *Low cost*: Does the strategy require minimal resources?
- *Equity*: Does the strategy unfairly benefit some at the expense of other regions, generations, or economic classes?
- *Institutional feasibility*: Is the strategy acceptable to the public? Can it be implemented with existing institutions under existing laws?
- *Unique or critical resources*: Would the strategy decrease the risk of losing unique environmental or cultural resources?
- *Health and safety*: Would the proposed strategy increase or decrease the risk of disease or injury?
- *Consistency*: Does the strategy support other national state, community, or private goals?
- *Private versus public sector*: Does the strategy minimize governmental interference with decisions best made by the private sector?

In the implementation of these criteria when designing adaptation policy, Scheraga and Grambsch (1998) suggest use of the following fundamental principles: (i) the effects of climate change vary by region; (ii) the effects of climate change may vary across demographic groups; (iii) climate change poses both risks and opportunities; (iv) the effects of climate change must be considered in the context of multiple stressors and factors, which may be as important to the design of adaptive responses as the sensitivity of the change; (v) adaptation comes at a cost; (vi) adaptive responses vary in effectiveness, as demonstrated by current efforts to cope with climate variability; (vii) the systemic nature of climate impacts complicates the development of adaptation policy; (viii) maladaptation can result in negative effects that are as serious as the climate-induced effects that are being avoided; and (ix) many opportunities for adaptation make sense whether or not the effects of climate change are realized.

Examples of adaptation include defending against floods through better flood defenses, and changing patterns of land use, such as avoiding more vulnerable areas for housing.

2.4 CONCLUSIONS

Climate change is one of the most significant challenges on the global scale. It directly affects flood-generating processes and flooding. Mitigation and adaptation are identified as the two approaches for dealing with climate change. The discussion on mitigation and adaptation portfolios has a global as well as a national/regional dimension. It should be recognized that mitigation and adaptation are very different regarding time frame and distribution of benefits. Both approaches can be related to sustainable development goals, but differ according to the direct benefits, which are global and long term for mitigation, while being local and shorter term for adaptation.

Adaptation can be both reactive (to experienced climate change) and proactive, while mitigation can only be proactive in relation to benefits from avoided climate change occurring over centuries. The IPCC (2007) also points out that there can be conflicts between adaptation and mitigation in relation to the implementation of specific national policy options. A common approach of many regional and national developing country studies on mitigation and adaptation policies has been to focus on the assessment of context-specific vulnerabilities to climate change.

Schipper and Burton (2008) suggest that research on adaptation should focus on assessing the social and economic determinants of vulnerability in a development context. The focus of the vulnerability assessment according to this framework should be on short-term impacts, i.e., should try to assess recent and future climate variability and extremes, economic and non-economic damage, and the distribution of these. Based on this, adaptation policies should be addressed as a coping strategy against vulnerability and potential obstacles. The role of various stakeholders and the public sector should also be considered.

Hydrological processes of relevance to flooding take place at a local scale and not at the global scale. Therefore, reaching general conclusions about flooding in future climates is loaded with various uncertainties. Many processes that play a role in flood generation are unresolved in models at the global level. At the river-basin scale, land use effects are more important than changes in the meteorological inputs of future climates. The general agreement of scientists is that the warming of the atmosphere will increase the capacity of the atmosphere to hold water, and this warming will also accelerate many of the processes involved in the redistribution of moisture in the atmosphere and will increase excess rainfall. However, making a pronouncement that all flooding will increase with changing climate is not possible at the current time.

Climate change studies agree that rainfall, and rainfall intensity, will change, indicating that the most prevalent flood-generating processes are likely to increase. Warming will also alter many other aspects of the global water cycle: increasing evaporation, changing land cover and soil properties, changing precipitation patterns and intensity and also affecting the processes involved in surface storage of water including snowpack generation, snowmelt, river ice breakup, and glacial melt. All these processes are active in flood generation and will result in changes in flooding patterns at the local scale and also in the spatial distribution of flooding processes on a larger sale as winters decrease in length and winter precipitation changes (Whitfield, 2012).

It is important to point out that the full range of possible climate outcomes, including flood impacts that remain highly uncertain, should be considered. Without taking these uncertain events into consideration, decision-makers will tend to be more willing to accept prospective future risks rather than attempt to avoid them through abatement. Flood management practice documents that, when faced with the risk of major damage, humans may make their judgment based on the consequences of the damage rather than on probabilities of events. The IPCC (2007) concludes that it is not clear that climate change flood impacts have a low probability; they are just very uncertain at present, and it is suggested that these uncertainties are taken into consideration in integrated assessment models, by adjusting the climate change flood damage estimates. The adjustments suggested include using historical data for estimating the losses of extreme floods, valuing ecosystem services, subjective probability assessments of monetary damage estimates, and the use of a discount rate that decreases over time in order to give high values to future generations. In this way the issues of climate change flood risk management have an element of decision-making under uncertainty, due to the complexity of the environmental and human systems and their interactions.

2.5 EXERCISES

2.1. Show, in graphical form, the impact of storm location on the shape of the flood hydrograph. Sketch the shape of the flood hydrograph corresponding to storm locations 1 and 2 (rainfall area shaded).

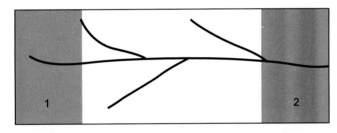

2.2. Figure below illustrates the two common urban hydrologic systems. For both systems (A and B) sketch the shapes of the natural and modified flood hydrographs.

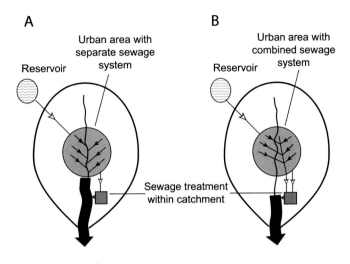

2.3. What are the most significant impacts of climate change in your country? In answering this question use as many examples as you can.

2.4. What are the flood-generating processes affected by climate change?

2.5. Describe in your own words the difference between climate change mitigation and adaptation?

2.6. Describe one major flood event in your country.

   a. What were the causes of the event?
   b. What were the main consequences?

2.7. List the potential adaptation strategies that may reduce the risk of flooding due to climate change and variability?

3 Risk management as adaptation to climate change

Several definitions of adaptation are available in the climate change literature. The following are some of the examples as summarized by Smit *et al.* (2000):

> Adaptation to climate is the process through which people reduce the adverse effects of climate on their health and well-being, and take advantage of the opportunities that their climatic environment provides.

> Adaptation involves adjustments to enhance the viability of social and economic activities and to reduce their vulnerability to climate, including its current variability and extreme events as well as longer-term climate change.

> The term adaptation means any adjustment, whether passive, reactive or anticipatory, that is proposed as a means for ameliorating the anticipated adverse consequences associated with climate change.

> Adaptation to climate change includes all adjustments in behavior or economic structure that reduce the vulnerability of society to changes in the climate system.

> Adaptability refers to the degree to which adjustments are possible in practices, processes or structures of systems in response to projected or actual changes of climate. Adaptation can be spontaneous or planned, and can be carried out in response to or in anticipation of change in conditions.

According to some of the typologies considered, adaptation can be planned or spontaneous; passive, reactive, or anticipatory, etc. According to the IPCC (2007), adaptation "has the potential to reduce adverse impacts of climate change and to enhance beneficial impacts, but will incur costs and will not prevent all damages."

The tendency of systems (e.g., natural, social, and engineering) to adapt is influenced by certain system characteristics. These include terms such as "sensitivity," "vulnerability," "resilience," "susceptibility," and "adaptive capacity," among others. The occurrence, as well as the nature of adaptations, is influenced by these. As Smit *et al.* (2000) point out, there is some overlap in the concepts captured by these terms. It may be argued that sensitivity, vulnerability, and adaptability capture the broad concepts.

Definitions of terms that describe system characteristics that are relevant to adaptation include:

Sensitivity: the degree to which a system is affected, either adversely or beneficially, by climate-related stimuli.

Vulnerability: the degree to which a system is susceptible to, or unable to cope with, adverse effects of climate change, including climate variability and extremes. It is a function of the character, magnitude, and rate of climate variation to which a system is exposed, its sensitivity, and its adaptive capacity.

Resilience: the degree to which a system rebounds, recoups, or recovers from a stimulus.

Responsiveness: the degree to which a system reacts to stimuli.

Adaptive capacity: the ability of a system to adjust to climate change (including climate variability and extremes) to moderate potential damage, to take advantage of opportunities, or to cope with the consequences.

Adaptability: the ability or capacity of a system to adapt to (to alter to better suit) climatic stimuli.

Adaptation is often the result of interactions between climatic and other factors. It varies not only with respect to its climatic stimuli but also with respect to other, non-climate conditions. It is important to highlight that the relationship between a changed climate system (e.g., higher temperatures, altered precipitation regime, etc.) and impacts on various systems is not necessarily linear. The role of adaptation (whether reactive or anticipatory, spontaneous or planned, etc.) is crucial for assessments of potential impacts of climate change.

Adaptation to climate change is a challenge that is complex and involves increasing risk. Efforts to manage these risks involve many decision-makers, conflicting values, competing objectives and methodologies, multiple alternative options, uncertain outcomes, and debatable probabilities (Noble *et al.*, 2005). Adaptation occurs at multiple levels in a complex decision environment, and is generally evaluated as better–worse, not right–wrong, based on multiple criteria. Identifying the best adaptation response is difficult.

Risk management techniques help to overcome these problems. Risk management offers a decision-making framework that assists in the selection of optimal strategies (according to various criteria) using a systems approach that has been well defined and generally accepted in public decision-making. In the context of adapting to climate change, the risk management process offers a framework for identifying, assessing, and prioritizing climate-related risks, and developing appropriate adaptation responses.

As a method for implementing adaptation to climate change, the risk management approach offers considerable benefits (Noble *et al.*, 2005).

- Vulnerability assessment is a central element of risk management. Vulnerability assessment is increasingly useful for guiding adaptation, since it helps reveal local- and larger-scale system vulnerabilities for which adaptation measures may be necessary to prevent serious adverse consequences. Future climate scenarios, based on the outputs of GCMs, will continue to provide valuable information, but the vulnerability-based approach is critical for helping identify specific risks and potential impacts that reflect the interests and values of people affected.

- Unlike "adapting," the concept of "managing risks" seems, from many perspectives, much more clear. Risk management is a familiar concept, especially in disaster management, whereas the notion of "adapting" is still poorly understood by many.

- Risk management provides a means for addressing uncertainties explicitly. Uncertainties exist in respect to uncertain future climate conditions and other aspects of climate change adaptation decision-making. Without a risk management view, decision-makers often receive uncertain responses to their question "what are we adapting to?"

- Risk management is relatively easy to apply in practice. In Canada, for example, many organizations have developed and accepted generic risk management procedures, and gained first-hand experiences in using risk management techniques (Canadian Standards Association, 1997 – reaffirmed in 2009). Increasingly, these are being applied to manage climate-related risks.

Climate change policy, strategy, and implementation already use the language and terminology of adaptation with increasing emphasis on the need for adaptation in the face of changing average climate and climate and weather extremes (Schipper and Burton, 2008). Increasing demand exists for assessment and promotion of flood (and other) disaster risk management practices that can contribute to climate change adaptation. This requires increasing synergy, merging, and complementarity between these two currently and still largely differentiated practices. The major aim of this book is to present a set of risk management tools as a practical approach to climate change adaptation as it relates to the risk of flooding.

Figure 3.1 The Venn diagram of integrated disaster management (after Simonovic, 2011).

3.1 FLOOD RISK MANAGEMENT DECISION PROCESS

The flood risk management community has developed key concepts and methods for managing, reducing, and transferring or sharing flood risk (GTZ, 2002; Sayers *et al.*, 2002; Kelman, 2003; GTZ, 2004; Levy and Hall, 2005; WMO, 2009). These concepts must evolve in order to take into account the ways changing climate may affect management schemes and challenges. This section presents the conceptual flood risk management framework as the basis for the management of flood risk due to climate change. Two main risk management approaches are presented in the following two sections – (i) a probabilistic approach and (ii) a fuzzy set approach – paving the path for the remaining chapters of the book.

Flood risk management, in its most general form, follows the disaster management cycle that has four closely integrated components: mitigation, preparedness, response, and recovery (Figure 3.1). Most traditional presentations of the integrated disaster management cycle involve graphing these four phases as independent components that follow each other. In this text, the Venn diagram presentation is introduced to capture the idea of integrated disaster management where each component of the cycle has some overlapping activities with other components (Simonovic, 2011).

In this book, flood risk management is viewed as a component of integrated disaster management as defined by Simonovic (2011). Managing, in everyday language, means handling or directing with a degree of skill. It also can be seen as exercising executive, administrative, and supervisory duties. Managing can be altering by manipulation or succeeding in accomplishing. Management is the act or art of managing.

> Integrated disaster management is an iterative process of decision-making regarding prevention of, response to, and recovery from a disaster (Simonovic, 2011).

This process provides a chance for those affected by a disaster to balance their diverse needs for protecting lives, property, and the environment, and to consider how their cumulative disaster-related actions may affect the long-term sustainability of the affected region. The guiding principles of the process are systems view, partnerships, uncertainty, geographic focus, and reliance on strong science and reliable data.

It is well documented in the flood risk management literature (GTZ, 2002; Sayers *et al.*, 2002; Kelman, 2003; GTZ, 2004; Levy and Hall, 2005; WMO, 2009) that many factors influence implementation of integrated disaster management activities that will minimize loss of life, injury, and/or material damage. It seems that governments, businesses, organizations, and individuals do not implement large-scale disaster management decisions that would enable them to avoid long-term losses from hazards (Simonovic, 2011). Making integrated disaster management a reality requires overcoming many social, political, and financial obstacles, as well as changing many human behaviors. One of the most important factors that influences flood risk management (as a component of integrated disaster management) is the decision-making process itself.

Decisions to (i) make an initial commitment of resources to taking precautionary measures and (ii) continue allocation of resources to follow through on them can be made at personal, organizational, or government levels. The decision-making process differs with the different level of decision-making. In the most general way, *decision-making* is defined as *the process of making an informed choice among the alternative actions that are possible*. The process may involve establishing objectives, gathering relevant information, identifying alternatives, setting criteria for the decision, and selecting the best option.

Flood risk management as a component of integrated disaster management involves complex interactions within and between the natural environment, human population (actions, reactions, and perceptions), and built environment (type and location). A different thinking is required to address the complexity of flood risk management. Mileti (1999, page 26) strongly suggested adaptation of a global systems perspective.

Systems theory is based on the definition of a system – in the most general sense as a collection of various structural and nonstructural elements that are connected and organized in such a way as to achieve some specific objective through the control and distribution of material resources, energy, and information (see Section 1.5). The basic idea is that all complex entities (biological, social, ecological, or other) are composed of different elements linked by strong interactions but a system is greater than the sum of its parts. This is a different view from the traditional analytical scientific model based on the law of additivity of elementary properties that views the whole as equal to the sum of its parts. Because complex systems do not follow the law of additivity, they must be studied differently.

A systemic approach to problems focuses on interactions among the elements of a system and on the effects of these interactions. Systems theory recognizes multiple and interrelated causal factors, emphasizes the dynamic character of processes involved, and is particularly interested in a system change with time – be it a flood, flood evacuation, or a flood disaster-affected community. The traditional view is typically linear and assumes only one, linear, cause-and-effect relationship at a particular time. A systems approach allows a wider variety of factors and interactions to be taken into account.

Using a systems view, Simonovic (2011, page 49) states that disaster losses are the result of interaction among three systems and their many subsystems:

- Earth's physical systems (the atmosphere, biosphere, cryosphere, hydrosphere, and lithosphere);
- human systems (e.g., population, culture, technology, social class, economics, and politics); and
- constructed systems (e.g., buildings, roads, bridges, public infrastructure, housing).

All of the systems and subsystems are dynamic and involve constant interactions between and among subsystems and systems. All human and constructed systems and some physical ones affected by humans are becoming more complex with time. This complexity is what makes national and international flood disaster problems difficult to solve. The increase in the size and complexity of the various systems is what causes increasing susceptibility to disaster losses. Climate change, changes in size and characteristics of the population, and changes in the constructed environment interact with changing physical systems to generate future exposure and define future disaster losses. The world is becoming increasingly complex and interconnected, helping to make flood disaster losses greater (Homer-Dixon, 2006).

3.1.1 Six-step procedure for flood risk management

The flood risk management process is a process for dealing with uncertainty within a public policy environment. It comprises a "systemic approach to setting the best course of action under uncertainty by identifying, understanding, acting on and communicating risk issues" (Simonovic, 2011) and by managing

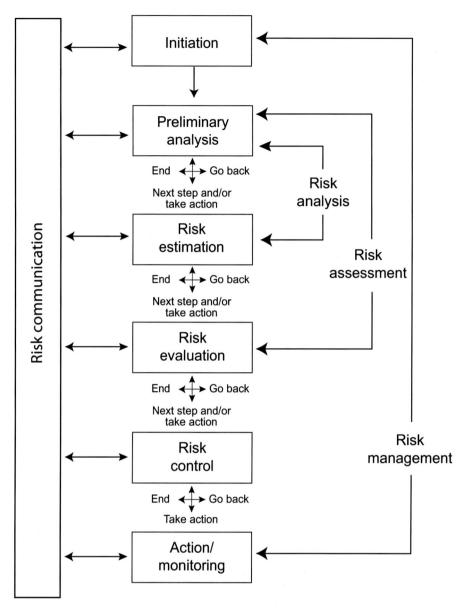

Figure 3.2 Steps in the process of decision-making under uncertainty – a model (modified after Canadian Standards Association, 1997, reaffirmed in 2009).

the flood risk in such a way that it is reduced to acceptable levels or accommodated by other actions.

Before the 1980s, risk management was an analysis tool used almost exclusively by financial institutions and the insurance industry. Since then, its use across engineering, health science, environmental science, and other disciplines has become increasingly common (Noble *et al.*, 2005). Risk management methodology is now of primary importance in deriving optimal decisions related to real or perceived impacts on human life, property, or the environment. Risk management has also become an important component of government policy analysis.

The generic risk management framework developed by the Canadian Standards Association (CSA) (1997, reaffirmed in

2009) is adopted for use in this textbook. The CSA Guideline, illustrated in Figure 3.2, lays out the general steps of the risk management process for the identification, analysis, evaluation, and control of risks and potential risks, including risks to health and safety. It offers a pragmatic and evolutionary approach to guide the development of strategies to avoid, reduce, control, or otherwise manage real and perceived risks. It also assists in setting priorities and balancing the effectiveness and costs of complex risk control strategies. Importantly, the process is iterative and allows for the inclusion of new information when it becomes available.

The center column prescribes a risk management process with standard steps and commitment to self-evaluation, not proceeding until a step has been satisfactorily accomplished. The left-hand bar prescribes continuing two-way interaction with the public. That interaction seeks to focus the process on public concerns and make its conclusions as credible as possible. Fischhoff and Morgan (2009) suggest that the Carnegie Mellon risk ranking approach could offer a scientifically sound approach to realizing this philosophy.

Each step of the process is separated by a decision diamond with three potential outcomes: end, go back, or next step and/or take action. The decision to end recognizes that the risk management decision process need not be continued further. This decision can be reached because: (i) concerns about the risk no longer exist; (ii) all associated risks are considered acceptable by stakeholders; (iii) the existing management options are judged to be sufficient and the residual risk is considered to be acceptable; or (iv) the source of hazard that generates the risk is eliminated.

The go back decision provides for the repetition of one (or more) previous steps in order to improve the accuracy and completeness of the risk information, to update the data or modify the assumptions. In this way, the process provides for continuous improvement.

The next step and/or take action decision involves proceeding to the next step. Immediate action may be necessary during emergency management where timing is critical. In such cases, the decision-making process may not involve all of the steps in a formal manner, but the steps are reviewed to determine that an emergency exists and that the particular actions are necessary.

STEP 1: INITIATION

The initiation step includes: (i) definition of the problem and associated risk issue(s); (ii) identification of the risk management team; (iii) assignment of responsibilities, authorities, and resources; and (iv) identification of potential stakeholders. The purpose of this step is to establish the administrative details of the process and link the process and the decision-maker in terms of time frame, reporting, resources for the process, etc.

Defining the problem or opportunity and the associated risk issues is often the most difficult step in the process and also the most important. The issues must be specifically defined, documented, and dealt with one at a time. There is often a need to prioritize issues. It may be beneficial to consult with stakeholders during this stage to validate or define the scope of the issues.

The role of the risk management team is to provide technical expertise and advice to the decision-maker. The team must have access to the knowledge bases needed to resolve issues. The team guides the process to its termination and must be prepared to change its composition as necessary throughout the decision process. It is necessary to provide the risk management team with both the authority to do its job and the resources necessary for accomplishing its objectives.

Stakeholders are defined in the process as anyone who can affect, is affected by, or believe they might be affected by, a decision or activity. In assembling the list of stakeholders it is important to think as broadly as possible and to include parties both inside and outside the problem domain.

STEP 2: PRELIMINARY ANALYSIS

The preliminary analysis includes: (i) definition of the scope of the decision; (ii) identification of hazards using risk scenarios; (iii) beginning of stakeholder analysis; and (iv) beginning of the development of a risk information library. The purpose of this step is to define the basic dimensions of the risk problem and then undertake an analysis and evaluation of potential risks. This step will identify one of the following courses of action: (i) a risk exists and action should be taken immediately; (ii) there is a need to undertake a more detailed analysis before taking any action; and (iii) the analysis should end, as it has been determined the risk is not an issue.

The decision-maker and risk management team determine the scope of the decisions to be made with due regard for the needs, issues, and concerns of the stakeholders. The decision-maker should anticipate that the problem or opportunity may extend beyond the initially determined scope of the decision framework and that new stakeholders may need to be included.

The flood hazards that generate risk include natural hazards, economic hazards (inflation, depression, changes in tax, etc.), technical hazards (system or equipment failure, water pollution, etc.), and human hazards (errors or omissions by poorly trained people, acts of sabotage or terrorism, etc.). Identification of hazards using risk scenarios starts by defining a risk scenario as a sequence of events with an associated frequency and consequence. The risk scenarios may be simple or quite complicated. There are various tools for identification of risk scenarios: failure mode and effect analysis, analysis of historical data, fault-tree analysis, event-tree analysis, hazard and operational studies, and others.

Stakeholder analysis provides the decision-maker with a profile of stakeholders. Through this process, stakeholders and their needs are determined together with risk issues and concerns. Stakeholder analysis also provides the risk management team with the context for a risk communication strategy.

At this step, the structure of the risk information library is initiated. Through the risk management process, this structure is further developed as additional information is incorporated into the risk information library. The library includes information necessary for making risk management decisions.

STEP 3: RISK ESTIMATION

The risk estimation includes: (i) definition of methodology for estimating frequency and consequences; (ii) estimation of frequency of risk scenarios; (iii) estimation of consequences of risk scenarios; and (iv) further refining of stakeholder analysis. The purpose of this step is to estimate the frequency and consequences associated with each risk scenario selected for the analysis.

This step is quite technical and requires making decisions on the appropriateness of use of historical data, models, professional judgment, or a combination of methods. Explicit definition of applied methods is required to avoid conflict between technical experts and laypersons. The choice of method will reflect the accuracy needed, cost, available data, the level of expertise on the team, and the acceptability of the method to stakeholders. Validation of the methods should ensure that they are appropriate for the stated objectives, that the analysis can be reproduced, that the analysis is not sensitive to the way data or results are formatted and that all assumptions and uncertainties are acknowledged and documented.

The purpose of frequency analysis is to determine how often a particular scenario might be expected to occur over a specified period of time. These estimates often rely on available historical data, where judgment about the future is based on what has occurred in the past. In cases when the data are not available or the past cannot be used to predict the future, as in the case of climate change, other methods may be used.

Consequence analysis involves estimating the impact of various scenarios on everyone and everything affected by the hazard. Flood consequences are often measured in financial terms, but they can also be measured by other factors: numbers of injuries or deaths, numbers of wildlife affected, impact on quality of life, and others. The benefit of measuring consequences in financial terms is that it provides a common measure for comparing dissimilar conditions.

Needs, issues, and concerns of stakeholders are further refined in this step. Permanent communication with stakeholders helps ensure that the stakeholder analysis continues to present a true profile of the stakeholders. The stakeholder analysis has to be updated as new data become available.

STEP 4: RISK EVALUATION

The risk evaluation includes: (i) estimation and integration of benefits and costs; and (ii) assessment of acceptability of risk to stakeholders. The main objective of risk evaluation is to determine the acceptability of the risks, as estimated in the previous steps, in terms of needs, issues, and concerns of stakeholders. One of three conclusions will result from the risk evaluation exercise: (i) the risk associated with the hazard is acceptable at its current level; (ii) the risk associated with the hazard is unacceptable at any level; or (iii) the hazard might be acceptable but risk control measures should be evaluated.

Up to this step in the process, only the risk has been considered. Prior to continuing the process, the benefits from flooding and any operational costs (other than risk) should also be considered. This information is then integrated into the risk information library. Both benefits and costs are evaluated in terms of the needs, issues, and concerns of the stakeholders. It is important here to consider more than the hard, financial, benefits and costs associated with the flood hazard. So-called soft benefits and soft costs should also be considered. They may include, for example, the comfort people are afforded knowing that a trusted agency is managing the flood risk. Another example is the increased trust developed between decision-makers and other stakeholders as a result of an effective communication program. Greater trust means less anxiety on the part of stakeholders. An example of a soft cost could be a reduction in quality of life or adverse effect of flooding on lifestyle. There may be a number of indirect benefits and costs associated with flood hazard that may not be readily recognized – for example, employment opportunities after the flood or other spin-off benefits.

It should be noted that there are a number of factors, other than expected value of flood loss, that affect stakeholder acceptance of risk. In this context the perception of risk is an important issue that affects the flood risk management process (Slovic, 2000; Simonovic, 2011). A major part of the risk management confusion relates to an inadequate distinction between three fundamental types of risk: (i) objective risk (real, physical), R_o, and objective probability, p_o, which is the property of real physical systems; (ii) subjective risk, R_s, and subjective probability, p_s. Probability is here defined as the degree of belief in a statement. R_s and p_s are not properties of the physical systems under consideration (but may be some function of R_o and p_o); and (iii) perceived risk, R_p, which is related to an individual's feeling of fear in the face of an undesirable possible event, is not a property of the physical systems but is related to fear of the unknown. It may be a function of R_o, p_o, R_s, and p_s.

Because of the confusion between the concepts of objective and subjective risk, many characteristics of subjective risk are believed to be valid also for objective risk. Therefore, it is almost universally assumed that the imprecision of human judgment is equally prominent and destructive for all flood risk evaluations and all risk assessments. This is perhaps the most important misconception that blocks the way to more effective societal disaster risk management. The ways society manages disaster risks appear to be dominated by considerations of perceived and subjective risks, while it is objective risks that kill people, damage the environment, and create property loss (Simonovic, 2011).

STEP 5: RISK CONTROL

Risk control includes: (i) identifying feasible risk control options; (ii) evaluation of risk control options according to various criteria (effectiveness, cost, etc.); (iii) assessment of stakeholders' acceptance of proposed action(s); (iv) evaluation of options for dealing with residual risk; and (v) assessment of stakeholders' acceptance of residual risk. The main objective of risk control is to determine the effectiveness of flood risk control options before they have been applied. The costs, benefits, and risks associated with the proposed flood risk control measures are considered.

In the consideration of feasible risk control options, six broad strategies for controlling risk are available: (i) avoiding exposure, (ii) reducing the frequency of loss, (iii) reducing the consequence of loss, (iv) separation of exposure, (v) duplicating assets (including redundancy), and (vi) transferring the control of losses to some other party. Evaluation of risk control options should be done in terms of their effectiveness in reducing losses, the cost of implementation, and the impact of control options on other stakeholder objectives.

Before risk control strategies are decided upon, they should be communicated to the stakeholders in order to assess their acceptability. A proposed option may appear acceptable to the decision-maker, but may not be acceptable to other stakeholders. The complete evaluation should be done in terms of the needs, issues, and concerns of all affected stakeholders.

Flood risk that cannot be totally eliminated by risk control measures must be financed, that is, someone has to pay for the loss. At this stage two basic risk financing techniques are available: retention and transfer. The responsibility for paying for the loss may be retained by the responsible organization (in the case of flooding, different levels of government) or transferred to some other organization. An example of this type of transfer arrangement is the purchase of commercial insurance where available.

STEP 6: ACTION

This step includes: (i) development of an implementation plan; (ii) implementation of chosen control, financing, and communication strategies; (iii) evaluation of the effectiveness of the risk management decision process; and (iv) establishment of a monitoring process. The main objective of this step is to implement the measures resulting from the risk management process and establish the monitoring program.

Prior to implementation of any of the chosen flood risk control measures, it is important to develop an implementation plan. This plan should consider training requirements, staffing requirements, job shifting or new positions, and financing requirements.

Monitoring is a key function of the risk management program in order to detect and adapt to changing circumstances, to ensure that the risk control is achieved, to ensure proper implementation, and to verify the correctness of assumptions.

After having undergone the extensive decision process, it is prudent to evaluate the effectiveness of the flood risk management process in satisfying the objectives of the decision-maker. This facilitates continuous improvement in the decision process itself.

3.2 APPROACHES TO FLOOD RISK MANAGEMENT AS ADAPTATION TO CLIMATE CHANGE

All strategic activities of flood risk management and adaptation require the use of reliable methodologies that allow an adequate estimation and quantification of potential losses and consequences to the natural systems, human population, and built environment. There are a wide range of approaches for integrating data with the flood risk management process. Inductive approaches model risk through weighting and combining different hazard, vulnerability, and risk control variables. Deductive approaches are based on the modeling of historical patterns of materialized risk (i.e., disasters, or damage and loss that have already occurred). An obstacle to inductive modeling is the lack of accepted procedures for assigning values and weights to the different vulnerability and hazard factors that contribute to flood risk. Deductive modeling will not accurately reflect risk in contexts where disasters occur infrequently or where historical data are not available. In spite of this weakness, deductive modeling offers a shortcut to flood risk indexing and can be used to validate the results from inductive models.

In the area of adaptation to climate change, there are a large number of studies that deal with impact and adaptation assessment of climate change (Schipper and Burton, 2008). These have become increasingly sophisticated in the past few years, but few have been able to provide robust information for decision-makers and flood risk managers (Dessai and Hulme, 2003). According to Schipper and Burton (2008) this occurs because of: (i) the wide range of potential impacts of climate change – issue of uncertainty; (ii) the mismatch of resolution between GCMs and local adaptation measures – issue of scale; (iii) impact assessments not designed to consider a range of adaptation options; and (iv) adaptation considered as an assumption rather than explored as a process.

The majority of adaptation studies have taken a prediction-oriented top-down approach (Dessai and Hulme, 2003) that considers a range of scenarios of world development, whose greenhouse gas emissions serve as input to GCMs, and whose output serves as input to impact models. In the next section an example is presented that follows the top-down approach (Peck et al., 2010, 2011; Eum et al., 2011). Some studies do not consider adaptation, while others assume arbitrary adaptation.

The second group of adaptation studies includes adaptation based on observation or trying to model adaptation with a bottom-up approach. Anticipatory adaptation strategies are then considered within a certain decision-making framework based on the physical impacts of climate change on flooding being examined with some consideration for the context.

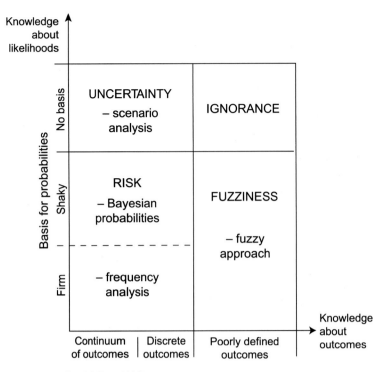

Figure 3.3 Risk, uncertainty and ignorance (after Stirling, 1998).

The flood risk management community in the context of integrated water resources management uses two approaches to deal with decision-making under uncertainty: a probabilistic approach and a fuzzy set approach (Simonovic, 2009, 2011). In flood risk management, probability is used based on historical records, for example, to determine the magnitude of the 100-year flood. This probability is called objective (classical, frequentist) because it is determined by long-run observations of the occurrence of an event. In contrast, climate change probabilities are subjective because they are based on the degree of belief that a person has that it will occur, given all the relevant information currently known to that person.

Stirling (1998) observed that in general risk assessment there is recognition of the intrinsic subjectivity of fundamental assumptions and the consequent necessity for active participation in analysis by all interested and affected parties. Despite this, there is considerable inertia in the implementation of these insights in formal policy-making and regulatory procedures. He stated that the issue seems often to be seen as a need for better "communication" and "management" than for better analysis, with attention devoted as much to the classification of different public perspectives as to techniques for direct stakeholder participation.

The newly emerging consensus on risk policy has very deep and robust theoretical roots, which reach right to the core of established methodologies. It represents basically coming to terms with two fundamental constraints that have been established, and at the same time neglected, for many years.

First, there is the problem of *ignorance* (see Figure 3.3). Put simply, it is a condition under which it is possible neither to resolve a discrete set of probabilities (or a density function) along a scale of outcomes (as is possible under *risk* proper – in the bottom left of the grid in Figure 3.3), nor even to define a comprehensive set of outcomes (as under *uncertainty* – in the top left of Figure 3.3). Ignorance, in this sense, lies in the top right hand corner of the grid in Figure 3.3. Ignorance arises from many familiar sources, including incomplete knowledge, contradictory information, conceptual imprecision, divergent frames of reference, and the intrinsic complexity of many natural and social processes. Under such circumstances, there often exists no basis for the many sophisticated techniques of probability theory. No matter how well informed, judgments concerning the extent to which "we don't know what we don't know" are subjective and value driven.

Second, it is impossible both democratically and consistently to aggregate individual preferences in a plural society (Arrow, 1963). The derivation of any single social preference ordering (or aggregate social welfare function) will violate at least one of a minimal set of economic conditions in the characterization of individual choice.

Stirling (1998) concluded that problems posed in risk management are as much a matter of analytical rigor as they are of policy legitimacy. Acknowledgement of the complexity and subjectivity of risk analysis need not be taken to imply the rejection of the clarity and rigor of quantitative techniques. If we cannot meaningfully speak of risk as a discrete scalar number, then we might see it rather as a vector with as many elements as there are relevant dimensions. Straightforward techniques such as simulation, optimization, and multi-objective analysis may go some way toward addressing these apparently intractable problems.

3.2.1 Probabilistic approach

Probability is a concept widely accepted and practiced in flood risk management. To perform operations associated with probability, it is necessary to use sets – collections of elements, each with some specific characteristics. Boolean algebra provides a means for evaluating sets. In probability theory, the elements that make up a set are outcomes of an experiment. Thus, the universal set Ω represents the mutually exclusive listing of all possible outcomes of the experiment, and is referred to as the *sample space* of the experiment. In examining the outcomes of rolling a dice, the sample space is $S = (1, 2, 3, 4, 5, 6)$. This sample space consists of six items (elements) or sample points. In probability concepts, a combination of several sample points is called an *event*. An event is, therefore, a subset of the sample space. For example, the event of "an odd outcome when rolling a dice" represents a subset containing sample points 1, 3, and 5.

Associated with any event E of a sample space S is a probability, shown by $\Pr(E)$ and obtained from the following equation:

$$\Pr(E) = \frac{m(E)}{m(S)} \tag{3.1}$$

where $m(.)$ denotes the number of elements in the set (.).

The probability of getting an odd number when tossing a dice is determined by using $m(odd\ outcomes) = 3$ and $m(sample\ space) = 6$. In this case, $\Pr(odd\ outcomes) = 3/6 = 0.5$. Note that Equation (3.1) represents a comparison of the relative size of the subset represented by the event E and the sample space S. This is true when all sample points are equally likely to be the outcome. When all sample points are not equally likely to be the outcome, the sample points may be weighted according to their relative frequency of occurrence over many trials or according to expert judgment. In disaster management practice we use three major conceptual interpretations of probability.

Classical interpretation of probability (equally likely concept) In this interpretation, the probability of an event E can be obtained from Equation (3.1), provided that the sample space contains N equally likely and different outcomes, i.e., $m(S) = N$, n of which have an outcome (event) E, i.e., $m(E) = n$. Thus $\Pr(E) = n/N$. This definition is often inadequate for flood risk management applications. For example, if failures of a pump to start in a flood-affected area are observed, it is unknown whether all failures are equally likely to occur. Nor is it clear whether the whole spectrum of possible events is observed. That case is not

similar to rolling a perfect dice, with each side having an equal probability of 1/6 at any time in the future.

Frequency interpretation of probability In this interpretation, the limitation on knowledge about the overall sample space is remedied by defining the probability as the limit of n/N as N becomes large. Therefore, $\Pr(E) = \lim_{N=\infty}(n/N)$. Thus, if we have observed 2000 starts of a pump in which 20 failed, and if we assume that 2000 is a large number, then the probability of the pump failing to start is $20/2000 = 0.01$. The frequency interpretation is the most widely used classical definition in flood risk management today. However, some argue that because it does not cover cases in which little or no experience (or evidence) is available, or cases where estimates concerning the observations are intuitive, a broader definition is required. This has led to the third interpretation of probability.

Subjective interpretation of probability In this interpretation, $\Pr(E)$ is a measure of the degree of belief one holds in a specified event E. To better understand this interpretation, consider the probability of improving an evacuation system by making a plan change. The manager believes that such a change will result in a performance improvement in one out of three evacuation missions in which the plan is used. It would be difficult to describe this problem through the first two interpretations. That is, the classical interpretation is inadequate since there is no reason to believe that performance is as likely to improve as to not improve. The frequency interpretation is not applicable because no historical data exist to show how often a plan change resulted in improving the evacuation. Thus, the subjective interpretation provides a broad definition of the probability concept.

3.2.2 A fuzzy set approach

One of the main goals of integrated flood disaster risk management is to ensure that flood protection performs satisfactorily under a wide range of possible future flooding conditions (Simonovic, 2011). This premise is particularly true of large and complex flood risk management systems. Flood risk management systems include people, infrastructure, and environment. These elements are interconnected in complicated networks across broad geographical regions. Each element is vulnerable to natural hazards or human error, whether unintentional, as in the case of operational errors and mistakes, or from intentional causes, such as a terrorist act.

The sources of uncertainty are many and diverse, as was discussed earlier, and as a result they provide a great challenge to integrated flood risk management. The goal to ensure failsafe protection system performance may be unattainable. Adopting high safety factors is one way to avoid the uncertainty of potential failures. However, making safety the first priority may render the system solution infeasible. Therefore, known uncertainty sources must be quantified.

The problem of flood system reliability has received considerable attention from statisticians and probability scientists. Probabilistic (stochastic) risk analysis has been used extensively to deal with the problem of uncertainty in flood risk management. A prior knowledge of the PDFs of both resistance and load, and their joint PDF, is a prerequisite of the probabilistic approach. In practice, data on previous flood disasters are usually insufficient to provide such information. Even if data are available to estimate these distributions, approximations are almost always necessary to calculate flood risk. Subjective judgment by a flood risk decision-maker in estimating the probability distribution of a random event – the subjective probability approach – is another approach to dealing with a lack of data. The third approach is Bayes's theory, where an expert judgment is integrated with observed information. The choice of a Bayesian approach or any subjective probability distribution presents real challenges. For instance, it is difficult to translate prior knowledge into a meaningful probability distribution, especially in the case of multi-parameter problems. In both subjective probability and Bayesian approaches, the degree of accuracy is strongly dependent on a realistic estimation of the decision-maker's judgment.

Until recently the probabilistic approach was the only approach for flood risk analysis. However, it fails to address the problem of uncertainty that goes along with climate change, human input, subjectivity, a lack of history and records. There is a real need to convert to new approaches that can compensate for the ambiguity or uncertainty of human perception.

Fuzzy set theory was intentionally developed to try to capture judgmental belief, or the uncertainty that is caused by a lack of knowledge. Relative to probability theory, it has some degree of freedom with respect to aggregation operators, types of fuzzy sets (membership functions), and so on, which enables it to be adapted to different contexts. During the past 40 years, fuzzy set theory and fuzzy logic have contributed successfully to technological development in different application areas such as mathematics, algorithms, standard models, and real-world problems of different kinds (Zadeh, 1965; Zimmermann, 1996). More recent disaster literature shows a slow introduction of fuzzy set theory in disaster management. Altay and Green (2006) report about 5% of research effort using fuzzy sets in their review. Among many examples, Chongfu (1996) provides an excellent introduction to the fuzzy risk definition of natural hazards with emphasis on urban hazards. More recently, Karimi and Hullermeier (2007) presented a system for assessing the risk of natural disasters, particularly under highly uncertain conditions, i.e., where neither the statistical data nor the physical knowledge required for a purely probabilistic risk analysis are sufficient. The theoretical foundation of this study is based on employing fuzzy set theory. The likelihood of natural

hazards is expressed by fuzzy probability. Moreover, uncertainties about the correlation of the parameters of hazard intensity, damage, and loss, i.e., vulnerability relations, have been considered by means of fuzzy relations. The composition of the fuzzy probability of hazard and the fuzzy vulnerability relation yields the fuzzy probability of damage (or loss). The system has been applied for assessing the earthquake risk in Istanbul metropolitan area. Akter *et al.* (2004) and Akter and Simonovic (2005) used a fuzzy set approach to multi-objective selection of flood protection measures in the Red River basin (Manitoba, Canada). They derived a methodology for integrating preferences of a large number of stakeholders through the use of fuzzy expected value. Ahmad and Simonovic (2007) and Simonovic and Ahmad (2007) presented an original procedure for spatial flood risk assessment using a fuzzy set approach. Three fuzzy risk indices – reliability, resilience, and robustness – were developed using fuzzy sets for flood risk mapping.

Shortly after fuzzy set theory was first developed in the late 1960s, there were a number of claims that fuzziness was nothing but probability in disguise. Probability and fuzziness are related, but they are different concepts. Fuzziness is a type of deterministic uncertainty. It describes the *event class ambiguity*. Fuzziness measures the *degree to which* an event occurs, not whether it occurs. At issue is whether the event class can be unambiguously distinguished from its opposite. Probability, in contrast, arises from the question of *whether or not* an event occurs. Moreover, it assumes that the event class is crisply defined and that the law of non-contradiction holds – that is, that for any property and for any definite subject, it is not the case both that the subject possesses that property and that the subject does not possess that property. Fuzziness occurs when the law of non-contradiction (and equivalently the law of excluded middle – for any property and for any individual, either that the individual possesses that property or that the individual does not possess that property) is violated. However, it seems more appropriate to investigate fuzzy probability for the latter case than to completely dismiss probability as a special case of fuzziness. In essence, whenever there is an experiment for which we are not capable of "computing" the outcome, a probabilistic approach may be used to estimate the likelihood of a possible outcome belonging to an event class. A fuzzy theory extends the traditional notion of a probability when there are outcomes that belong to several event classes at the same time, but to different degrees. Fuzziness and probability are orthogonal concepts that characterize different aspects of human experience. Hence, it is important to note that neither fuzziness nor probability governs physical processes in nature. These concepts were introduced by humans to compensate for our own limitations.

Let us review two examples that show a difference between fuzziness and probability. *Russell's paradox.* That the laws of non-contradiction and excluded middle can be violated was pointed out by Bertrand Russell with the tale of the barber.

Russell's barber is a bewhiskered man who lives in a town and shaves a man if and only if the man does not shave himself. The question is, who shaves the barber? If he shaves himself, then by definition he does not. But if he does not shave himself, then by definition he does. So he does and he does not. This is a contradiction or *paradox*. It has been shown that this paradoxical situation can be numerically resolved as follows. Let *S* be the proposition that the barber shaves himself and *not-S* the proposition that he does not. Since *S* implies *not-S* and vice versa, the two propositions are logically equivalent, i.e., $S = not\text{-}S$. Fuzzy set theory allows for an event class to coexist with its opposite at the same time, but to different degrees, or in the case of paradox to the same degree, which is different from zero or one.

Misleading similarities. There are many similarities between fuzziness and probability. The largest, but superficial and misleading, similarity is that both systems quantify uncertainty using numbers in the unit interval [0,1]. This means that both systems describe and quantify the uncertainty numerically. The structural similarity arising from lattice theory is that both systems algebraically manipulate sets and propositions associatively, commutatively, and distributively. These similarities are misleading because a key distinction comes from what the two systems are trying to model. Another distinction is in the idea of observation. Clearly, the two models possess different kinds of information: fuzzy memberships, which quantify similarities of objects to imprecisely defined properties; and probabilities, which provide information on expectations over a large number of experiments.

3.3 AN EXAMPLE: CLIMATE CHANGE-CAUSED FLOOD RISK TO MUNICIPAL INFRASTRUCTURE, CITY OF LONDON (ONTARIO, CANADA)

The following example illustrates the top-down approach for flood risk assessment to municipal infrastructure due to the climate change. It provides an original approach to incorporating climate change into flood risk analysis and combines elements of probabilistic and fuzzy set approaches. The study was conducted for the City of London (Peck *et al.*, 2011) as one of the first steps in the implementation of the municipal climate change adaptation strategy.

3.3.1 Background

The City of London is located in southwestern Ontario, Canada, within the Upper Thames River basin (Figure 3.4). The City is the 10th largest in Canada, with a population of approximately 352,400 and an area of 42,057 ha. The City is characterized by

the Thames River, which flows south through the city where the branches meet at a location locally known as the Forks. The city has a well-documented history of flooding dating back to the 1700s, with the worst recorded flood event occurring in 1937. This flood was destructive of both life and property; five deaths were recorded and over 1,100 homes experienced significant flood damage. Fanshawe Dam on the North branch of the Thames is used to control downstream flooding. The City of London has a high-density urban core located at the Forks, which is largely protected by a series of dykes.

This case study measures the impact of changing flood patterns within the city, due to climate change, on its municipal infrastructure (Peck *et al.*, 2011). Since infrastructure is designed to last for many years, and is based on codes that use historical data, the changing climate has the potential to have a serious impact on current and future infrastructure. Study into how different infrastructure elements react to varying flooding conditions and the compounding effects of multiple failures or partial failures is an integral component of flood risk management. Table 3.1 shows an overview of the types and quantities of data being considered for the infrastructure within the City of London. Interviews were conducted with experts across the infrastructure categories at the City of London to better understand each system and gather input for the risk analysis. The departments and divisions involved in this process include: Risk Management Division, Wastewater and Drainage Engineering, Planning and Development – Building, Transportation Planning and Design, Water Operations Division, Water Engineering Division, Pollution Control Operations, Environmental Programs and Customer Relations, and Corporate Security and Emergency Management Division.

The study considers transportation infrastructure (including bridges, culverts, and arterial roads), buildings (residential, commercial, and industrial), critical infrastructure (fire stations, emergency management services (EMS), police stations, hospitals, schools, and pollution control plants (PCPs)), flood protection structures, sanitary and storm networks, and the drinking water distribution network. Each of these infrastructure elements has different failure mechanisms under flood loading.

Table 3.1 *Infrastructure summary: City of London*

Infrastructure	Quantity
Bridges and culverts	216
Arterial roads	520 km
Buildings	>3000 (within the floodplain considered)
Sanitary/storm pipe network	>1327 km
Pollution control plants	6
Stormwater management facilities	100

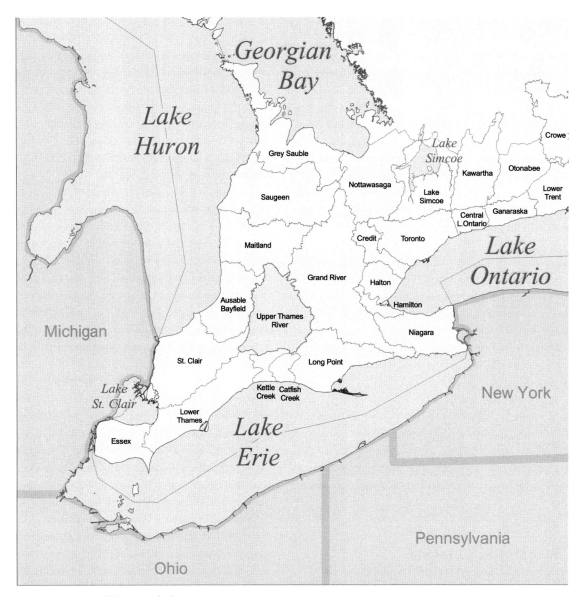

Figure 3.4 The Upper Thames River watershed.

TRANSPORTATION

Arterial/primary roads are considered in the study. Roads are a critical network in the event of any disaster as they allow for evacuation and rescue access for emergency services. The effect of flooding on roads has been well documented in regions such as the Gulf Coast of the USA where hurricanes make flooding common. Primary failure mechanisms for an inundated roadway include scour of the embankments and subsoil (washout) and rutting. Other failures include total collapse due to extreme scour and surface wear from debris impact. One of the most common impacts of flooding on a roadway is that it decreases its design life. Therefore, while the road may not experience catastrophic failure, it will become more susceptible to damage and likely require repair at an earlier date than planned or budgeted for. The degree

to which the road is damaged depends largely on the velocity and turbulence of the floodwater as well as the road surface material. An inundated road also becomes a danger to human safety and must be closed, therefore causing it to experience functional failure. This can hamper emergency access for fire, police, and ambulance services.

The City of London's Transportation Master Plan (TMP) indicates that the primary mode of transport for citizens is through the use of private vehicles. This demonstrates the importance of keeping the roadways in good condition in order to maintain a high level of productivity in the city. There are over 520 km of primary and arterial roadways within the City of London. Due to the size of this network and its importance for the city's day-to-day operations, emergency response, and budgeted investment, it is crucial that the transportation division be prepared for an increase in the frequency of extreme flood events. This study provides a unique way of measuring the expected impact of a flood on a roadway. The inundation depth is used as a key variable in predicting the structural and functional losses that are expected during the flood.

BRIDGES AND CULVERTS

The bridges documented in the Bridge Management System operated by the city are included in the flood risk assessment process. The study includes footbridges and culverts. The main failure mechanisms of a bridge during a flood are washout due to embankment and/or pier scour from the fast moving water. Other major flood damage can be the overturning of the structure due to the forces of water and possible build-up of ice and/or debris creating a damming effect. Debris may also be expected to contribute to the damage of non-structural elements such as railings, conveyance cables, and streetlights. Similarly to a road, the functionality of a bridge will be compromised should the deck become submerged and the bridge is closed.

Culverts are designed for some overhead. However, if this is exceeded the culvert can experience the same damage as previously described for roads and bridges. In addition, the loss of function can be extended to account for the water that exceeds the culvert's design capacity. This will cause damming action behind the culvert, increasing the inundation depth upstream.

There are 117 culverts, 99 bridges, and 8 footbridges within the city. These structures are an integral part of the transportation network and must be prepared to cope with the increased flood load due to climate change. The majority of the bridges are in good condition, as indicated by the Bridge Management System. This study examines not only the current condition of the structure and how that impacts its ability to withstand increased loads due to flow increase, but also its structural design, such as capacity (culverts) and elevation (bridges) in relation to the modified design floods.

CRITICAL FACILITIES

Critical facilities are defined in this study as the buildings that provide essential or emergency services and include: hospitals, emergency management services (EMS), fire, police, pollution control plants (PCPs), and schools. Many of these services are especially important during a flood event and so they are studied separately from the rest of the building infrastructure.

All critical facilities may experience the same failure mechanisms with respect to structural and equipment damage. Any costly equipment that is below the inundation depth may be lost and the building envelope itself may be compromised in the event of large inundation or extreme foundation scour, depending on the velocity and depth of flooding. The functionality of the critical facility may also be affected depending on its proximity to the floodplain.

Hospitals must have accessibility from many different routes for ambulances and possible evacuation in case of a large flood disaster. In addition, the location of the personnel who staff the hospital will impact the operations at the building even if the hospital itself is not submerged. The inflow of patients during a flood disaster event may also increase, and the hospitals should be prepared to deal with this influx of patients (as well as any from hospitals that may need to evacuate due to flooding inundation).

Similarly to hospitals, fire fighting, EMS, and police infrastructure will lose functionality if major access roads are cut off due to flooding. The location of the personnel will also affect the operations of the infrastructure, especially if many are located near, or cut off by, the floodwaters. An increase in demand for emergency services must be expected during the disaster flood event. Fire Services London manages all of the fire stations for the city. None of the stations are within the existing floodplain. However, some stations are in close proximity to high risk areas. Therefore, they may experience an increased demand for their services during and immediately following a flood disaster event. The main stress on the system will be the increased demand and reduced access due to flooded roads, which will increase the response time. Thames Emergency Medical Services manages the emergency services for the City of London through the use of six locations. None of these locations are in the floodplain; however, similarly to the fire stations, some will be affected due to their proximity to the areas at risk.

Pollution control plants are generally located in low-lying areas near the river due to the nature of their design and function. Thus they are highly vulnerable to flooding. A PCP may experience partial to full failure depending on the extent of inundation it experiences. For example, if the secondary treatment system is inundated, the plant may still run primary treatment and bypass the secondary system. This is not the ideal operation, but it is better than allowing raw sewage to pass untreated into the system. However, if the outlets become submerged such that the water is backing up into the plant, a full bypass may be necessary. This means that, for the duration of the high water levels, the sewage will be discharged directly to the river. Any electrical equipment that becomes inundated during a flood can also be lost or damaged by the water. It is documented that 4 ft of water is enough to short out all electronics. In addition, damage can be expected to include pump stations, exposed sewer mains, and washout, silt, and debris interfering with manholes and mains. The study uses inundation depth to measure flood damage and so a combination of visual inspection with structural details from each plant is used to determine the degree of failure that can be expected for those plants that fall within the flood scenarios.

London has six PCPs: Southland, Greenway, Vauxhall, Oxford, Adelaide, and Pottersburg. Together the plants have the capacity to handle approximately 298 ML per day. Currently Pottersburg, Adelaide, and Greenway experience difficulty discharging during extreme flow events. Emergency overflows are in place to manage the discharge in addition to a bypass at Pottersburg. Due to the location of the plants within the floodplain, access during a flood event is a concern. Vulnerable aspects of the plant include the tanks, clarifiers, and electrical equipment.

The final type of critical facilities considered is schools. It is assumed that schools will be closed and/or evacuated in the event that they are inundated (therefore experience total failure of functionality). Structurally, a school will be affected in the same way as any other building of similar construction that experiences inundation. This structural damage is related to the velocity and depth of the floodwaters, the age and condition of the building envelope and foundation, and the surrounding infrastructure (debris damage). The proximity of the school to the floodplain (if it is not within the floodplain) determines the level of its functionality loss based on school bus access, walking access, and the level of safety. Schools may also experience loss of contents such as computers, desks, and books if the building becomes inundated.

STORMWATER MANAGEMENT SYSTEM

The stormwater system consists of a network of sewers, manholes, outlets, and stormwater management facilities (SWMFs). There exist over 1,300 km of storm gravity sewers, 6.7 km of combined sewers, 18,472 storm access holes and 100 SWMFs. Hydraulic modeling of the stormwater network is not conducted in this case study. Floodwaters affect stormwater management by overwhelming the system. The pipe networks can become unable to handle the extreme volume of water causing it to back up the pipes and flood the roadways out of the manholes and inlets. In the case of combined sewer systems the sewage may back up and through basement drains causing major damage to buildings. If SWMFs are inundated fully, they will be no longer be able to provide storage or treatment for the area and will therefore lose their function for the duration of the inundation. Extensive hydraulic modeling is needed to fully understand and predict the response of the entire system to extreme flooding scenarios.

There are two major flooding mechanisms that may affect the city. The first is flooding due to the river overtopping its banks. This type of flooding may occur due to large and intense (a large amount of precipitation within a relatively short period of time) storm events leading to increase in flow within the river which then overtops its banks. The flooding of the Thames and its tributaries is the type of flooding considered in this case study. The infrastructure that is affected is therefore within the floodplains or in close proximity.

The second type of flooding may occur due to a large amount of rainfall that overwhelms the SWMFs but does not cause the river to flood. The problem may be compounded by urbanization and land use change, which lead to a reduction of natural runoff and rainfall absorption. As land use changes from agricultural and rural to developed and urban, more stormwater infrastructure is required to manage the large volumes of rainfall runoff. When an extreme storm event occurs, both flooding types combine, increasing the amount of infrastructure affected. As the stormwater system becomes overwhelmed, more water is discharged to the already full river, amplifying the flooding. The extent of the analysis conducted in the case study includes overlay of the final flood risk map over the stormwater network to identify key areas of the intersection of high risk with vulnerable storm infrastructure.

FLOOD CONTROL STRUCTURES

There are many flood control structures at work within the City of London. Due to the position of the city around the Thames River, and the propensity of the Thames to flooding, these control structures are important in the management of water levels for both safety and recreation in the city. The city's largest dam, Fanshawe, is located on the North Thames branch at the northeast end of the city. It is an embankment dam with a concrete spillway that controls a drainage area of 1,450km^2 at its outlet. The total storage volume is approximately 35.6 $\times 10^6$ m^3. The hydraulic modeling done for this study begins at the outlet of Fanshawe dam; as such the dam is not included in the case study.

There is an extensive network of dykes designed to protect the city from flood damage. As of 2006, there were approximately 5.5 km of dykes around sections of the north, south, and main Thames River. Along the south branch are the Clarence and Ada-Jacqueline dykes. Along the main branch are Riverview, Byron, and Coves. The West London Dyke (WLD) is the largest. It runs along the north and main branches at the Forks. Finally, the Broughdale Dyke is on the north branch. The WLD was recently repaired and some parts were replaced to bring it up to acceptable conditions and the 250-year regulatory flood levels.

The majority of the dykes within the city are earthen fill dykes. Ada-Jacqueline was repaired in the 1990s using riprap and portions of Broughdale are composed of gabion. WLD is constructed using reinforced concrete panels and has been restored and replaced in some sections with the flood wall. The Coves contains a flap gate that is used as a stormwater management release structure. Recent vegetation and erosion studies done on the dyke network have indicated that the main vulnerabilities of the system are due to erosion from the river. This causes undercutting, which can lead to failure. Failure may also be caused by overtopping or breeching of the dyke.

BUILDINGS

Due to the intensive urbanization of the city, with the densest development occurring around the Forks area of the Thames, flooding has the potential to have devastating effects on residential and business properties. The study identifies all buildings that experience any level of inundation in each climate scenario. The level of inundation is defined in the stage–damage curves as the level of the water above the first floor entrance. The amount of damage sustained by a building during a flood is typically measured using stage–damage curves. These curves are used in the study for the calculation of flood risk due to climate change. The foundation of the building is critical in determining its response to inundation. In addition, the age, structure type, and condition of the building all play an important role. This study assumes that all buildings will be evacuated in the event of a flood and therefore only structural and functional impacts are considered. The data were provided by the Municipal Property Assessment Corporation. Factors taken into consideration when evaluating the risk to buildings include age, design, value, inundation level, and total area inundated.

There are approximately 3,014 buildings affected by the modified floodplains (due to climate change) of which the majority (2,823) are residential. The average residential building value is \$95,177 in 2008 C\$. The most common building type is property code 301 – single family detached, not on water. The next most common type is property code 370, residential condominium. Together these two categories account for 85% of the affected properties. The average age of the residential structures is 52 years. Flooding not only impacts the physical building structure, it can also cut off access to commercial industries causing business disruption and economic damage. These damages are taken into account in this case study.

3.3.2 Overview of the risk assessment methodology

The Public Infrastructure Engineering Vulnerability Committee (PIEVC) established by Engineers Canada recently conducted an assessment of the vulnerability of Canadian public infrastructure to changing climatic conditions (Engineers Canada, 2008; Simonovic, 2008). The major conclusion of the assessment was that failures of public infrastructure due to climate change will become common across Canada.

The integrated risk assessment procedure developed for this case study includes: (i) climate modeling; (ii) hydrologic modeling; (iii) hydraulic modeling; and (iv) infrastructure risk assessment. Figure 3.5 provides a visual overview of the risk assessment methodology. There is a vertical interconnectivity between all the steps in the methodology. Output from each step is used as input into the next step. The climate modeling approach based on the use of GCM data together with a weather generator (WG) is used to provide precipitation data for a set of climate change scenarios. Precipitation data are transformed into flow data using the hydrologic model of the watershed. Flow information is processed through hydraulic analysis to obtain the extension and depth of flood inundation. Quantitative and qualitative risk calculations are

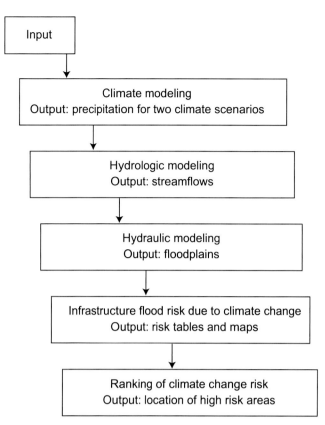

Figure 3.5 Infrastructure risk assessment to climate change.

performed in the next step to generate a detailed spatial distribution of flood risk to municipal infrastructure due to climate change.

The methodology in this study is specific to the flood hazard – identified as the most critical climate change impact in the Upper Thames River basin – but the general approach and methodology may also be applied to other hazards.

3.3.3 Climate modeling

An original inverse impact modeling approach (Simonovic, 2010) is used in the case study for assessing the vulnerability of river basin hydrologic processes to climate forcing. The approach consists of the following four steps:

Step 1 Identification of critical hydrologic exposures that may lead to local failures of water resource systems in a particular river basin. Critical exposures are analyzed together with existing guidelines and management practices. The vulnerable components of the river basin are identified together with the risk exposure. The water resource risk is assessed from three different viewpoints: risk and reliability (how often the system fails), resiliency (how quickly the system returns to a satisfactory state once a failure has occurred), and vulnerability (how significant the likely consequences of a failure may be). This step is accomplished in collaboration with local water authorities.

Step 2 In the next step, the identified critical hydrologic exposures (such as floods and droughts) are transformed into corresponding critical meteorological conditions (e.g., extreme precipitation events, sudden warming, prolonged dry spells). A hydrologic model is used to establish the inverse link between hydrologic and meteorological processes. Reservoir operation, floodplain management, and other anthropogenic interventions in the basin are also included in the model. In this study, the US Army Corps of Engineers (USACE) Hydrologic Engineering Center Hydrologic Modeling System (HEC-HMS) is used to transform inversely extreme hydrologic events into corresponding meteorological conditions. HEC-HMS is a precipitation–runoff model that includes a large set of mix-and-match methods to simulate river basin, channel, and water control structures.

Step 3 A WG is used to simulate the critical meteorological conditions under present and future climatic scenarios. The WG produces synthetic weather data that are statistically similar to the observed data. Since the focus is mainly on extreme hydrologic events, the generator reflects not only the mean conditions, but also the statistical properties of extreme meteorological events. The K-nearest neighbor (K-NN) algorithm is used to perform strategic resampling to derive new daily weather data with altered mean or variability. In the strategic resampling, new weather sequences are generated from the historical record based on prescribed conditioning criteria. For a given climatic variable, regional periodical deviations are calculated for each year and for each period.

Step 4 In the final stage, the parameters of the WG are linked with GCM and an ensemble of simulations reflecting different future climatic conditions is generated. The frequency of critical meteorological events causing specific water resources risks is then assessed from the WG outputs.

The main impacts of climatic change, obtained by earlier studies in the Upper Thames River basin (Cunderlik and Simonovic, 2005, 2007; Simonovic, 2010), include: (i) flooding that will occur more frequently in the future, regardless of the magnitude of floods; and (ii) that low flow conditions in the Upper Thames River basin will remain the same as currently observed. Therefore, the changing climatic conditions are expected to increase flood damage.

Figure 3.6 illustrates the main findings of earlier studies for one station in the basin (Byron). The solid line represents the flood flow frequency that corresponds to the historical data and the dashed line shows the flood flow frequency for the upper bound of climate change (CC_UB). Simple comparison between the two climatic conditions shows that the 100-year flood (historical data) becomes an event with a return period of 32 years under changed climate. If a comparison is made between the magnitude

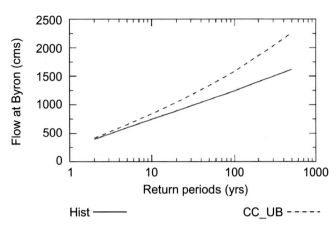

Figure 3.6 Flood flow analyses under climate change: the Upper Thames River basin.

of the 100-year flood under historic and climate change conditions, 1,286 m³/s and 1,570 m³/s respectively, an increase of more than 20% is observed.

Starting from the conclusions of the previous work, this case study examines the increase of flood risk due to climate change and its impact on the municipal infrastructure. Practical implementation of the inverse approach to the problem of high flows includes the use of a WG and an event-based hydrologic model (discussed in the following section). For flood frequency analysis, a large number of rainfall events need to be simulated by the WG. These events are used as inputs into the hydrologic model. The outputs of the hydrologic model (peak flows) are used in frequency analysis.

One of the main contributions of the work conducted in the case study is the assessment of climate change impacts for a range of climate scenarios (Simonovic, 2010). A general suggestion is to consider the assessment of impacts for two extreme climate scenarios that will define the lower (CC_LB) and upper bound (CC_UB) of potential climate change. Two climate scenarios are derived by integrating a WG, which perturbs and shuffles historical data, with inputs from GCMs. The CC_LB is obtained by perturbing and shuffling locally observed data with the assistance of a WG tool. Through the perturbation process, it allows the extreme (minimum and maximum) generated values to be outside of the historic range. In this way the character of the lower bound scenario reflects the existing conditions (greenhouse gas emissions, land use, population, etc.) and their potential impact on the development of future climate. The CC_UB is derived by perturbing and shuffling historical data and combining them with the input from the GCMs. The choice of GCM is made on the basis that the CC_UB should represent the most critical impact of climate change for the basin under consideration. Selection of the range of potential climate change through the use of two scenarios compensates for the existing level of uncertainty present in global modeling of climate change for a watershed. It is noted in the literature that the global models offer various predictions of future climate as a consequence of (i) the selected global model, (ii) the selected global model simulation scenario, and (iii) the spatial and temporal resolution of the selected global model. It is important to point out that both climate scenarios are equally likely as well as the range of climatic conditions between the two.

The WG used in this study (Yates *et al.*, 2003) modified by Sharif and Burn (2006) is based on the K-NN algorithm. The K-NN algorithm, further modified by Simonovic (2010) with p variables and q stations proposed has the following steps:

(1) Calculation of regional means of p variables (x) across all q stations for each day in the historic record:

$$\overline{X}_t = \left[\overline{x}_{1,t}, \overline{x}_{2,t}, \ldots, \overline{x}_{p,t} \right] \qquad \forall\, t = \{1, 2, \ldots, T\} \quad (3.2)$$

where

$$\overline{x}_{i,t} = \frac{1}{q} \sum_{j=1}^{q} x_{i,t}^{j} \qquad \forall\, i = \{1, 2, \ldots, p\} \qquad (3.3)$$

(2) Compute the potential neighbors of size $L = (w + 1) \times N - 1$ days long for each variable p with N years of historical record and selected temporal window of size w. All days within that window are selected as potential neighbors to the current feature vector. Among the potential neighbors, N data corresponding to the current day are eliminated in the process to prevent the possibility of generating the same value as that of the current day.

(3) Compute the regional means for all potential neighbors selected in step (2) across all q stations for each day.

(4) Compute the covariance matrix, C_t, for day t using the data block of size $L \times p$.

(5) Select randomly a value of the first time step for each variable p from all current day values in the record of N years.

(6) Compute the Mahalanobis distance expressed by Equation (3.4) between the mean vector of the current days ($\overline{X}_t$) and the mean vector of all nearest neighbor values ($\overline{X}_k$), where $k = 1, 2, \ldots, L$.

$$d_k = \sqrt{\left(\overline{X}_t - \overline{X}_k\right) C_t^{-1} \left(\overline{X}_t - \overline{X}_k\right)^{\mathrm{T}}} \qquad (3.4)$$

where T represents the transpose matrix operation, and C^{-1} represents the inverse of the covariance matrix.

(7) Select the number of $K = \sqrt{L}$ nearest neighbors out of L potential values.

(8) Sort the Mahalanobis distance d_k from smallest to largest, and retain the first K neighbors in the sorted list (they are referred to as the K nearest neighbors). Then, use a discrete probability distribution giving higher weights to closest neighbors for resampling out the set of K neighbors.

The weights are calculated for each k neighbor using the following Equations (3.5) and (3.6):

$$w_k = \frac{1/k}{\sum\limits_{i=1}^{K} 1/i} \qquad (3.5)$$

where $k = 1, 2, \ldots, K$. Cumulative probabilities, p_j, are given by:

$$p_j = \sum_{i=1}^{j} w_i \qquad (3.6)$$

Note that Sharif and Burn (2006) used the cumulative probability of K neighbors according to Equation (3.6) while Yates *et al.* (2003) used just a probability for each of K neighbors according to Equation (3.5).

(9) Generate a random number $u(0,1)$ and compare it to the cumulative probability p_j to determine the nearest neighbor of the current day. If $p_1 < u < p_K$, then day j for which u is closest to p_j is selected. On the other hand, if $u < p_1$, then the day corresponding to d_1 is selected, and if $u = p_K$, then the day corresponding to d_K is selected. Once the nearest neighbor is selected, the weather of the selected day is used for all stations in the region. By this characteristic of the K-NN algorithm, therefore, cross-correlation among variables in the region is preserved. In this step, the improved K-NN algorithm provides a reasonable method that can randomly select one among K neighbors because it uses the cumulative probability. However, in the algorithm by Yates *et al.* (2003) the first nearest neighbor may be selected in most cases because it selects the one of K nearest neighbors for which u is closest to the probability of each neighbor.

(10) This step was added by Sharif and Burn (2006) to generate variables outside the range of historical data by perturbation. They suggested the optimal bandwidth (λ) that minimizes the asymptotic mean integrated square error (AMISE) for a Gaussian distribution. In the univariate case, the optimal bandwidth becomes:

$$\lambda = 1.06\, K^{-1/5} \qquad (3.7)$$

where K represents the number of nearest neighbors. In addition, they suggested that a new value can be achieved from a value with mean $x_{i,t}^{j}$ and variance $(\lambda \sigma_i^{j})^2$, i.e., the perturbation process is conducted according to Equation (3.8):

$$y_{i,t}^{j} = x_{i,t}^{j} + \lambda \sigma_i^{j} z_t \qquad (3.8)$$

where $x_{i,t}^{j}$ is the value of the weather variable obtained from the original K-NN algorithm; $y_{i,t}^{j}$ is the weather variable value from the perturbed set; z_t is the normally distributed random variable with zero mean and unit variance,

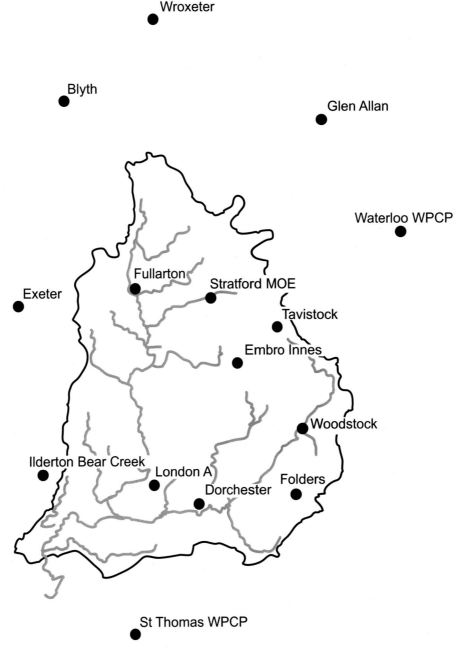

Figure 3.7 Schematic map of meteorological stations in the basin.

for day t. To prevent the negative values for bounded variables (i.e., precipitation), the largest acceptable value of $\lambda_a = x_{*,t}^j / 1.55\sigma_*^j$ is employed, where * refers to a bounded weather variable (Sharif and Burn, 2006). If the value of the bounded weather variable computed previously is still negative, then a new value of z_t is generated.

Sharif and Burn (2006) have developed the K-NN algorithm to generate three variables ($p = 3$: precipitation, maximum temperature, and minimum temperature). If there are more meteorological

variables available in the basin for use with the WG model, the calculation of Mahalanobis distance expressed by Equation (3.4) becomes quite demanding. Therefore, a modified WG model is integrated with principal component analysis (PCA) to decrease the dimensions of the mean vector of the current day ($\overline{X}_t$) and the mean vector of all nearest neighbor values ($\overline{X}_k$) in step (6) by employing only the first principal component.

(6a)   Calculate the eigenvector and eigenvalue for the covariance matrix (C_t).

(6b) Find the eigenvector related to the largest eigenvalue that explains the largest fraction of the variance described by the p variables.

(6c) Calculate the first principal component with the eigenvector found in step (6b) using Equations (3.9) and (3.10):

$$PC_t = \overline{\mathbf{X}}_t \mathbf{E} \qquad (3.9)$$

$$PC_k = \overline{\mathbf{X}}_k \mathbf{E} \qquad (3.10)$$

where PC_t and PC_k are the values of the current day and the nearest neighbor transferred by the eigenvector from step (6b), respectively; and $\mathbf{E}$ is the eigenvector related to the largest eigenvalue.

After calculating PC_t and PC_k with the one-dimensional matrix (Equations (3.9) and (3.10)), the Mahalanobis distance is computed using:

$$d_k = \sqrt{(PC_t - PC_k)^2 / \mathrm{Var}(\mathbf{PC})} \quad \forall k = \{1, 2, \ldots, K\} \quad (3.11)$$

where $\mathrm{Var}(\mathbf{PC})$ represents the variance of the first principal component for the K nearest neighbors.

This case study employs daily precipitation, maximum temperature, and minimum temperature for 15 stations in the basin (Figure 3.7) for the period from 1964 to 2006 ($N = 43$) to generate two climate scenarios using the WG model described above. Among the 15 stations used in this study, only three locations are selected to show the comparison of WG: (1) Stratford for illustrating the characteristics of the northern part of the basin; (2) London for the southwestern part; and (3) Woodstock for the southeastern part of the basin. For the application of WG, the temporal window of 14 days ($w = 14$) and 43 years of historical data are used resulting in 644 days as the potential neighbors ($L = (w + 1) \times N - 1 = 644$) for each variable.

The CC_UB uses the CCSRNIES B21 scenario provided by the Canadian Climate Impacts Scenarios group at the University of Victoria (www.cics.uvic.ca/, last accessed November 10, 2010). The GCM time slice of 2040–2069 is used to represent climate conditions for the 2050s. To include the impact of climate change in the WG model, the observed historical data are modified by adding (in the case of temperature) or multiplying (in the case of precipitation) the average change between the reference scenario and the future climate scenario to the regional observed historical data at each station. The monthly change for the CC_UB scenario for precipitation and temperature variables is shown in Table 3.2. The CC_UB scenario shows the increase in precipitation during the period from January to September with a significant increase during the spring season, from March to June.

The climate modeling phase results in 200 years of daily values for three meteorological variables: precipitation, maximum temperature, and minimum temperature, for the CC_LB and CC_UB scenarios. All variables generated by the WG model are compared with the observed historical data for verification purposes.

Table 3.2 *Monthly changes in precipitation and temperature between the historic and the CC_UB scenarios*

Month	CC_UB scenario Precipitation (% change)	Temperature (difference in °C)
Jan	0.1767	4.43
Feb	0.0638	3.29
Mar	0.1507	4.52
Apr	0.2284	5.78
May	0.2414	4.50
Jun	0.1855	3.32
Jul	0.0503	3.59
Aug	0.0788	4.09
Sep	0.0427	2.11
Oct	−0.1151	3.11
Nov	−0.1555	4.64
Dec	−0.031	1.43

CLIMATE MODELING RESULTS: PRECIPITATION

The WG model is first used with 43 years of observed data (1964 to 2006) to simulate the lower bound of potential climate change. The underlying assumption in this scenario is that neither mitigation nor adaptation measures will be introduced into the social-economic-climatic system and the future state of the system will be the consequence of already existing conditions within the system (concentration of greenhouse gases, population growth, land use, etc.). Figure 3.8 shows the comparison between the generated and observed precipitation data for the CC_LB scenario. Then, the historic data are modified using information from the GCM (shown in Table 3.2) to provide the WG input for the CC_UB scenario. Figure 3.9 shows the precipitation comparison for the CC_UB scenario.

The synthetic data generated using the WG model are shown using the box plot while dots represent the percentile values of the observed data corresponding to the minimum, 25th, 50th, 75th, and maximum value (from the bottom up). The median value of the observed historical data is shown as the solid line. The results confirm that the WG regenerates well the percentile values of the observed data. In addition, due to the implementation of the perturbation step in the WG algorithm the generated data include values outside of the observed minimum and maximum values.

Detailed results of climate modeling also include maximum and minimum temperature data for two climate scenarios. These data are used in extensive analyses of climate change impacts and timing of flood events (Eum and Simonovic, 2010).

The precipitation results obtained by the WG for two climate scenarios are used as the input into the hydrologic analysis.

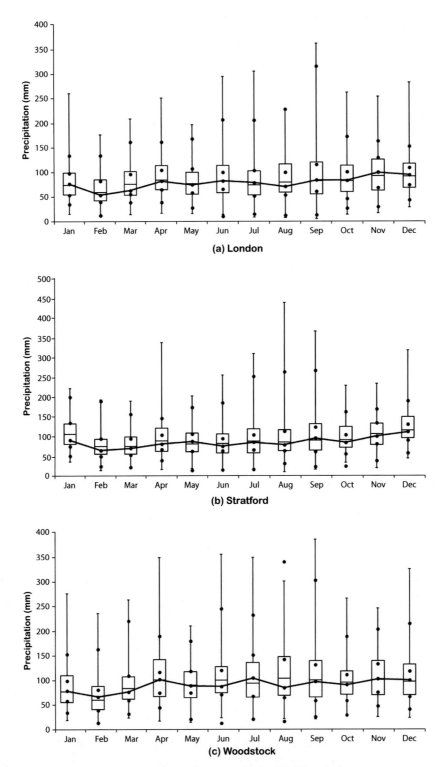

Figure 3.8 Comparison of the generated and the observed precipitation values for the CC_LB scenario.

3.3.4   Hydrologic modeling

The meteorological variables generated by the WG model are used, as input data for a hydrologic model, to further assess the impacts of climate change on the hydrologic conditions in the

basin. This study selected the HEC-HMS version 3.3 due to flexible time and spatial scales, ease of use, short set-up time, etc. (USACE, 2008). Despite the fact that HEC-HMS is not "the best" model for use in urban watersheds, reasons such as (i) availability

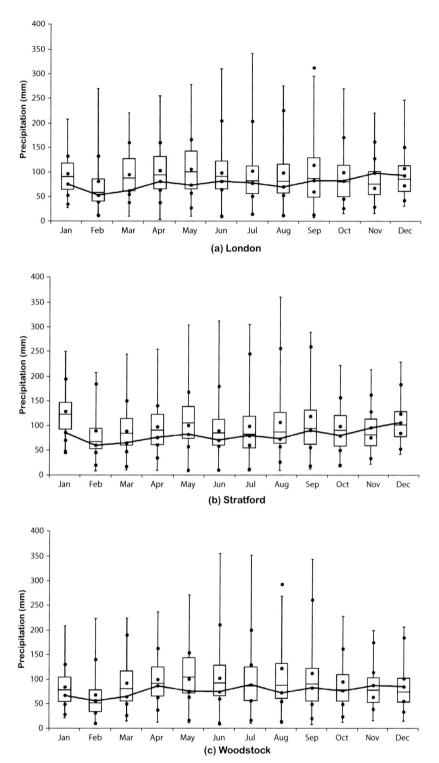

Figure 3.9 Comparison of the generated and the observed precipitation values for the CC_UB scenario.

of the calibrated model for the whole Upper Thames River basin, (ii) limited modeling time and resources available for the study, and (iii) limited flow data for most sub-watersheds within the City of London boundaries, led to the selection of this model for use in the study.

The WG model generates daily precipitation and temperature variables at 15 stations within the Upper Thames River basin. However, the HEC-HMS requires extreme precipitation data with at least hourly resolution. In addition, the spatial resolution of model input data has to be adjusted too. The meteorological input

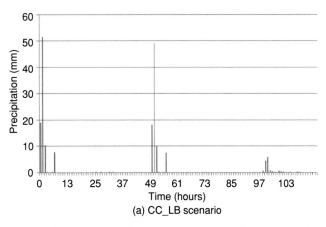

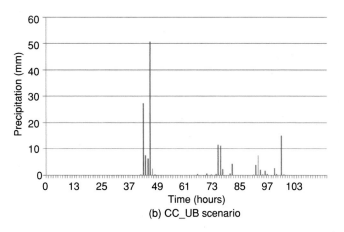

Figure 3.10 Hyetographs of an annual extreme event for London station.

data (precipitation) are available at 15 stations within the basin and required for each sub-basin in the Upper Thames River basin. Therefore, the temporal disaggregation and the spatial interpolation schemes are implemented to provide the necessary input data. The spatial interpolation, based on the inverse distance method and the location information for 15 measurement stations, is applied to obtain the meteorological data for each sub-basin. The disaggregation procedure based on the method of fragments is implemented to convert daily data into hourly.

This study generates daily data for 200 years and any given year contains a number of events. The main objective of the hydrologic analysis is to perform the flow frequency analysis of extreme annual flood events. Therefore, the 5-day annual extreme events that produce the largest annual events (200 events altogether) for the entire basin are selected. Figure 3.10 shows one example of the extreme flood hyetographs at London station for both climate change scenarios.

Cunderlik and Simonovic (2005, 2007) have developed the HEC-HMS model with 34 sub-basins for the Upper Thames River basin and successfully applied the model to assess climate change impacts in the basin. For the assessment of the vulnerability of municipal infrastructure to climate change within the City of London, more detailed description of the hydrologic conditions within the City of London is required. The nesting procedure of additional sub-basins (for improved spatial resolution within the city boundaries) into original model structure is implemented. There are four major sub-basins in the city: Medway Creek, Stoney Creek, Pottersburg Creek, and Dingman Creek, which are divided into 5, 6, 4, and 16 sub-watersheds, respectively. In addition to four sub-basins, this study also delineates the main river basin within the city. At the end, the complete HEC-HMS model used in this study (Eum and Simonovic, 2010) consists of 72 sub-basins, 45 reaches, 49 junctions, and 3 reservoirs (Figure 3.11).

An hourly rainfall event from July 5 to July 16 of 2000 that covered almost the entire basin is used for model calibration.

There are no measured streamflow data available for the Stoney Creek and the Pottersburg Creek during the July 2000 event. The Stoney Creek is also affected by backwater from the North Thames River, which further complicates the selection of proper measurement data for calibration. Following the recommendation from the Upper Thames River Conservation Authority (UTRCA), the flood event of October 4 to 7, 2006, is used for calibration of the Stoney Creek sub-basins. This event is not affected by the backwater impact.

A calibrated hydrologic model, HEC-HMS, is used to convert climate input into flow data within the City of London. The annual extreme precipitation events for each of 200 years and both climate scenarios (total of 400 flood events) are selected and used as input into the HEC-HMS model. For each flood event, the streamflow values are calculated for each sub-basin and each control point. Each hydrologic simulation run is done using a 5-day time horizon. The simulation results provide the essential hydrologic information for each sub-basin and each control point for two climate scenarios and 200 years. Within the City of London 171 locations of interest are identified – mostly representing input profiles for the hydraulic analysis (Eum and Simonovic, 2010).

A frequency analysis is used to relate the magnitude of extreme events to their frequency of occurrence. The results of the hydrologic analyses (using the HEC-HMS model) in this study are used as input into the Hydrologic Engineering Center's (USAE) River Analysis System (HEC-RAS), which calculates the extent and depth of flood inundation for two regulatory flood return periods, 100 and 250 years. Flood frequency analysis of the hydrologic model output is conducted to provide the input for the hydraulic analysis. In this study, the method of L moments and three extreme event probability distributions are used: Gumbel, generalized extreme value (GEV), and Log-Pearson type III (Eum and Simonovic, 2010). Previous studies done in the basin and recommendations provided by Environment Canada suggest the use of the Gumbel distribution. Therefore, the Gumbel distribution is used in this study in order to provide for the comparison

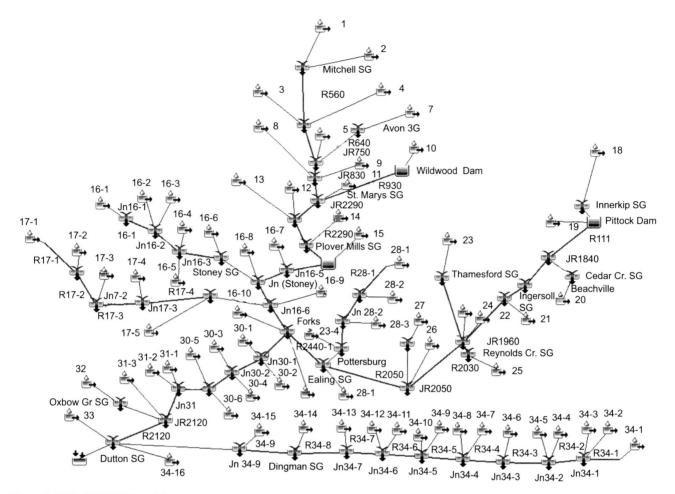

Figure 3.11 The HEC-HMS model structure.

of the results of flood frequency analysis under climate change with those used in the current flood plain management by the City of London. Figure 3.12 shows the results of the flood frequency analysis for two climate scenarios at North Thames, South Thames, and Forks locations. Similar results are obtained for all 171 locations. As expected, the results of flood frequency analysis for the CC_UB scenario show an increase in both flood frequency and flood magnitude when compared with the CC_LB scenario. The difference between the two scenarios identifies the range of climate change flood impacts that may be expected at each location.

Two return period flow values for two climate scenarios are then provided as input into the hydraulic analysis.

3.3.5 Hydraulic modeling

The traditional process of floodplain mapping based on hydraulic calculations of water surface elevations is adopted for the purpose of climate change flood risk assessment to municipal infrastructure for the City of London (Sredojevic and Simonovic, 2010). Standard computer software, HEC-RAS (USACE, 2006), is used

for hydraulic modeling and computation of water elevation in the basin. HEC-GeoRAS, an extension of ArcGIS (USACE, 2005), is used for the preparation of spatial data for input into a HEC-RAS hydraulic model and the generation of geographic information system (GIS) data from the output of HEC-RAS for use in floodplain mapping.

The climate modeling provides meteorological data (precipitation) for hydrologic analysis. The HEC-HMS hydrologic model (USACE, 2008) is used to transform the climate data generated by the WG model into flow data that are required for hydraulic analysis. The methodology used in hydraulic analysis consists of three steps: (i) pre-processing of geometric data for HEC-RAS, using HEC-GeoRAS; (ii) hydraulic analysis in HEC-RAS; and (iii) post-processing of HEC-RAS results and floodplain mapping, using HEC-GeoRAS.

The first step in the pre-processing stage is the creation of a digital terrain model (DTM) of the river system in a triangulated irregular network (TIN) format. The TIN also serves for the delineation of floodplain boundaries and calculation of inundation depths. The pre-processing starts with the development of the geometry data file for use with HEC-RAS. The geometry data file contains

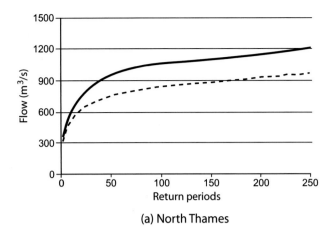

(a) North Thames

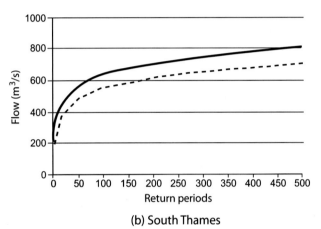

(b) South Thames

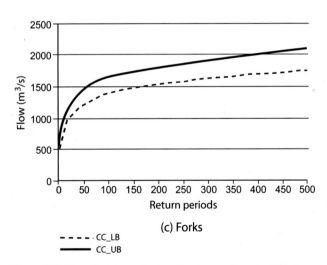

(c) Forks

- - - - CC_LB
——— CC_UB

Figure 3.12 Flood frequency for two climate scenarios at selected locations.

important information about cross sections, hydraulic structures, river bank points, and other physical attributes of river channels. The pre-processing is done by using the HEC-GeoRAS to generate the following geometric data: *River, Banks, Flowpaths, XsCutLines, Bridges, Levees,* and *InlineStructures.*

After completion of the pre-processing stage, the hydraulic analysis is performed using the HEC-RAS modeling program for the computation of water surface profiles. The analysis starts by importing geometric data (GIS layers) generated in the previous stage. The hydraulic analysis is performed using flow data for two climate scenarios (CC_LB and CC_UB). For both climate scenarios, steady flow data are used for flow return periods of 100 and 250 years. Flow information is entered at all 171 locations for which the frequency analyses were completed as the ending step of the hydrologic analysis. The HEC-RAS model is executed for the subcritical flow profiles. Two water surface profiles (100 and 250) are generated within the boundaries of the City of London for both climate scenarios: 100 CC_LB, 100 CC_UB, 250 CC_LB, 250 CC_UB.

The post-processing of the water surface profiles is performed using the same maps that were used for the pre-processing of geometry data. Floodplain mapping is performed within the limits of the bounding polygon using the water surface elevations generated by the HEC-RAS. An example map is shown in Figure 3.13.

A summary of the results is presented in Table 3.3. It is clear that the climate change conditions result in significant increases in flooded area. Figures 3.14, 3.15, and 3.16 show the comparison of floodplain areas for the two climate scenarios and 100-year return period at the location close to the University of Western Ontario (North Thames), the Forks (confluence of the North and the South Thames), and the Broughdale Dyke (North Thames), respectively. Similarly, Figures 3.17, 3.18, and 3.19 show the comparison of floodplain areas for the two climate scenarios and current conditions (UTRCA scenario) and 250-year return period at the location close to the University of Western Ontario (North Thames), the Forks (confluence of the North and the South Thames), and the Broughdale Dyke (North Thames), respectively.

3.3.6 Risk calculation

A comprehensive risk assessment has been undertaken to better understand climate change-caused flood impacts on municipal infrastructure and provide a measurement of risk as the basis for the development of climate change adaptation options. The main goals of the assessment are (i) to provide the level of risk to infrastructure that may be affected by flooding; and (ii) to prioritize areas of high infrastructure risk for future climate change adaptation planning decisions. The areas in the City of London identified as high risk areas are flagged for further socio-economic studies. The methodology is based on an integrated risk index for each infrastructure element considered in the study. The risk index allows for comparison among various locations that may be flooded. Each risk level for a particular location provides the source of risk (type of infrastructure that may be affected) and relative contribution of each source to the overall risk.

Table 3.3 *Comparison of flooded areas for two climate scenarios*

| River/Creek | Flooded area (m²) | | | | | |
| | 100-year return period | | | 250-year return period | | |
	CC_LB	CC_UB	Difference	CC_LB	CC_UB	Difference
Main Thames River	2,717,208	3,228,637	511,429	3,189,657	3,342,766	153,109
North Thames River	4,951,784	6,327,229	1,375,445	6,144,150	6,497,384	353,234
South Thames River	2,676,651	2,885,980	209,329	2,886,324	3,128,588	242,264
Medway Creek	1,143,686	1,170,080	26,394	1,219,177	1,242,106	22,929
Stoney Creek	974,141	1,008,950	34,809	1,030,558	1,104,061	73,503
Pottersburg Creek	2,853,112	3,063,310	210,198	3,069,149	3,283,552	214,403
Mud Creek	72,339	123,697	51,358	124,241	226,260	102,019
Dingman Creek	7,750,220	8,011,897	261,677	8,302,463	9,061,872	759,409
Total	**23,139,141**	**25,819,780**	**2,680,639**	**25,965,719**	**27,886,589**	**1,920,870**

Figure 3.13 The floodplain depth grid and bounding polygon for the CC_UB 250-year flood at one location along the North Thames. See color plates section.

RISK INDEX

Risk is defined as the product between a hazard and vulnerability in the context of this case study (Peck *et al.*, 2011). This study measures vulnerability, which is defined by Engineers Canada in the context of engineering infrastructure and climate change as "the shortfall in the ability of public infrastructure to absorb the negative effects, and benefit from the positive effects, of changes in the climate conditions used to design and operate infrastructure" (Engineers Canada, 2008).

An original risk measure termed the risk index, R, is defined. This index is calculated for each infrastructure element and incorporates quantitative and qualitative data to address both objective

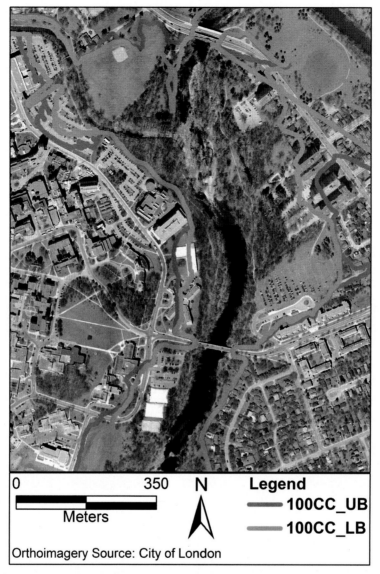

Figure 3.14 Comparison of floodplain boundaries for 100 CC_LB and 100 CC_UB at the location of the University of Ontario (North Thames). See color plates section.

and subjective types of uncertainty. The mathematical expression of the risk index is:

$$R_{ke} = P \times \sum_{i=1}^{3}(D_i \times IM_i) \qquad (3.12)$$

where P is the probability of occurrence of the hazard event (dimensionless); D_i is the economic loss for each impact category, i ($); IM_i is the impact multiplier (fraction of damage sustained for each impact); e is the infrastructure element; k is the infrastructure type from 1 to 6 (building, bridge, barrier, critical facility, PCP, and road); and i is the impact category, from 1 to 3, representing function, equipment/contents, and structure. For a 100-year flood event, the probability, P, of occurrence in any given year is 1 in

100, or 1%. Similarly, the probability of a 250-year event is 1 in 250 or 0.4%.

The risk index is tabulated and normalized for each infrastructure element across each of the five scenarios (two climate scenarios – 100 CC_LB, 100 CC_UB, 250 CC_LB, 250 CC_UB – and one scenario describing current conditions – 250 UTRCA). These values are then combined and displayed spatially using GIS in the form of risk maps. Risk is portrayed geographically by dissemination areas (DA) classification, consistent with the Statistics Canada method of representing data. There are 527 DAs within the City of London (Figure 3.20). Each DA is defined by Statistics Canada as "a small, relatively stable geographic unit comprised of one or more adjacent dissemination blocks," available online at Statistics Canada (2011).

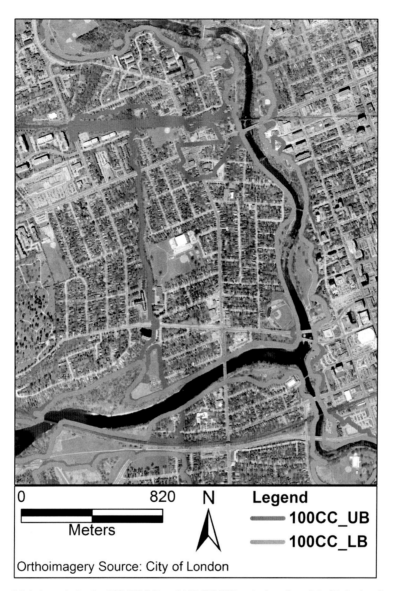

Figure 3.15 Comparison of floodplain boundaries for 100 CC_LB and 100 CC_UB at the location of the Forks (confluence of the North Thames and the South Thames). See color plates section.

The risk index is used to aid in the prioritization of areas of infrastructure at risk. Equation (3.13) shows the calculation of risk to a DA for all infrastructure elements of interest (bridges, buildings, barriers, roads, critical facilities, and/or PCPs):

$$Rq = \sum_{e=1}^{m} R_{eq} \qquad (3.13)$$

where q is the DA, and m is the number of infrastructure elements of interest.

IMPACT MULTIPLIERS, IM_i
The second element of the risk equation represents the impact to the infrastructure as a result of its interaction with the flood hazard. The damage is both direct (such as loss of structural integrity and

components) and indirect (such as a loss of functionality). Damages resulting from flooding are extremely varied and include losses ranging from inconvenience to structural damage to death. This study considers three variables as a measure of these consequences: the loss of function (IM_1), loss of equipment (IM_2), and loss of structure (IM_3). Each of these factors (termed impact multipliers) is measured as a percentage loss and calculated using both quantitative and qualitative information. They are incorporated into the risk index as demonstrated by expanding Equation (3.12) as shown below:

$$R_e = P \times (D_1 \times IM_1 + D_2 \times IM_2 + D_3 \times IM_3) \quad (3.14)$$

The quantitative data include the ability of the infrastructure to withstand direct damage due to flooding in addition to actual

Figure 3.16 Comparison of floodplain boundaries for 100 CC_LB and 100 CC_UB at the location of the Broughdale Dyke (North Thames). See color plates section.

inundation. The qualitative data include information gathered through interviews relating to the decision-makers' expertise and experience. This includes the condition of the infrastructure and how that may affect its response to flooding. It is important to note that the measure of the impact multiplier may be different across the varying infrastructure types; however, they are consistent across any one particular infrastructure type.

Loss of function (IM_1) The loss of function impact multiplier, IM_1 measures the degree to which the infrastructure has lost its functionality. This is defined in the case study as the degree to which the infrastructure no longer functions at an acceptable level

relative to that for which it was originally designed, as a result of flooding. The value of IM_1 is an integer belonging to [0,1] where 0 denotes no loss of function and 1 denotes total loss of function.

The transportation infrastructure (roads, bridges, culverts, and footbridges) is designated as having an IM_1 equal to 1 once they are inundated. Buildings and critical facilities are assigned an IM_1 of 1 if they are inundated or if all possible access routes are blocked due to flooding. Flood protection structures have an IM_1 value of 1 once their design capacity has been reached.

Partial loss of function may occur in the case of critical infrastructure such as fire stations, EMS, hospitals, and schools if

Figure 3.17 Comparison of floodplain boundaries for 250 CC_LB, 250 CC_UB, and 250 UTRCA at the location of the University of Ontario (North Thames). See color plates section.

some, but not all, of the access routes are blocked by floodwaters. The methodology assigns a fractional value of IM_1 depending on the number of incoming or outgoing major routes and the number of routes that are flooded. The relationship used to calculate IM_1 for fire stations and EMS buildings is

$$IM_1 k = \frac{(n - m)}{n} \qquad (3.15)$$

where $k = 4$ (critical facility types); n is the total number of major access routes; and m is the number of routes obstructed by floodwaters. In the case of schools and hospitals, the loss of function multiplier is calculated based on the total number of access routes within one intersection from the building.

For PCPs, the loss of function multiplier is 0 or 1. IM_1 is 1 if the danger flow or elevation danger point as indicated by the City of London Flood Plan is exceeded, or the outlet invert is inundated up to the plant elevation.

Loss of equipment (IM_2) The second impact multiplier, IM_2, estimates the percentage of equipment lost as a direct result of inundation. Equipment is defined as contents or non-structural components of the infrastructure. In the case of residential buildings this would be the housing contents or anything that would be expected to be taken in a move. Transportation infrastructure (roads, bridges, culverts, and footbridges) and flood protection structures (dykes) do not have an IM_2 component. Buildings and

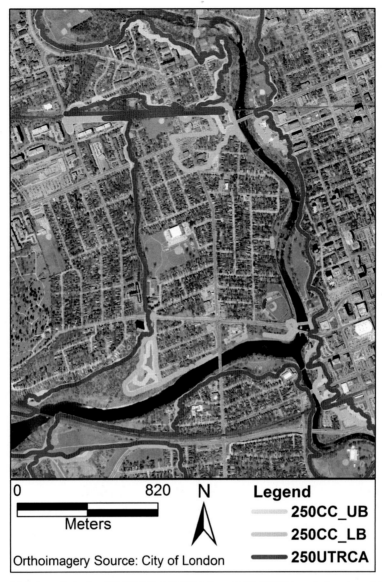

Figure 3.18 Comparison of floodplain boundaries for 250 CC_LB, 250 CC_UB, and 250 UTRCA at the location of the Forks (confluence of the North Thames and the South Thames). See color plates section.

critical facilities have equipment values estimated using methods based on building type and value and are estimated as 30% of the total structure's value. The equipment values for PCPs are estimated based on the City of London's 2010 Wastewater Budget.

Loss of structure (IM_3) The final impact multiplier, IM_3, measures the percentage structural loss of the infrastructure. This is the degree to which the structural integrity is compromised as a result of flooding. The flood depth was used in the calculation of IM_3 in addition to the infrastructure element's condition, age, capacity, and other knowledge gained during interviews with experts in each area. IM_3 is a measure of both quantitative and qualitative structural loss. The methodology uses an innovative approach in the incorporation of qualitative and subjective data with the quantitative measures. The qualitative portion uses fuzzy set theory to allow for subjectivity and differences of opinion with respect to the condition of the infrastructure, its failure mechanisms, and its response to flooding.

The deterministic element of IM_3 is calculated using stage–damage curves. These curves use the inundation depth as input to estimate the percentage damage (LS) to the infrastructure (both structural and contents) as a result of flood inundation.

The qualitative element of IM_3 is used to quantify the subjective uncertainty associated with potential failure of the infrastructure system. Assessment of subjective uncertainty is conducted with the assistance of experts for various types of infrastructure. The qualitative component of IM_3 allows for the measure of partial

Figure 3.19 Comparison of floodplain boundaries for 250 CC_LB, 250 CC_UB, and 250 UTRCA at the location of the Broughdale Dyke (North Thames). See color plates section.

failure as well as the impact of the structure's current conditions on its response to flooding as perceived by experts in the field. This measurement is termed the fuzzy reliability index (El-Baroudy and Simonovic, 2004; Simonovic, 2009). The premise for the combination of the fuzzy reliability index with the quantitative structural loss measure is that the condition of the infrastructure will affect the amount of structural damage sustained by the infrastructure during a flood. The condition of the infrastructure is not quantified by the stage–damage curves and therefore the input of those who are the most familiar with infrastructure may provide for a more accurate assessment of the risk. The condition of the infrastructure is measured using fuzzy analysis through interviews performed with experts within the city (Peck *et al.*, 2011).

Once combined with a flood event, the condition of the infrastructure will affect its structural loss measure. Therefore, to calculate IM_3 the fuzzy risk index and the deterministic measure must be combined. An increase in the compatibility measure indicates a decrease in risk to the particular infrastructure (e.g., a bridge that is considered not to be in an acceptable state with respect to its condition will experience higher damage than a bridge that is considered to be in an acceptable state). To represent this inverse relationship in the calculation of the loss of structure impact multiplier (IM_3), the following equation is used:

$$IM_3(CM) = \begin{cases} 1, & CM = 0 \\ Min\left(1, LS \times \dfrac{1}{CM}\right), & CM > 0 \end{cases} \quad (3.16)$$

Figure 3.20 Dissemination areas for the City of London, Ontario, as of 2007.

where IM_3 is the loss of structure impact multiplier used in Equation (3.12); CM is the compatibility measure obtained through the fuzzy compatibility analysis (Peck *et al.*, 2011); and LS is the percentage loss of structure from the stage–damage curves ($LS \leq 1$).

Therefore, in this study when CM is 0, the structure is deemed to be completely unsafe, or experiencing a total loss ($IM_3 = 1$). The stage–damage curves are assumed to represent the damage to a structure at a completely acceptable state. As such, for CM less than 1, the risk to the infrastructure will increase proportionally. A CM value of 1 (completely safe) will yield $IM_3 = LS$.

ECONOMIC LOSS

Economic loss refers to the potential monetary damage incurred by an infrastructure element as a result of a flood event. It is a value that is applied which provides a higher weight to those structures that are more expensive to repair or replace. This is in favor of the City of London's priority of protecting and investing in the infrastructure that could potentially cause the most interference as a result of a flood event. The economic loss factor is different for each piece of infrastructure. There is an associated economic loss value for each type of impact multiplier (IM_1, IM_2, IM_3) as shown in Equation (3.14). These may be referred to as monetary losses due to loss of infrastructure function (D_1), monetary losses associated with infrastructure's equipment (D_2), and

finally the monetary loss incurred by damage to the infrastructure itself (D_3).

3.3.7 Analysis of results

The implementation of the methodology to the City of London offers, first, insight into the spatial risk for each climate scenario individually that results from an aggregation of all the risk indices. The second set of results provides insight into the risk to the specific infrastructure elements and their individual contributions to the aggregated risk. The third and final set of results provides the comparison of risk between each event scenario (the 100 CC_LB to 100 CC_UB, 250 CC_LB to 250 CC_UB, and 250 UTRCA to 250 CC_UB) and examines the contribution of climate change to the overall risk to London's infrastructure.

For all risk maps, the risk increases from 0 to 1, with 1 being the darkest color and 0 being white or very light grey. The maps are shown for the entire city, with the Forks located at the mid-left of the image. The major roads are labeled for comprehension. The risk maps of the entire city for the five climate scenarios are shown in Figure 3.21.

In the 100 CC_LB scenario, the PCPs and barriers dominate the risk index. The reason for this is that the PCPs have vulnerable equipment of very high value and are in a highly hazardous area, being located along the river. The risk factor for each barrier

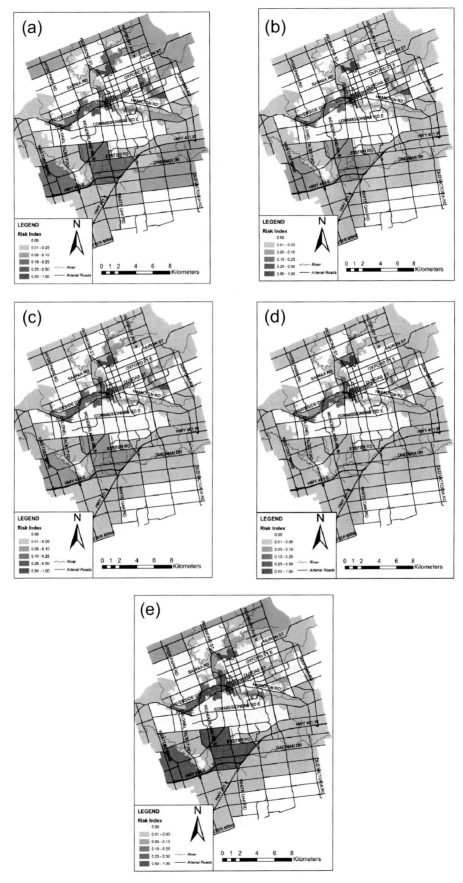

Figure 3.21 Risk to all infrastructure for (a) 100 CC_LB, (b) 100 CC_UB, (c) 250 CC_LB, (d) 250 CC_UB, and (e) 250 UTRCA. See color plates section.

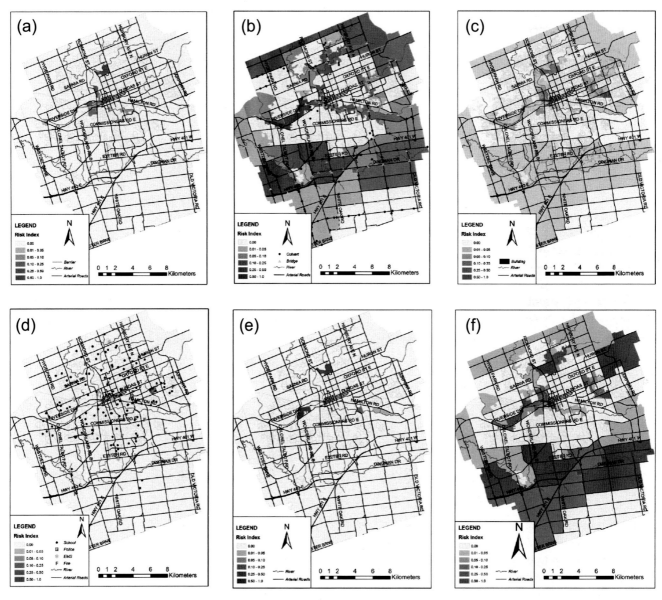

Figure 3.22 100 CC_UB infrastructure risk indices for (a) barriers, (b) bridges, (c) buildings, (d) critical infrastructure, (e) PCPs, and (f) roads. See color plates section.

is based on the consequences of a breach. Therefore, any areas behind a barrier that fails will have a high risk index. For the 100 CC_LB scenario, the highest risk is along the south bank of the Main Thames branch. Due to the high risk at the PCP, the majority of the DAs have risk indices below 0.25. There is a high concentration of risk along the Dingman West branch, due to the size of the DAs and the number of culvert and bridge structures. More detail is examined in the specific infrastructure maps. For the 100 CC_UB scenario, the spatial distribution of risk is very similar to the 100 CC_LB. However, the extent of risk is increased to cover more DAs and the distribution of risk is more even. That is, the risk is more evenly spread across the

city. This is likely due to the fact that under the 100 CC_UB some barriers fail, which leads to high risk being spread across the city.

For the 250 CC_LB, there is a sharp contrast in risk distribution. A few areas containing PCPs and dykes have very high risk values, while the remaining DAs at risk have a much lower risk index in comparison. This demonstrates that the 250 CC_LB is dominated by PCPs and dyke failure. Conversely, the 250 CC_UB has a more even distribution of risk since the PCP and dyke failures are more closely matched with the number of buildings that become catastrophically damaged due to the flood. Finally, the 250 UTRCA scenario shows a very high concentration of risk in

Figure 3.23 Greenway PCP under 100 CC_UB inundation.

the Dingman West area and along the Forks and Main Thames branch. The remainder of the DAs at risk are at a very low risk index in comparison to the others. The reason for this uneven distribution is not immediately apparent and must be examined further in the detailed infrastructure risk maps.

The second set of results shows the specific infrastructure across each climate scenario, enabling further analysis on the contributions of each element to the overall risk. The presentation in Figure 3.22 includes results for all infrastructure types for the 100 CC_UB scenario. The presentation of results for other scenarios is available in Peck *et al.* (2011). Figure 3.22 shows maps for the roads, bridges and culverts, PCPs, critical infrastructure, and barriers for the 100 CC_UB scenario.

By examining the risk indices for each infrastructure type separately, conclusions can be drawn on the contributions of each element to the overall risk. For example, from Figure 3.21(a), it is evident that a high risk area exists along the south bank of the Main Thames branch just downstream of the Forks. By examining Figure 3.22, it is seen that the same area has risk contributed by barriers, bridges, buildings, and a PCP. Some risk is

contributed by roads but it is not as strong as the other categories. In addition, the maps can be used to determine characteristics of one type of infrastructure. One example, shown in Figure 3.23, is a magnified view of Greenway PCP under the 100 CC_UB flood scenario. Here it is possible to see what processes are inundated, which is used in the calculation of the loss of function multiplier.

Another case shows that the highest risk to roads in the 100 CC_UB scenario occurs mainly in the Dingman creek area. Further research into this area can determine the exact causes – such as the poor pavement quality in the area. In this way, the maps can be used to prioritize areas of high risk that should be examined in greater detail.

The next set of results shows the contribution of climate change to the risk to infrastructure across the city. The two maps (Figure 3.24) show the percentage difference between the lower bound and upper bound climate scenarios for both return periods.

For the 100-year flood, the risk map shows that there is a large increase in the risk due to climate change. Many DAs along the north, south, and Medway branches display increases of at least

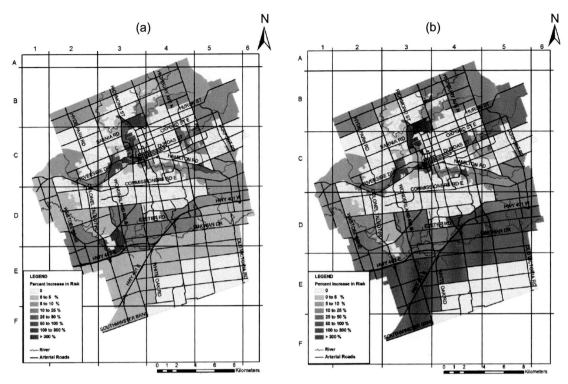

Figure 3.24 Percentage increase in risk index between the lower and upper bound, 100- and 250-year return events. See color plates section.

100% and, in many cases, of over 300%. Dingman West does not appear to be affected strongly by climate change; however, it does undergo a slight increase in risk of between 5% and 25%. On the other hand, the 250-year flood shows a more evenly distributed, but higher increase overall. There is a large increase in risk due to climate change in the Dingman area as well as along the North Thames. There are fewer critical (more than 300% change) areas here than in the change between the 100-year flood scenarios, but the overall change in risk is higher.

The comparison between the 250 UTRCA and 250 CC_UB scenarios is shown in Figure 3.25. The overall trend of the map does show an increase in risk between the two scenarios. The majority of the added risk occurs in the Pottersburg area, Dingman area, and west London. The reason for the large increase in risk in the Pottersburg area is that the current UTRCA floodplain does not accurately model the damming effect of a culvert and rail embankment located west of Clarke road, south of Hamilton Road. During the modeled 250 CC_UB scenario, the culvert reaches capacity and the rail embankment acts as a dam, causing the water to back up and flood the Pottersburg area.

The results of integrated risk analyses are very useful for quickly identifying and prioritizing areas at high risk. Further inspection of the detailed infrastructure maps leads to more information on the specific infrastructure elements that contribute to the overall risk. Areas on the map that have been identified as high risk should be studied further to determine specific courses

of action and policies to enact. The results are of high importance to the City of London. They indicate the most at-risk areas that result from the changes in flood patterns due to climate change that can be expected in the future. These results are used to form recommendations on engineering, operations, and policy for the city (Peck *et al.*, 2011), which have been relayed to the city council via formal reports and a workshop.

3.4 CONCLUSIONS

Adaptation to climate change is defined in the context of flooding as a risk management process. A six-step methodology for risk management is presented as the process to be used in adaptation to climate change. The two main approaches to flood risk management are identified as probabilistic and fuzzy set. The following two parts of the book provide detailed analytical methodologies for the implementation of these two approaches.

An example of the assessment of climate change-caused flood risk to municipal infrastructure is presented for the City of London (Ontario, Canada). It provides an original approach to incorporating climate change into flood risk analysis and combines elements of probabilistic and fuzzy set approaches. The study is conducted for the City of London (Peck *et al.*, 2011) as one of the first steps in the implementation of the municipal climate change adaptation strategy.

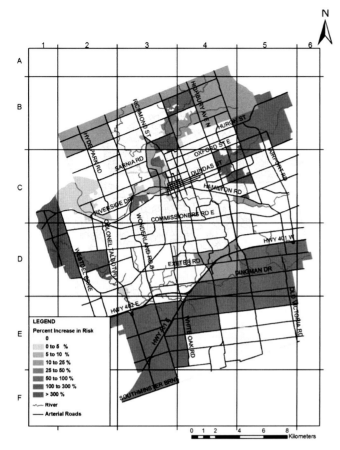

Figure 3.25 Percentage increase in risk between the 250 UTRCA and 250 CC_UB scenarios. See color plates section.

3.5 EXERCISES

3.1. Select one flood example that you are familiar with.

 a. Describe in your own words the integrated flood disaster management process that applies to the selected example.

 b. For the selected example present the details of a six-step flood risk management procedure.

 c. What is in your opinion the most critical step in the flood risk management procedure?

3.2. How is climate change addressed in the example selected in Exercise 3.1?

3.3. What are the two main approaches to flood risk management as adaptation to climate change? Present, in the form of an essay, a minimum of three examples of flood risk management using one, or both, of the approaches introduced in this section (maximum 500 words for each example).

3.4. Read the article by Stirling (1998) listed in the references.

 a. Illustrate your understanding of the concepts presented in Figure 3.3.

 b. Provide an example to support your answer to question (a).

3.5. How is climate change addressed in the City of London example presented in Section 3.3?

 a. Provide your understanding of the advantages and disadvantages of the approach based on the range of potential impacts of climate change (use of upper and lower bound of change).

 b. What are the other ways of incorporating the climate change impact in similar analyses?

3.6. Describe the risk assessment methodology used in the City of London example presented in Section 3.3.

3.7. Read the article by Sharif and Burn (2006) listed in the references.

 a. What is the main advantage of the WG tool presented in the article?

 b. Explain the perturbation process.

 c. Is there, in your opinion, a way to improve the tool presented in the article?

3.8. The City of London example uses 100-year and 250-year floods as regulatory floods in the region.

 a. Find the definitions of these two regulatory floods from the documentation that applies to Ontario (Canada). (Tip: search the web information available for the local conservation authority.)

 b. Are you familiar with similar regulations in your region? Discuss the regulations that apply to your region.

3.9. How would you apply/modify the risk calculation procedure for the example in your region?

 a. Describe the example.

 b. Identify the key elements of risk.

 c. Show your modified risk calculation procedure.

Part II
Flood risk management: probabilistic approach

4 Risk management: probabilistic approach

In flood risk management probability is used based on historical records, for example, to determine the magnitude of the 250-year flood. This probability is called objective (as pointed out in Section 3.2) because it is determined by long-run observations of the occurrence of an event.

4.1 MATHEMATICAL DEFINITION OF RISK

At a conceptual level, we defined risk (in the Definitions) as a significant potential unwelcome effect of system performance, or the predicted or expected likelihood that a set of circumstances over some time frame will produce some harm that matters. More pragmatic treatments view risk as one side of an equation, where risk is equated with the probability of failure or the probability of load exceeding resistance. Other symbolic expressions equate risk with the sum of uncertainty and damage, or the quotient of hazards divided by safeguards (Samuels, 2005; Singh *et al.*, 2007; Simonovic, 2009, 2011).

Let us start with a risk definition based on the concept of load and resistance, terms borrowed from structural engineering. In the field of flood risk management these two variables have a more general meaning. For example, the flood rate can represent the load, whereas the levee height can be considered as the resistance. Load l is a variable reflecting the behavior of the system under certain external conditions of stress or loading. Resistance r is a characteristic variable that describes the capacity of the system to overcome an external load. When the load exceeds the resistance ($l > r$) there should be a failure or an incident. A safety or reliability state is obtained if the resistance exceeds or is equal to the load ($l \leq r$).

Mathematically, risk may be expressed as the probability of exceeding a determined level of economic, social, or environmental consequence at a particular location and during a particular period of time. Convolution is a mathematical operation on two functions f and g, producing a third function that is typically viewed as a modified version of one of the original functions. Convolution can be used to describe the mutual conditioning of hazard

and vulnerability. They are mutually conditioning situations and neither can exist on its own.

Most commonly in flood management, risk has been defined as the probability of a system failure. Risk involves uncertainty as well as loss or damage. The risk of flooding involves the probability of occurrence of the flood as well as the damage that might result from the flood event. To understand the linkage between hazard and risk it is useful to consider the source–pathway–receptor–consequence (SPRC) framework of Sayers *et al.* (2002) shown in Figure 4.1. This is, essentially, a simple conceptual model for representing systems and processes that lead to a particular consequence. For a risk to arise there must be a hazard that consists of a "source" or initiator event (e.g., high water level), a "receptor" (e.g., flood plain properties), and a "pathway" between the source and the receptor (e.g., flood levee). A hazard does not automatically lead to a harmful outcome, but identification of a hazard does mean that there is a possibility of harm occurring, with the actual harm depending upon the exposure to the hazard and the characteristics of the receptor. The bottom part of Figure 4.1 links the SPRC framework to the load–resistance concept of flood risk definition.

Equation (1.1) in Section 1.3 was used as the most general mathematical definition of flood risk. However, a drawback of this definition is that it equates the risk of a high probability/low consequence event with that of a low probability/high consequence event. In real life these two events may not result in the same risk. Therefore, an improved measure of flood risk is given by:

$$Risk = Hazard\ uncertainty$$
$$+ Consequence\ of\ the\ system's\ vulnerability \quad (4.1)$$

To illustrate the definition let us consider the following simple example.

Example 1
Most people are interested in large-scale flooding. What is the risk of the largest flood on record occurring next summer in London, Ontario?

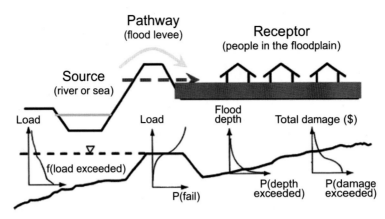

Figure 4.1 Source–pathway–receptor–consequence framework for the assessment of flood risk (after Sayers *et al.*, 2002).

Solution

We would like to know whether the coming summer in London will be the wettest on record and the consequences if it is. Thus, risk (assuming large flooding events are harmful) in this case is the probability of occurrence of the largest flood next summer in London and the resulting consequences in terms of flood damage to property in London, crop damage, damage to infrastructure, loss of business, cost of emergency management, and so on.

Flood risk is conditional and conditions are often implied by the context and are not explicitly stated. For example, the risk of death from a flood is relatively small in the developed countries, but its value will significantly differ from one place to another and from one country to another, depending on the climate, flood protection measures, warning issued, communication, rescue operations, people's perception, etc. In contrast, the risk of death from a flood of the same magnitude can be relatively large in the developing world. The Indian Ocean tsunami of December 26, 2004, caused an unimaginable loss of human life (over 230,000 people) in Indonesia and surrounding countries. The very large tsunami of March 11, 2011, following the Tohoku earthquake in Japan, caused loss of human life in the range of 30,000 – incomparable to the 2004 event in the Indian Ocean. The difference is due to advanced warning systems, better infrastructure, and the receptivity of Japanese people to warning.

Hazard is a situation or occurrence of an event that could, in particular circumstances, lead to harm. Hazard can be considered as a latent danger or an external risk factor of an exposed system. This can be mathematically expressed as the probability of occurrence of an event of a certain intensity at a particular location and during a determined period of exposure. Thus, hazard is a source of risk and risk includes the chances of conversion of that source into actual loss. For example, it is not advisable to cross a river in a rubber boat during flooding conditions, because the chances of drowning are relatively significant even for an expert swimmer. But if one attempts to cross the river by motorboat, equipped with a powerful engine, rugged body, and life jackets, etc., the risk of drowning is considerably smaller. Thus safeguards help reduce risk. Mathematically, one can write (Kaplan and Garrick 1981):

$$Risk = Hazard/Safeguards \qquad (4.2)$$

Equation (4.2) uses division rather than subtraction. As safeguards tend to zero, even a small hazard can lead to a high value of risk; as safeguards increase, the risk becomes smaller. In day-to-day life this equation is seen to work. For example, after the tragic experience of Hurricane Katrina in 2005 in New Orleans and along the Mississippi Gulf Coast, people in western Louisiana and Texas heeded the warnings about Hurricane Rita and the result was significantly lower loss of human life.

Example 2

Consider a detention pond for local flood control in an urban area. What could the risk be from this detention structure?

Solution

The detention dam may be overtopped and breached. As a result, the dam breach may cause harm to people in the urban area. The risk would be the probability of specified damage or harm in a given period. For water control structures, hazard from failure depends on the size of the structure. Therefore, decisions about the recommended design load (or design flood) are based upon the size of the structure and its hazard potential. Widely used recommendations of the USACE regarding selection of spillway design flood for a dam are given in Table 4.1. A typical classification of reservoirs according to size and the hydraulic head is given in Table 4.2. The hazard potential classification of reservoirs is given in Table 4.3.

Characterization of risk helps establish its specific context. To characterize risk, two basic elements are necessary: (i) probability of occurrence of hazard; and (ii) extent of damage (governed by the vulnerability of a given system). The concept of triplet definition of risk introduced in the Definitions is widely applied in flood risk management: (i) What can happen – what can go

Table 4.1 *Recommendations for selection of design flood*

Hazard	Size	Design flood
Low	Small	50–100 years
	Intermediate	100 years – 0.5 PMF
	Large	0.5 PMF – PMF
Significant	Small	100 years – 0.5 PMF
	Intermediate	0.5 PMF – PMF
	Large	PMF
Large	Small	0.5 PMF – PMF
	Intermediate	PMF
	Large	PMF

PMF: probable maximum flood

Table 4.2 *Reservoir classification by size*

Type of dam	Storage capacity (10^6 m^3)	Hydraulic head	Inflow design flood
Small	0.5–1.0	7.5–12	100 year
Intermediate	10–60	12–30	SPF
Large	>60	>30	PMF

SPF: standard project flood
PMF: probable maximum flood

Table 4.3 *Reservoir hazard potential classification*

Category	Loss of life	Economic loss
Low	None expected	Minimal
Significant	Few	Appreciable
High	More than a few	Excessive

Table 4.4 *Illustrative list of scenarios for levee failure*

Scenario or event	Probability	Consequences
Structural failure		
Levee overtopping	0.001	Failure of levee
Piping	0.002	Excessive erosion
Failure of levee drainage	0.0004	May lead to failure of levee
Sloughing on levee slopes	0.0009	Localized damage to levee body
Earthquake of magnitude >8 on Richter scale occurs	0.0001	Extensive damage to the levee
Performance failure		
Flood bigger than design flood occurs	0.0002	Large water levels damage levee
Others	0.00003	Unknown

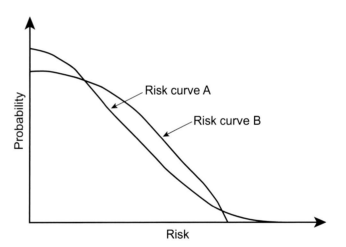

Figure 4.2 Risk curves.

wrong during a given event? (ii) How likely is it to happen – what is the probability of the event occuring? (iii) If it does happen, what are the consequences? (Simonovic, 2009 after Kaplan and Garrick, 1981). These three questions are answered by preparing a list of scenarios in which the answers to the questions are arranged as a triplet:

$$R = [s_i, p_i, x_i], \quad i = 1, 2, \ldots, N \quad (4.3)$$

where s_i is the scenario or hazardous event identification, p_i is the probability of that event or scenario, x_i is a measure of loss and represents the consequences of that event or scenario, and R is the risk. The scenario list can be arranged in the form of a table. An example table for a failure of a levee is shown in Table 4.4.

If this table contains all the possible scenarios, it is the estimation of risk. In real life, the list of scenarios can be extensive. Kaplan and Garrick (1981) suggest that a category "others," including all the scenarios that have not been thought of, may be

added to the list for completeness. The problem of assigning a probability to this category remains to be solved.

The triplet definition of risk, Equation (4.3), suggests that hazard can be defined as a subset of risk, a set of doubles:

$$H = [s_i, x_i], \quad i = 1, 2, \ldots, N \quad (4.4)$$

If the consequences are arranged in order of increasing severity of damage:

$$x_1 \leq x_2 \leq \ldots \leq x_N \quad (4.5)$$

then the second column of Table 4.4 can be accumulated and a smooth curve relating x and p can be plotted as shown in Figure 4.2, known as a risk curve.

4.2 CLASSIFICATION OF RISK

Singh *et al.* (2007) classify risk into three categories: (i) risks for which statistics of identified casualties are available; (ii) risks for which there may be some evidence, but where the connection between suspected cause and damage cannot be established; and (iii) estimates of probabilities of events that have not yet occurred. Additionally, there are risks that are unforeseen.

All systems have a probability of failure no matter how small it is and the complete avoidance of all risk is not possible. The objective is to reduce the probability to an acceptable level of individual and societal risk. An engineering approach to quantify risk begins with a physical appreciation of possible failure mechanisms or modes and their analysis. This entails quantification of the reliability of the components and examination of the systematic failure to establish the overall vulnerability of the complete system, based on experience, and verified by analysis, testing, and inspection.

An examination of past events helps with an understanding of failure modes. Consider, for example, the case of river levees for flood control and their failure modes. In light of the potential for major consequences involved in levee failures, it is inappropriate to wait for disasters to occur such that a body of case histories can be built to form a basis for management of floods. Therefore, an anticipatory approach based on judgment and experience is required. Such an approach can follow a systems view (discussed in Section 1.5) of using risk estimation through methods based on a systematic decomposition of a complex system into its component subsystems and the use of predictive techniques and modeling. The system failure mechanisms can be analyzed and risk can be estimated by aggregating together models of individual subsystems. This method requires a wide range of data on past failures and knowledge about the various processes that could occur.

There are other more traditional methods in widespread use that are essentially deterministic in nature. A deterministic method can be illustrated by the employment of a factor of safety, which can be defined as the ratio of design load to design resistance. A multitude of factors affect both design load and design resistance. In practice, therefore, there will be a distribution of load and resistance, with their mean and variance values. If the mean load is smaller than the mean resistance, then it can be shown that failure will occur where the upper end of the load distribution encounters the lower end of the resistance distribution. This leads to definitions of safety factors and safety margins in *probabilistic* form.

Although a deterministic approach incorporates the concept of the variability of load and resistance, it implies that there is a level of probability of failure that is acceptable for design purposes, and that level can be quantified. In contrast, the probabilistic approach includes the low-probability events in the assessment. By necessity, sufficient data must be available. In terms of decision-making, the deterministic approach incorporates implicit value judgments as to an acceptable standard of practice and is derived from an

extension of past practice and experience, which may be inadequate to deal with rapidly changing conditions. The probabilistic approach, on the other hand, describes hazard in terms of risk of failure and its associated consequences. Thus, it enables an acceptable decision to be made based on a design process and the needed judgments to be made.

The risk of failure and its consequences are significantly influenced by management. Management of safety involves the training of staff to observe, record, and report; not to panic; to systematically react to the onset of a potential flood disaster; and to organize evacuation and rescue procedures in the event of a flood disaster.

Risk can be negative and positive. The concept of risk is widely used in a society that is future oriented, which sees the future to be managed. We wish to minimize a multitude of risks, such as those related to human health, flooding, environmental pollution, disease epidemic, drug addiction, wildfires, and social violence. This indeed is the basis of insurance, introduced as a mechanism to spread the risk (Skipper and Kwon, 2007).

The notion of risk is inseparable from the concepts of probability and uncertainty. Risk is not the same as hazard or danger. It relates to hazards that are assessed in terms of future possibilities. One classification distinguishes between two types of risk: *external* and *manufactured* (Singh *et al.*, 2007). External risk is the risk that originates from outside, or from nature. Manufactured risk is the risk that is created by our actions and can occur in a situation that we have very little experience of confronting. According to this classification, the flood risks due to climate change are manufactured risks, influenced by intensifying globalization. This is the risk created by the very impact of our developing knowledge upon the world. Flooding risk from land use change, such as urbanization, can be categorized as a manufactured risk too. From the earliest days of human civilization up to the threshold of modern times, risks were primarily due to natural or external sources: floods, famines, earthquakes, tsunamis, etc. Recently, the focus has shifted from what nature does to us to what we have been doing to nature. This marks the transition from the predominance of external risk to that of manufactured risk. Therefore, natural phenomena, such as floods, droughts, and extreme weather, are not entirely natural; rather, they are being influenced by human activities, as suggested by their unusual features. As manufactured risk expands, there is a new riskiness to risk. The very idea of risk is tied to the possibility of calculation. However, in many cases we simply do not know what the level of risk is and we could not know for sure until it is too late. Thus, one way to cope with the rise of manufactured risk is to employ the precautionary principle. It presupposes action about issues even though there is insufficient scientific evidence about them. This leads to the concept of risk management: the balance of risk and danger.

Another classification of risk, introduced in Section 3.1.1, leads to risk management confusion due to an inadequate distinction between three fundamental types of risk (Slovic, 2000; Simonovic, 2009, 2011):

- *Objective risk* (real, physical), R_o, and objective probability, p_o, which is a property of real physical systems.
- *Subjective risk*, R_s, and subjective probability, p_s. Probability is here defined as the degree of belief in a statement. R_s and p_s are not properties of the physical systems under consideration (but may be some function of R_o and p_o).
- *Perceived risk*, R_p, which is related to an individual's feeling of fear in the face of an undesirable possible event, is not a property of the physical systems but is related to fear of the unknown. It may be a function of R_o, p_o, R_s and p_s.

Because of the confusion between the concepts of objective and subjective risk, many characteristics of subjective risk are believed to be valid also for objective risk. Therefore, it is almost universally assumed that the imprecision of human judgment is equally prominent and destructive for all flood risk evaluations and all risk assessments. This is perhaps the most important misconception that blocks the way to more effective societal flood risk management. The ways society manages risks appear to be dominated by considerations of perceived and subjective risks, while it is objective risks that kill people, damage the environment and create property loss.

4.3 RISK CRITERIA

Natural hazards, such as floods, have always been a source of concern to planners, designers, operators, and managers of engineering systems. Because of their destructive nature, they are assigned a major role in design considerations. There are many phenomena for which it is difficult to assess the degree of risk. By increasing the factor of safety one can reduce the risk, but there is a natural reluctance to pay the often exorbitant cost that is associated with the safety factor. Even with an increase in safety, there will always be some risk because it is not possible to build a system that is so strong that it can withstand all conceivable disasters. Thus, in reality flood risk cannot be eliminated entirely, but it can be reduced to an acceptable level.

A risk criterion is a qualitative and quantitative assessment of the acceptable level of risk with which to compare the assessed risk. A risk criterion is employed to balance the risk of loss against the cost of increase in safety. There are several probabilistic risk criteria that are employed in flood risk management (modified after Borgman, 1963): (i) return period, (ii) encounter probability, (iii) distribution of waiting time, (iv) distribution of total damage, (v) probability of zero damage, and (vi) mean total damage. The choice of the most appropriate criterion is a matter of decision-making judgment. These criteria emphasize different risk aspects and can be derived from mathematical risk models.

We have already pointed out that risk cannot be perceived simply as a one-dimensional objective concept, such as the product of the probabilities and consequences of any event. Risk perception is inherently multi-dimensional and subjective, with a particular risk or hazard meaning different things to different people and different things in different contexts. So we can say that risk is a social construct.

Statistical estimation of a risk may involve developing probabilities of future events based on a statistical analysis of past historical events. One of the key steps in risk assessment is defining *acceptable risk*. The acceptability depends on the context in which to assess risk and the availability of financial resources. It may be added that the notion of an acceptable probability is highly subjective. Consider a design that can give protection against a flood event x up to a certain magnitude x_d. The design is acceptable if the exceedance probability, $Prob(x \geq x_d) = 1 - Prob(x_d)$, is smaller than the acceptable value P_{Acc} (Plate, 2002). Vrijling *et al.* (1995) define the acceptable value of risk for a person that reflects their preference as:

$$P_{Acc} = \frac{\beta_i \times 10^{-4}/year}{v_{ij}} \quad (4.6)$$

where β_i can range from 0.01 for a high risk for an action that gives no benefit to the person to 10 for a risky activity that brings high satisfaction to the person; and v_{ij} is the vulnerability of an individual to an event x_i. For a larger region or a nation, the acceptable probability, according to Vrijling *et al.* (1995), becomes:

$$P_{Acc} = \frac{10^{-3}/year}{n^2} \quad \forall n \geq 10 \; casualties \quad (4.7)$$

A problem with the idea of acceptable risk is that the acceptability or not of a risk can only be expressed together with the associated costs and benefits. A decision-maker will choose the optimum mixture of risk, cost, and benefit and might be willing to take a higher risk only if it is associated with either lower cost or more benefit. The relationships (4.6) and (4.7) implicitly assume that risks are linearly comparable, which is known not to be true. If we look at Figure 4.2 we cannot say whether the risk is higher in the case of curve A or curve B.

After the flood hazard events have been identified and their data collected, an assessment of risk can be made. In the case of flooding, unwanted situations arise if the water level exceeds a critical value.

Example 3

City X is located on the banks of River Y. The river flow follows an extreme value distribution with a mean of 256.8 m^3/s and standard deviation of 78.2 m^3/s. The rating curve of River Y at a gaging site near X is given by $Q = 97.03 \times (h - 214.5)^{0.64}$. The water authority is asked by a developer Z to approve construction of an office building on a plot of land near the river at an elevation of 223.0 m. Find the risk of flooding at this plot every year. If the acceptable flood risk for approval of a development is 5%, will the developer be granted a permit? If the risk is greater than 5%, what will be the necessary increase in ground elevation to provide for granting a permit?

Solution
Parameters of the extreme value distribution can be calculated from the provided mean and variance (see for example Singh *et al.*, 2007, page 214) as follows:

$$\alpha = \frac{1.282}{78.2} = 0.0164 \tag{4.8}$$

$$u = 256.8 - \frac{0.5772}{0.0164} = 221.605 \tag{4.9}$$

From the river rating curve, the flow at stage 223.0 m will be 381.7 m³/s. So, the probability that the flow in a given year will exceed a value of $q = 381.7$ m³/s will be:

$$P[Q \geq q] = 1 - F_Q(q) = 1 - e^{-e^{[0.0164(q-221.605)]}}$$
$$P[Q \geq 381.7] = 1 - F_Q(381.7) = 1 - e^{-e^{[0.0164(381.7-221.605)]}}$$
$$P[Q \geq 381.7] = 0.0695 \approx 7\% \tag{4.10}$$

There is about 7% risk that the office building will be under water in any given year. Since the acceptable risk is 5%, the water authority will not issue a permit.

If the developer wants to build the office building at this very site, it will have to raise the plinth level by soil filling. In order to achieve the acceptable risk level of 5%, the flow calculated from (4.10) should be $q_{5\%} = 403$ m³/s. Using the rating curve we find that the safe building elevation is 223.75 m. So raising the ground elevation by 0.75 m will be required in order to secure the development permit at the selected location.

4.4 PROBABILISTIC RISK MODELING

Fundamental to probabilistic risk modeling is the assessment of uncertainties. These cover a range of types. This section presents the discussion of statistical independence-based risk modeling and return period modeling modified after Singh *et al.* (2007).

4.4.1 Statistical independence-based risk modeling

Consider a flood time series where at each integer value of time, say, $1, 2, 3, \ldots$, along the time axis t, flood events occur. The flood time series may be water elevation, flood discharges, and so on. Each flood event has a flow value, denoted by X, which measures the degree of danger the event creates. It is assumed that X is a random variable with the distribution function $F(x)$ defined as the probability that X will be less than or equal to x (a spec value of X) for one of the selected future flood events. In mathematical notation:

$$F(x) = P[X \leq x] \tag{4.11}$$

where $P(.)$ denotes the probability that the flood event within parentheses will not occur. It is assumed that the flow values of flood events are statistically independent and have the same distribution $F(x)$. For statistical independence, the time scale for measuring X may be in years. If the flood flow of a river at a given gaging station, for example, represents the time series, then X may represent the instantaneous maximum annual flood flow. At the end of the first year ($t = 1$), the flow value of the flood event would be the maximum flow for that year. Similarly, if the time series is represented by the water elevation at a particular location, then the largest elevation during a year would be one value of X for that year.

4.4.2 Return period modeling

In flood risk modeling and hydraulic design in general, the return period is commonly used, because of its simplicity. However, it is prone to misinterpretation and misuse. The return period $T(x)$ is defined as

$$T(x) = \frac{1}{1 - P[X \leq x]} = \frac{1}{P[X \geq x]} \tag{4.12}$$

where $T(x)$ represents the average time between flood events having values equal to or exceeding x. It does not mean that the flood event will certainly take place. The flood events occurring in the time interval $n - 1 < t \leq n$ are plotted at $t = n$, not at their actual time of occurrence in the interval. $T(x)$ will be slightly greater than the real average time between the flood events with flow of x. The magnitude of the difference will depend on the time scale used. For example, the return period for time measured in years will be different from the return period for time measured in decades.

Consider a random variable $W(x)$ that denotes the time interval between two successive exceedances of x. Then, $W(x)$ can be referred to as the time between events. If exceedance of x has just occurred, we can compute the probability that the next exceedance will occur n time units away. Because of the assumption of statistical independence, what occurs during one time unit has no effect on the probabilities of future occurrences. This means that the same result would be obtained if any integer on the time axis was selected as the initial point, irrespective of whether an exceedance has just occurred. If $W = n$, this means that there must have been $n - 1$ flood events without an exceedance of x (probability of each $= q$) followed by an exceedance (probability $= p$). Here the following relationships apply:

$$p = 1 - F(x) \quad \text{and} \quad q = F(x) = 1 - p \tag{4.13}$$

Thus,

$$P(N = n) = q^{n-1}p = [F(x)]^{n-1}[1 - F(x)], \quad n = 1, 2, \ldots$$
$$P(N = n) = q^{n-1}(1 - q) = q^{n-1} - q^n \tag{4.14}$$

The expected or theoretical average of n can be expressed as:

$$T = 1P(N = 1) + 2P(N = 2) + 3p(N = 3) + \cdots$$

$$T = E[N] = \sum_{n=1}^{\infty} nP(N = n) = \sum_{n=1}^{\infty} nq^{n-1} - \sum_{n=1}^{\infty} nq^n \quad (4.15)$$

which results in:

$$T = \frac{1}{1 - q} = \frac{1}{1 - F(x)} \quad (4.16)$$

The quantity T is known in hydrologic literature as the average return period. On average, a flood above a level x will occur once every T years. It should be noted that the distribution function of N is a geometric progression:

$$P(N \le n) = p(1 + q + q^2 + q^3 + \cdots + q^{n-1}) = p\frac{1 - q^n}{1 - q}$$

$$= 1 - q^n \quad (4.17)$$

Since $q = F(x) = 1 - 1/T$, Equation (4.17) can be expressed in terms of the return period as:

$$P(N \le n) = 1 - \left(1 - \frac{1}{T}\right)^T \quad (4.18)$$

We can also find the variance of N as follows:

$$Var[N] = E[N]^2 - [E(N)]^2, \quad 0 < p < 1 \quad (4.19)$$

or

$$Var[N] = \frac{1 + q}{(1 - q)^2} - \frac{1}{(1 - q)^2} = \frac{q}{(1 - q)^2} = \frac{q}{p^2} \quad (4.20)$$

Let us illustrate these concepts with an example.

Example 4

Find the variance of the return period if $q = 0.9$ and 0.99.

Solution
When $q = 0.9$, $p = 1 - 0.9 = 0.1$. Hence $T = 10$ and variance can be calculated using Equation (4.20) as:

$$Var[T] = \frac{q}{p^2} = \frac{0.9}{0.01} = 90 \quad (4.21a)$$

In the second case when $q = 0.99$, $p = 1 - 0.99 = 0.01$ and $T = 100$:

$$Var[T] = \frac{q}{p^2} = \frac{0.99}{0.0001} = 9900 \quad (4.21b)$$

This example shows that $E[N]$ is not the best measure. Therefore, it is better to calculate the probability of no exceedance within a given period of n as:

$$P(N > n) = \alpha$$

$$P(N > n) = \sum_{k=n+1}^{\infty} P(N = k) = \sum_{k=n+1}^{\infty} q^{k-1}(1 - q) \quad (4.22)$$

which yields

$$\alpha = q^n = [F(x)]^n \quad (4.23)$$

Equation (4.23) has three variables: α, n, and x. Recall that:

$$T = \frac{1}{1 - q} \quad \text{or} \quad q = 1 - \frac{1}{T} \quad (4.24)$$

$$\alpha = \left(1 - \frac{1}{T}\right)^n = \left(1 - \frac{1}{T}\right)^T \quad \text{for} \quad n = T \quad (4.25)$$

and therefore:

$$\lim_{T \to \infty} \left(1 - \frac{1}{T}\right)^T = e^{-1} \cong 0.368 \quad (4.26)$$

According to Equation (4.26), P (at least one exceedance within a large return period) is:

$$P = 1 - 0.368 = 0.632 \quad (4.27)$$

Example 5

For the 100-year flood (with a return period of 100 years), find the probability of at least one exceedance within the return period.

Solution
Given $T = 100$, then using Equation (4.25) we find:

$$\alpha = \left(1 - \frac{1}{T}\right)^n = \left(1 - \frac{1}{100}\right)^{100} = 0.99^{100} = 0.366 \quad (4.28)$$

and

$$P(A) = 1 - 0.366 = 0.634 \quad (4.29)$$

4.4.3 Flood risk and return period

The link between flood risk and return period is of significant importance in planning, design, and operation of flood protection infrastructure. As in the previous section, let $p = 1 - F(x)$ and $q = F(x) = 1 - p$ be the probability that a value of the random variable X will be equal to or greater than x. The probability that x will occur in the next year by definition is $p = 1/T$. The probability that x will not occur in the next year is $q = 1 - p = 1 - 1/T$. The probability that x will be equaled or exceeded in any n successive years is given by $(1 - 1/T)^n$. The probability that x will occur for the first time in n years is:

$$q^{n-1}p = \left(1 - \frac{1}{T}\right)^{n-1}\frac{1}{T} \quad (4.30)$$

The probability that x will occur at least once in the next n years is the sum of the probabilities of its occurrence in the first, second, . . . and nth years and is therefore:

$$p + pq + pq^2 + \cdots + pq^{n-1} \quad (4.31)$$

So, the risk, or probability that the event will occur once is:

$$R = 1 - q^n = 1 - \left(1 - \frac{1}{T}\right)^n \quad (4.32)$$

Table 4.5 *Encounter probabilities E_1 [$= 1 - (1 - 1/T)^L$] for estimated life L and return period T_1*

L	T_1																	
	5	10	15	20	25	30	40	50	60	80	100	120	160	200	250	300	400	500
1	0.200	0.100	0.067	0.050	0.040	0.033	0.025	0.020	0.017	0.012	0.010	0.008	0.006	0.005	0.004	0.003	0.002	0.002
2	0.360	0.190	0.129	0.098	0.078	0.066	0.049	0.040	0.033	0.025	0.020	0.017	0.012	0.010	0.008	0.007	0.005	0.004
3	0.488	0.271	0.187	0.143	0.115	0.097	0.073	0.059	0.049	0.037	0.030	0.025	0.019	0.015	0.012	0.010	0.007	0.006
4	0.590	0.344	0.241	0.185	0.151	0.127	0.096	0.078	0.065	0.049	0.039	0.033	0.025	0.020	0.016	0.013	0.010	0.008
5	0.672	0.410	0.292	0.226	0.185	0.156	0.119	0.096	0.081	0.061	0.049	0.041	0.031	0.025	0.020	0.017	0.012	0.010
6	0.738	0.469	0.339	0.265	0.217	0.184	0.141	0.114	0.096	0.073	0.059	0.049	0.037	0.030	0.024	0.020	0.015	0.012
7	0.790	0.522	0.383	0.302	0.249	0.211	0.162	0.132	0.111	0.084	0.068	0.057	0.043	0.034	0.028	0.023	0.017	0.014
8	0.832	0.570	0.424	0.337	0.279	0.238	0.183	0.149	0.126	0.096	0.077	0.065	0.049	0.039	0.032	0.026	0.020	0.016
9	0.866	0.613	0.463	0.370	0.307	0.263	0.204	0.166	0.140	0.107	0.086	0.073	0.055	0.044	0.035	0.033	0.025	0.020
10	0.893	0.651	0.498	0.401	0.335	0.288	0.224	0.183	0.155	0.118	0.096	0.080	0.061	0.049	0.039	0.032	0.025	0.020
12	0.931	0.718	0.563	0.460	0.387	0.334	0.262	0.215	0.183	0.140	0.114	0.096	0.072	0.058	0.047	0.039	0.030	0.024
14	0.956	*0.771*	0.619	0.512	0.435	0.378	0.298	0.246	0.210	0.161	0.131	0.111	0.084	0.068	0.055	0.046	0.034	0.028
16	0.972	0.815	0.668	0.560	0.480	0.419	0.333	0.276	0.236	0.182	0.149	0.125	0.095	0.077	0.062	0.052	0.039	0.032
18	0.982	0.850	0.711	0.603	0.520	0.457	0.366	0.305	0.261	0.203	0.165	0.140	0.107	0.086	0.070	0.058	0.044	0.035
20	0.988	0.878	0.748	0.642	0.558	0.492	0.397	0.332	0.285	0.222	0.182	0.154	0.118	0.095	0.077	0.065	0.049	0.039
25	0.996	0.928	0.822	0.723	0.640	0.572	0.469	0.397	0.343	0.270	0.222	0.189	0.145	0.118	0.095	0.080	0.061	0.049
30	0.999	0.958	0.874	0.785	0.706	0.638	0.532	0.455	0.396	0.314	0.260	0.222	0.171	0.140	0.113	0.095	0.072	0.058
35	0.999+	0.976	0.911	0.834	0.760	0.695	0.588	0.507	0.445	0.356	0.297	0.254	0.197	0.161	0.013	0.110	0.084	0.068
40	0.999+	0.985	0.937	0.871	0.805	0.742	0.637	0.554	0.489	0.395	0.331	0.284	0.222	0.182	0.148	0.125	0.095	0.077
45	0.999+	0.991	0.955	0.901	0.841	0.782	0.680	0.597	0.531	0.432	0.364	0.314	0.246	0.202	0.165	0.140	0.107	0.086
50	0.999+	0.955	0.968	0.923	*0.870*	0.816	0.718	0.636	0.568	0.467	0.395	0.342	0.269	0.222	0.182	0.154	0.118	0.095

The value of risk can be obtained directly from the probability of non-exceedance in n years. This equation can be used to calculate the probability that x will occur within its return period:

$$R_T = 1 - \left(1 - \frac{1}{T}\right)^T \tag{4.33}$$

For large T, it has already been shown that $P_T = 1 - e^{-1} \cong 0.63$. This indicates that the probability that x will occur within its return period is about 64%. Thus, a levee designed to withstand a flood with a 100-year return period has a 64% chance that this design flood will be exceeded before the end of the first 100-year period.

For flood infrastructure design purposes it might be desirable to specify some probability that the undesirable event would occur within the design period and calculate the required return period. If R is the risk that the event will occur within the design period, then using Equation (4.33) we can calculate the design return period T. The probability R is also called the encounter probability. Suppose a levee is built for a postulated life of L time units (say, years). The probability that an event with intensity x will occur during the life of the levee is the encounter probability, $E(x)$, and is a measure of risk. The probability of no exceedance during L time units is $[F(x)]^L$. Hence, the probability of one or more exceedances is:

$$E = 1 - [F(x)]^L \tag{4.34}$$

so the relationship between E and T is given by:

$$E = 1 - \left[1 - \frac{1}{T}\right]^L \tag{4.35}$$

Equations (4.34) and (4.35) have the same appearance because there are one or more exceedances of x in time L if $N \leq L$. Hence, $E = P(N \leq L)$. Table 4.5 shows values of the encounter probability for various values of the estimated life L and return periods T. Table 4.6 shows the return periods for various values of the encounter probability and estimated life. A comparison of the time between events and the encounter probability brings out several interesting properties. For example, a levee with a 50-year life has a better than even chance of encountering 50-year floods during its life. Indeed, the probability is 0.636. Thus, depending on the amount of risk the decision-maker is willing to take, a much higher return period flood will have to be used for a 50-year levee. As an example, for a 10% risk, a 475-year flood will have to be used.

Example 6

Suppose a flood control reservoir and corresponding dam are designed with a projected life of 25 years. The designer wants to take only a 10% chance that the dam will be overtopped within this period. What return period flood should be used?

Table 4.6 *Return periods T_1 for estimated life L and encounter probability E_1 [$= 1 - (1 - 1/T)^L$]*

L	E_1								
	0.02	0.05	0.10	0.15	0.20	0.30	0.40	0.50	0.70
1	50	20	10	7	5	3	3	2	1
2	99	39	19	13	9	6	4	3	2
3	149	59	29	19	14	9	6	5	3
4	198	78	38	25	18	12	8	6	4
5	248	98	48	31	23	15	10	8	5
6	297	117	57	37	27	17	12	9	6
7	347	137	67	44	32	20	14	11	6
8	396	156	76	50	36	23	16	12	7
9	446	176	86	56	41	26	18	13	8
10	495	195	95	62	45	29	20	15	9
12	594	234	114	74	54	34	24	18	10
14	693	273	133	87	63	40	28	21	12
16	792	312	152	99	72	45	32	24	14
18	892	351	171	111	81	51	36	26	15
20	990	390	190	124	90	57	40	29	17
25	1238	488	238	154	113	71	49	37	21
30	1485	585	285	185	135	85	59	44	25
35	1733	683	333	216	157	99	69	51	30
40	1981	780	380	247	180	113	79	58	34
45	2228	878	428	277	202	127	89	65	38
50	2475	975	475	308	225	141	98	73	42

Solution

Given $n = 25$ and $R = 10\% = 0.10$, we find from the Table 4.5 the return period $T = 238$ years. This is the return period of the flood that should be used in the design.

4.5 PROBABILISTIC TOOLS FOR FLOOD RISK MANAGEMENT

In flood risk management, probability is based on historical data. This probability is called objective, classical, or frequentist and deals with uncertainty that can be measured quantitatively. Climate change probabilities are subjective because they are based on the degree of belief that a person has that change will occur. The probabilistic approach still has its place in the consideration of climate change as discussed in Sections 3.1 and 3.2. Figure 3.3 adopted from Stirling (1998) defines the place for a probabilistic approach in the space bounded by the knowledge about likelihoods and knowledge about outcomes. The focus of this section of the book is on the presentation of three probabilistic tools: Monte Carlo simulation, evolutionary optimization, and multiobjective goal programming, that can be of value for flood risk management under climate change.

4.5.1 Monte Carlo simulation for flood risk management under climate change

Simulation models describe how a system operates, and are used to predict what changes will result from a specific course of action. Such models are sometimes referred to as cause-and-effect models. They describe the state of the system in response to various inputs, but give no direct measure of what decisions should be taken to improve the performance of the system. Therefore, simulation is a problem-solving technique (Simonovic, 2009). Simulation consists of designing a model of the system and conducting experiments with the model either for better understanding of the functioning of the system or for evaluating various strategies for its management. Nowadays most simulation experiments are conducted numerically using a computer. A mathematical model of the system is prepared and repeated runs of this model are taken. The results are available in the form of tables, graphs, or performance indices. They are analyzed to draw inferences about the adequacy of the model.

In the application of a probabilistic approach, the design of real-world systems is based on observed historical data. For example, the observed streamflow data are used in sizing a reservoir, designing a levee, determining capacity of a diversion channel, etc. However, in the consideration of climate change it is well known that the historical records and the observed pattern of data are not likely to repeat in the future (Milly *et al.*, 2008). Besides, the performance of a system critically depends on the extreme values of input variables and the historical data may not contain the entire future range of input variables. An important implication of these facts is that we may not get a complete picture of the system performance and risks involved when historical data are used in analysis. Thus, for instance, the risks of a flood risk management system failing during its economic life cannot be determined because this requires a very large sample of data, which are not commonly available.

BASICS OF MONTE CARLO SIMULATION

Let us consider a system for which some or all inputs are random, system parameters are random, initial conditions may be random, and boundary condition(s) may also be random in nature. The probabilistic properties of these are known. For analysis of such systems, simulation experiments may be conducted with a set of inputs that are synthetically (artificially) generated. The inputs are generated so as to preserve the statistical properties of the random variables. Each simulation experiment with a particular set of inputs gives an answer. When many such experiments are conducted with different sets of inputs, a set of answers is obtained. These answers are statistically analyzed to understand or predict the behavior of the system. This approach is known as Monte Carlo simulation (MCS). Thus, it is a technique to obtain statistical properties of the output of a system given the properties of the inputs and the system (Singh *et al.*, 2007). By using it, decision-makers can get better insight into the working of the system and can determine the risk of failure (e.g., chances of a reservoir flood storage being insufficient for flood attenuation, levee height being insufficient to prevent overtopping, etc.). Sometimes, MCS is defined as any simulation that involves the use of random numbers.

In MCS, the inputs to the system are transformed into outputs by means of a mathematical model of the system. This model is developed such that the important features of the system are represented in sufficient detail. The main steps in MCS are:

- assembling inputs;
- preparing a model of the system;
- conducting experiments using the inputs and the model; and
- analyzing the output.

Sometimes, a parameter of the system is systematically changed and the output is monitored in the changed circumstances to determine how sensitive it is to the changes in the properties of the system.

The main advantages of MCS are: (i) that it permits detailed description of the system, its inputs, outputs, and parameters; and (ii) savings in time and expenses. It is important to remember that the synthetically generated data are no substitute for the observed data but this is a useful pragmatic tool that allows extraction of detailed information from the available data.

The name of the technique comes from the use of roulette wheels, similar to those in use in the casino at Monte Carlo, in the early days to generate random numbers. The generation of random numbers forms an important part of MCS. During the initial days of mathematical simulation, mechanical means were employed to generate random numbers. Printed tables of random numbers were also in use for quite some time. The current approach is to use a computer-based routine to generate random numbers. Various techniques in use today are available in many statistics textbooks (for example Singh *et al.*, 2007, page 439).

MONTE CARLO SIMULATION, CLIMATE CHANGE, AND FLOOD RISK MANAGEMENT

Many studies dealing with climate change uncertainty take advantage of the MCS approach. Some of them are reviewed here to illustrate the utility of the approach in flood risk management under climate change.

Muzik (2002) examines how flood frequencies and magnitudes in a midsize subalpine watershed on the eastern slopes of the Rocky Mountains in Alberta, Canada, would change under doubled-CO_2 conditions. Given the limited spatial resolution of present GCMs and their uncertain performance at the regional scale, a first-order analysis is carried out, in which only rainfall intensity changes are considered to have the most significant impact on future floods. Estimates of storm rainfall increases are based on the literature survey, GCM projections for the study area, and transposition of southern climatic conditions. Two scenarios of likely most severe changes are selected: (i) a 25% increase in the mean and standard deviation of the Gumbel distribution of rainfall depth for storm durations from 6 to 48 hours; and (ii) a 50% increase in the standard deviation only. The HEC-1 watershed model and the soil conservation service runoff curve method for abstractions are used in MCS. Comparison of Monte Carlo-derived flood-frequency curves for the two scenarios with the present day curve shows that scenario 1 is more critical in terms of flood flow increases than scenario 2. Under scenario 1, the mean annual flood in the study watershed would increase by almost 80% and the 100-year flood would increase by 41%.

Prudhomme *et al.* (2003) present a rigorous methodology for quantifying some of the uncertainties of climate change impact studies, excluding those due to downscaling techniques, and applied to a set of five catchments in Great Britain. Uncertainties in climate change are calculated from a set of 25,000 climate scenarios randomly generated by MCS, using several GCMs, SRES-98 emission scenarios, and climate sensitivities. Flow series representative of current and future conditions are simulated using a conceptual hydrologic model. Generalized Pareto distributions are fitted to peak-over-threshold series for each scenario, and future flood scenarios compared to current conditions for four typical flood events. Most scenarios show an increase in both the magnitude and the frequency of flood events, generally not greater than the 95% confidence limits. The largest uncertainty can be attributed to the type of GCM used, with the magnitude of changes varying by up to a factor of 9 in Northern England and Scotland. The recommendation of the study is that climate change impact studies should consider a range of climate scenarios derived from different GCMs, and that adaptation policies do not rely on results from only very few scenarios.

The paper by Booij (2005) assesses the impact of climate change on flooding in the River Meuse using spatially and temporally changed climate patterns and a hydrologic model with three different spatial resolutions. This assessment is achieved by selecting a hydrologic modeling framework and implementing appropriate model components, derived in an earlier study, into the selected framework. Additionally, two other spatial resolutions for the hydrologic model are used to evaluate the sensitivity of the model results to spatial model resolution and to allow for a test of the model appropriateness procedure. Monte Carlo generations of a stochastic precipitation model under current and changed climate conditions are used to assess the climate change impacts. The average and extreme discharge behavior at the basin outlet is well reproduced by the three versions of the hydrologic model in the calibration and validation. The results show improvement with increasing model resolution. The model results with synthetic precipitation under current climate conditions show a small overestimation of average discharge behavior and a considerable underestimation of extreme discharge behavior. The underestimation of extreme discharges is caused by the small-scale character of the observed precipitation input at the sub-basin scale. The general trend with climate change is a small decrease of the average discharge and a small increase of discharge variability and extreme discharges. The variability in extreme discharges for climate change conditions increases with respect to the simulations for current climate conditions. This variability results from both the stochasticity of the precipitation process and the differences between the climate models. The total uncertainty in river flooding with climate change (over 40%) is much larger than the change with respect to current climate conditions (less than 10%). However, climate changes are systematic changes rather than random changes and thus the large uncertainty range is shifted to another level corresponding to the changed average situation.

The Suchen (2006) report on deterioration of existing flood defenses within the Thames Estuary (UK) states that flood risk is once again becoming increasingly important, exacerbated by the effects of climate change and growth of the exposed population and assets. This report utilizes risk assessment tools to review flood defense strategies including the Thames Estuary 2100 Project. A portfolio of defense, social, and regulatory initiatives as well as sustainable strategies is discussed subject to controlling future flood risks using various source–pathway–receptor frameworks (mentioned in Section 4.1) with MCS techniques. Failure mechanisms of earth embankments are also assessed using existing methods. An effective flood risk model used is based on the SPRC framework. Each component of the model can be quantified by: (i) source: fragility curves of defenses; (ii) path: flood inundation model; (iii) receptor: people and property (database); and (iv) consequence: resultant damage. This model incorporates defense reliability, changes in climate and flood plains by modeling the hydraulic loading on the system as a dependent variable but with the performance of each defense being independent. This setup enables use of MCS – sampling defense system-states under various loading conditions referencing fragility curves. Ultimately the outputs are expected annual damage (EAD) and spatial inundation.

Park (2008) presents an approach to assess climate change under uncertainty using MCS. He finds that in the absence of climate policy, the 95% bound on temperature change in 2100 is 5.79 °C. The stringent climate policies with aggressive emissions reductions over time significantly lower the temperature change, compared to the no policy case. His work is not directly related to flood risk management but offers a traditional implementation of MCS. This paper builds on previous estimates of uncertainty in future climate changes but with two important improvements: (i) It runs MCS with different probability distributions for uncertain parameters and scenarios, based on a DICE model output (DICE denotes a family of optimization global integrated assessment models of climate change). DICE links an optimal economic growth model to a description of anthropogenic climate change with the implied economic impacts. Economic output is described by a constant-returns-to-scale Cobb–Douglas production function with labor and capital as input factors. DICE maximizes a global welfare function by determining the optimal division of economic output over time into consumption, investment, and emissions abatement. (ii) It estimates uncertainty under three policy scenarios as well as a no policy case, to show the impact of the range of uncertainty in estimates for temperature change in 2100 and 2200. Using the Monte Carlo approach, the paper provides a more comprehensive picture of the relative likelihood of different future climates. To examine the relative contributions of uncertain climate parameters (climate sensitivity and radiative forcing), the author uses a reduced-form version of the climate model to generate the PDFs of temperature change by MCS. First, the probability distributions for the uncertain input parameters are defined. Two different probability distributions are used: uniform and triangular distributions. Radiative forcing is modeled using a log-normal distribution. Then, the output parameter (temperature change) is identified, and the simulation with 10,000 iterations is run.

To explore the potential for changes in future flood risk, Shaw and Riha (2011) employ a compound frequency distribution which assumes that annual maximum discharges can be modeled by combining the cumulative distribution functions of discharges resulting from annual maximum rainfall, annual maximum snowmelt, and occurrences of moderate rain on wet soils. A compound frequency distribution comprises univariate generalized extreme value (GEV) and gamma distributions. They find that a hypothetical 20% increase in the magnitude of rainfall-related stream discharge results in little change in 96th percentile annual maximum discharge. For the 99th percentile discharge, two water bodies in their study have a 10% or less increase in annual maximum discharge when annual maximum rainfall-related discharges increased by 20%, while the third water body has a 16% increase in annual maximum discharges. Their analyses are done using 50-plus years of historical discharge and meteorological data from three watersheds in different physiographic regions of New York State, USA. Additionally, in some of their cases, annual maximum discharges could be offset by a reduction in the discharge resulting from annual maximum snowmelt events. While only intended as a heuristic tool to explore the interaction among different flood causing mechanisms, the use of a compound flood frequency distribution suggests a case can be made that not all water bodies in humid, cold regions will see extensive changes in flooding due to increased rainfall intensities. Their work uses Monte Carlo analysis to assess whether the averaged 2-month runoff ratios of the warmest winters are statistically different from randomly drawing values from the full record of runoff ratios. In the Monte Carlo analysis, they generate 10,000

Table 4.7 *The discharge frequency data*

T_r (years)	Exceedance probability	Discharge (m^3/s)
500	0.002	898.8
200	0.005	676.1
100	0.010	538.5
50	0.020	423.0
20	0.050	298.8
10	0.100	222.5
5	0.200	158.4

Table 4.8 *The stage discharge data*

Discharge (m^3/s)	Stage (m)
898.8	8.32
676.1	7.57
538.5	6.70
423.0	5.80
298.8	4.76
222.5	4.00
158.4	3.24

unique combinations of semi-monthly runoff ratios and take each combination's average, sampling with replacement from all the possible monthly values. Each sample is as large as the number of warm winters in the record of the respective site (i.e., either four or six).

The following includes the detailed presentation of an application of the MCS for learning purposes.

Example 7

Suppose you are asked to design a flood protection levee. Determine the height of a levee for 100-year return period flood protection with the discharge frequency curve as provided in Table 4.7 and the stage discharge curve in Table 4.8. The freeboard value is 1 m.

In the design of a levee use the implicit probabilistic simulation procedure with log-normal distribution and given population properties for a 100-year return period in Table 4.9. The value of x in the table represents the discharge.

Find (a) the expected value of a levee height and (b) the 90th percentile (the height of a levee which will account for the flood stage value at or below 90%).

Solution

The Monte Carlo sampling method starts with a cumulative distribution function $F(x)$, which gives the probability P that the

Table 4.9 *Monte Carlo simulation input data for log-normal distribution*

	Mean	Standard deviation
100 year discharge (m^3/s)	$\mu = 538.5$	$\sigma = 100$
Stage (m)	$\mu = -6E - 06x^2 + 0.0134x + 1.2903$	$\sigma = 0.3$

variable X will be smaller than or equal to the distribution of an uncertain input variable x:

$$F(x) = P(X \le x) \qquad (4.36)$$

where $F(x)$ takes values from zero to one. The next step is to look at the inverse function $G(F(x))$ written as:

$$G(F(x)) = x \qquad (4.37)$$

The inverse function is used in the generation of random samples from the distribution. Thus, to generate a sample from an input probability distribution fitted to the uncertain variable, a random number (r) is generated between zero and one. This value is substituted into Equation (4.37) where $F(x)$ is equal to (r). The random number r is generated from the uniform $(0, 1)$ distribution to provide equal opportunity of an x value to be generated in any percentile range. The MCS process is automated with the use of a software package such as MATLAB (2011).

(a) The expected value of levee height. Using the expected value of the log-normal distribution we first find the expected value of discharge for the 100-year return period and then the corresponding expected value of stage. The equations below correspond to the expected value for the log-normal distribution. These equations are implemented with the MCS.

$$f(X) = \int_0^\infty \frac{1}{x\sqrt{2\pi\sigma_1^2}} e^{-(\ln x - \mu_1)^2/2\sigma_1^2} dx \qquad (4.38)$$

$$E(X) = \int_0^\infty x \frac{1}{x\sqrt{2\pi\sigma_1^2}} e^{-(\ln x - \mu_1)^2/2\sigma_1^2} dx \qquad (4.39)$$

where

$$\mu_1 = \ln\left(\frac{\mu^2}{\sqrt{\sigma^2 + \mu^2}}\right) \qquad (4.40)$$

$$\sigma_1 = \sqrt{\ln\left(\frac{\mu^2 + \sigma^2}{\mu^2}\right)} \qquad (4.41)$$

where μ and σ correspond to the normal distribution mean and standard deviation, respectively.

```
%MATLAB Lognormal distribution Monte Carlo Intensity Simulation

%number(n) of random number generated iterations
n=2000;
 %format of Lognormal distribution is Lognormal(m,v) where
%'mu' is equal to the discharge population mean and 'sigma' is equal to the discharge
population
%standard deviation
mu = 538.5;
sigma = 100;
%Transformation of 'mu' and 'sigma' to lognormal location, 'm' and shape,
%'s'parameters
m = log((mu^2)/sqrt(sigma^2+mu^2));
s = sqrt(log(sigma^2/(mu^2)+1));
 %X is the random variables generated using lognormal distribution
X = lognrnd(m,s,n,1);
%MX and STD are the expected value and standard deviation respectively
MX = mean(X);
STD = std(X);
%Percentiles
percentile=quantile(X,[.90]);
%%%%%%%%%%%%%%%%%%%%%%%%%%%%%
%%%%%%%%%%%%Summary of Results: comparing Population & Sample Distribution
%%%%%%%%%%%%%%%%%%%%%%%%%%%%%
z = (0:0.02:1000);
%lognormal pdf of population distribution
y = lognpdf(z,m,s);
subplot(2,1,1),plot(z,y), title ({'LogNormal Population pdf',;['mean:', num2str(mu),', Std:',
num2str(sigma)]}),xlabel('x'); ylabel('p');
%lognormal sample distribution from Monte Carlo simulation
subplot(2,1,2),hist(X,100),title ({'LogNormal Random Simulation',;['Expected value:',
num2str(MX),' m^3/s, 90th Percentile:', num2str(percentile),' m^3/s']});
%%%%%%%%%%%%%%%%%%%%%%%%%%%%%
%END of Program
```

Figure 4.3 MATLAB computer code for MCS of discharge.

In this example, 2000 trials of input combinations are evaluated through the use of a random number generator in an automated process. The MATLAB computer program was developed for the implementation of MCS and is provided in Figures 4.3 and 4.4.

Step 1. Use the discharge parameters for the log-normal distribution provided in Table 4.9 as inputs into the MATLAB MCS (code in Figure 4.3). We find that the 100-year expected discharge is $E(Q) = 539.93$ m^3/s. Figure 4.5 shows the detailed output of the Monte Carlo discharge simulation.

Step 2. Use the stage parameters for the log-normal distribution provided in Table 4.9 as inputs into the MATLAB MCS (code in Figure 4.4) to find the corresponding expected stage value. The mean stage value is first calculated using the mean stage–discharge function provided in Table 4.9:

$$\mu = -6.0(10^{-6})x^2 + 0.0134x + 1.2903 \quad (4.42)$$

where x is the expected discharge, $E(Q) = 539.93$ m^3/s, resulting in:

$$\mu = -6.0(10^{-6})(539.93)^2 + 0.0134(539.93) + 1.2903$$
$$= 6.763 \text{ m} \quad (4.43)$$

The mean stage value from Equation (4.43) and standard deviation given in Table 4.9 are used as inputs into the MATLAB MCS (code in Figure 4.4) yielding an expected flood stage value of $E(H) = 6.773$ m. Figure 4.6 shows the detailed output of stage MCS.

Step 3. Using the freeboard of 1 m and expected flood stage of 6.773 m the total expected height of the levee is:

$$H_t = H + H_f = 6.773 + 1 = 7.773 \text{ m} \quad (4.44)$$

(b) The 90% of levee height. This part of the exercise will also be solved using MCS.

Step 1. Given the input discharge values in Table 4.9 the MCS using code from Figure 4.3 yields a 90th percentile discharge of 664.01 m^3/s. Figure 4.5 shows the detailed output of discharge MCS.

Step 2. Use the discharge value of 664.01 m^3/s and substitute into the mean stage Equation (4.42). The result is a mean stage value of 7.56 m in addition to the standard deviation of 0.3 m provided in Table 4.9.

The values of mean and standard deviation are used as input for the 90th percentile stage MCS. Figure 4.6 shows the detailed output of stage MCS. The 90th percentile flood stage is $H = 7.16$ m.

```
%MATLAB Lognormal distribution Monte Carlo Intensity Simulation
%number(n) of random number generated iterations
n=2000;
%format of Lognormal distribution is Lognormal(m,v) where
%'mu' is equal to the stage population mean and 'sigma' is equal to the stage
population
%standard deviation
mu = 6.763;
sigma = 0.3
%Transformation of 'mu' and 'sigma' to lognormal location, 'm' and shape,
%'s'parameters
m = log((mu^2)/sqrt(sigma^2+mu^2));
s = sqrt(log(sigma^2/(mu^2)+1));
%X is the random variables generated using lognormal distribution
X = lognrnd(m,s,n,1);
%MX and STD are the expected value and standard deviation respectively
MX = mean(X);
STD = std(X);
%Percentiles
percentile=quantile(X,[.90]);
%%%%%%%%%%%%%%%%%%%%%%%%%%%%%
%%%%%%%%%%%%Summary of Results: comparing Population & Sample
Distribution
%%%%%%%%%%%%%%%%%%%%%%%%%%%%%
z = (0:0.02:20);
%lognormal pdf of population distribution
y = lognpdf(z,m,s);
subplot(2,1,1),plot(z,y), title ({'LogNormal Population pdf',;['mean:', num2str(mu),',
Std:', num2str(sigma)]}),xlabel('x'); ylabel('p');
%lognormal sample distribution from Monte Carlo simulation
subplot(2,1,2),hist(X,100),title ({'LogNormal Random Simulation',;['Expected value:',
num2str(MX),' m, 90th Percentile:', num2str(percentile),' m']});
%%%%%%%%%%%%%%%%%%%%%%%%%%%%
```

Figure 4.4 MATLAB computer code for MCS of stage.

Step 3. The design levee height that flood level will be at or below for 90% of the time given the freeboard addition of 1 m is:

$$H_t = H + H_f = 7.16 + 1 = 8.16 \, \text{m} \qquad (4.45)$$

The levee height of 8.16 m is conservative and provides a high safety level that may not be economically feasible.

The MCS code developed using MATLAB and shown in Figures 4.3 and 4.4 is included on the website: directory Probabilistic tools, sub-directory Simulation, sub-sub-directory Monte Carlo, files discharge.m and stage.m.

MONTE CARLO SIMULATION OF FUTURE FLOODING CONDITIONS IN THE OKAVANGO DELTA

This section illustrates the value of the Monte Carlo probabilistic simulation approach to the management of climate change-caused flood risk. The example is taken from Wolski *et al.* (2002) by permission of Professor Piotr Wolski of the Harry Oppenheimer Okavango Research Centre. This example utilizes a method based on the MCS for making predictions of future temporal flooding patterns for the Okavango Delta. The method is developed to allow for incorporation of upstream water abstraction and global climate change in the assessment of the effects on the flooding in the delta. The MCS is implemented to describe future conditions in probabilistic terms based on a large number of simulations that mimic the observed variability. The method uses an annual, lumped regression model of the delta's flooding and a series of satellite-derived flood images.

Introduction The Okavango Delta is a large wetland situated in northern Botswana that has developed on a near terminal alluvial fan of the Okavango River. The size of the inundated area of the wetland varies considerably, responding to discharge of the Okavango River, with local rainfall playing a less significant role. That variation is visible at various time scales (from months to years to decades to centuries).

The strongest variation results from seasonality of rainfall in the catchment of the Okavango River and the consequent seasonal

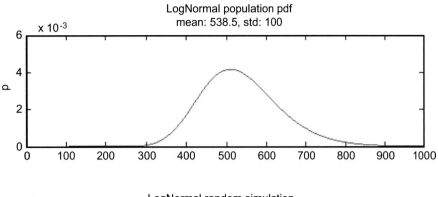

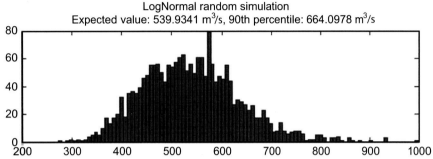

Figure 4.5 The output of Monte Carlo discharge simulation.

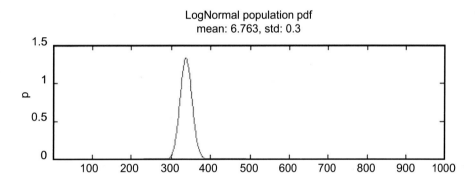

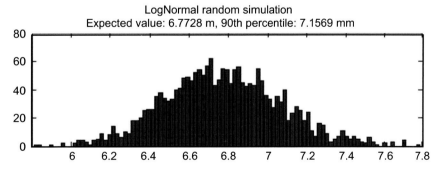

Figure 4.6 The output of Monte Carlo stage simulation.

changes in delta inflow (100 m³/s during the low flow season and up to 1,500 m³/s during the peak discharge period, with an annual mean discharge of 320 m³/s). An area of approximately 4,000–6,000 km² is permanently inundated during the low flow season (January–March), which expands to 9,000–12,000 km² during the flood season (July–September). Flood extent also varies between years responding to interannual differences in rainfall over the delta and discharge of the Okavango River.

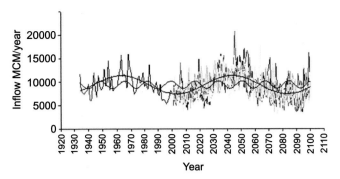

Figure 4.7 Sequence of observed and simulated annual inflow into the Okavango, with the long-term oscillation superimposed (after Wolski *et al.*, 2002).

Longer-term variations in inflow to the delta are also evident (see Figure 4.7). For example, the 1970s were characterized by a sequence of years with large floods, whereas the floods during the 1990s were much smaller.

Another important factor determining the dynamics of flooding, which acts in a much longer time frame, is the process of sedimentation. That process is responsible for a gradual, lateral shift of flow between the subsystems of the delta.

The dynamic nature of the flood distribution and size results in different regions of the Okavango Delta experiencing different temporal patterns of inundation. Inundation is the main determinant of the ecology of the delta. The three main units or ecoregions of the wetland that can be distinguished in the delta are: permanent swamp, seasonal floodplains, and occasional floodplains. They differ in duration and frequency of inundation, which determines vegetation structure and composition. Due to the long-term variation in flooding extent, the location and size of these units varies with time. Obviously, vegetation has had to adapt to these changing conditions. However, little is known about the dynamics, especially response time, of vegetation changes caused by changes in flooding patterns. The long-term change caused by global climate change also plays an important role in understanding the potential extent of inundation and its impact on the ecology of the Okavango Delta.

An original method is developed for making predictions of future temporal flooding patterns using models and historic data available for the Okavango Delta. The method allows for incorporation of upstream water abstractions and global climate change. It can thus be used for assessing the effects of anthropogenic interference or natural change on the flooding in the Okavango Delta.

General description of the method There are two tools used in modeling of flood size and distribution in the Okavango Delta: a regression model of maximum annual flood (Gumbricht *et al.*, 2004) and a time series of flood maps derived from satellite data.

The regression model relates hydrologic inputs, i.e., local rainfall and inflow from the Okavango River to the annual maximum size of the flooded area. Flood distribution maps show that over the last 15 years the spatial distribution of the flood varied relatively little between years; i.e., floods of similar size occurring in various years looked remarkably similar. This information can be used to translate the model-derived maximum size of the flooded area into a map of that flood.

Using the available tools it is possible to determine the future temporal flooding pattern in any part of the Okavango Delta, provided with predictions of model inputs, local rainfall, and flow in the Okavango River. The model provides information on maximum annual flood size only, and the duration of flooding cannot be predicted. The temporal flooding pattern is thus described in terms of flood frequency, the number of years in a longer time period during which a given area is inundated, and the duration of that inundation is disregarded.

To predict future rainfall and flow in the Okavango, an assumption is made that statistical description of variation in the future time series of rainfall and inflow will be the same as it was in the past (which does not hold for the climate change conditions). Each series is treated as being composed of high frequency random noise superimposed on low frequency deterministic trends. By generating a large number of realizations of future time series of rainfall and inflow and using them in the regression model a large number of possible (in probabilistic terms) future series of maximum flooded area can be obtained. For each of such series the flood sizes that would occur with given frequencies are then determined. Using all the realizations, the PDFs describing given associations of flood size and flooding frequency are developed. They provide how likely it is that a flood of a given size would occur with a given frequency. Finally, the frequency probability information is transferred to maps.

The following steps are involved in the implementation of the methodology:

Step1. Generate realizations of future rainfall and Okavango River discharge. Future changes such as upstream water abstraction and/or climate change are implemented at this stage by imposing them as deterministic factors.

Step 2. Translate each realization of rainfall and inflow into a time series of the size of flood inundation using the regression model.

Step 3. Obtain the probability of a given flood frequency being associated with a given size of flood inundation.

Step 4. Translate flood frequency probabilities into flood frequency probability maps.

Simulation scenarios Currently the Okavango River flows with its natural, unaltered regime, without any significant anthropogenic interference. The first simulation scenario assumes that such a regime will continue in the future.

Developments in the region are likely to occur, which will result in abstraction of water from the river and thus alter its flow regime. At the current time there is no water abstraction scheme scheduled to be implemented. However, there is a proposal to abstract approximately 100 MCM/year at Rundu. This abstraction is analyzed as the second simulation scenario.

Another factor that has an immediate effect on the hydrologic inputs to the Okavango Delta is climate change. Various GCMs exist that simulate future climatic conditions (rainfall and temperature) accounting for global warming. This study uses the results of the Hadley Centre's HadCM3 model (information available on line at www.metoffice.gov.uk/research/modelling-systems/unified-model/climate-models/hadcm3, last accessed June, 2011).

The implementation of the results of climate change models in the analysis is challenging due to the lack of a reliable rainfall–runoff model for the catchment (the model is in the process of development and one of the major obstacles is lack of rainfall data from Angola). The output of the GCM is expressed as reduction in rainfall in relation to present conditions. The method used in this study needs rainfall and discharge data for the Okavango River. Thus, there is a need to determine how the discharge in the Okavango will be affected by the simulated reduction in rainfall in the Okavango River catchment. The authors implemented a simpler solution in calculating reduction in flow by relating Okavango River discharge to rainfall from Mwinilunga in Zambia (a station that has a sufficiently long rainfall record (80 years) and is located in the same climatic zone as the Okavango catchment). From the resulting regression equation it was determined that reduction in rainfall by 100 mm would result in reduction in annual flow by 1,100 MCM. The HadCM3 model simulates the reduction in annual rainfall over the delta to be 60 mm in 2100 for one emission scenario, compared to contemporary conditions. For the catchment, the HadCM3 model predicts a reduction of 130 mm. The latter translates to a reduction in flow of 1,430 MCM. In order to incorporate these values into the procedure presented here, use the linear gradual increase from 0 in 2002 to the values mentioned above in 2100. This is the third simulation scenario.

In summary, the following scenarios have been simulated: (i) the natural flow regime (hereafter called "natural flow"); (ii) abstraction of 100 MCM at Rundu (hereafter called "abstraction"); and (iii) global climate change – progressive reduction in delta rainfall up to a maximum of 60 mm/year in 2100 and progressive reduction in the Okavango flow up to a maximum of 1,430 MCM/year in 2100 (hereafter called "climate change").

Model analyses The following is the description of the implementation of the presented methodology.

Step 1: Future rainfall and Okavango River discharge. The time series of the Okavango River discharge at Mohembo and rainfall over the delta, taken as an average of the Maun and Shakawe records, were subject to Fourier analysis in order to detect cyclicity. Two long-term cycles were found in both: an 18-year cycle, and a cycle of length exceeding 60 years. The length of the long-term cycle observed in records of the Okavango flow was taken as 80 years based on the 3,000 years of proxy data available for southern Africa.

Two cycles, 18 and 80 years, are deterministic components in the total variation in flow and it is assumed that they will continue in the future. The cyclicity was subtracted from the data and the resulting values were normalized and checked for normality of distribution with the Wilk–Shapiro test. Parameters of the normal distribution describing the data were then determined. Subtraction of cyclicity removed the autocorrelation in the data.

The MCS was performed to generate future sequences of rainfall and discharge. The MCS covered the period between 2003 and 2100. One thousand realizations were generated for each of the considered scenarios. The PDF of the rainfall and inflow series was based on de-trended and normalized data, and thus each of the generated series was scaled back to the original values and the trend (cyclicity) added. The effects of abstractions and global climate change were imposed on the generated rainfall and inflow by subtracting the anticipated reductions from the generated rainfall and inflow respectively. The simulated inflows are illustrated in Figure 4.7, which also shows the historic data and the 18- and 80-year cycles. Simulated results are presented by plotting only 10 simulations for clarity.

Obviously, the delta rainfall and inflow co-vary. The correlation coefficient between them is, however, relatively small: $r = 0.34$. The future sequences of rainfall and inflow are simulated as random independent variables, not correlated. In spite of this, the correlation coefficient between the simulated rainfall and inflow is similar to that in the original data. This effect can be attributed to the fact that the majority of the correlation in the original data resulted from the presence of similar patterns in the low frequency component (long-term trend or cyclicity), and these are preserved in the deterministic component of the simulated sequences.

Step 2. Development of time series of flood inundation size. The regression model of Gumbricht *et al.* (2004) was used to transform rainfall and inflow into time series of flood inundation size. The model operates on an annual basis. The maximum annual flood inundation area (A) is the dependent variable, while total annual discharge of the Okavango at Mohembo (Q), total annual rainfall (P) over the delta, evaporation (E), and the previous year's maximum flood inundation (L) are the explanatory variables.

$$A(km^2) = 0.96 \times Q(10^6 m^3) + 5.59 \times P(mm) + 0.148$$
$$\times L(km^2) - 747 \times E(mm) + 2811 \qquad (4.46)$$

The model has been calibrated on the flood inundation data from 1984 to 2000, which comprised flood inundations as big as 12,000 km² and as small as 4,500 km². The calibration process

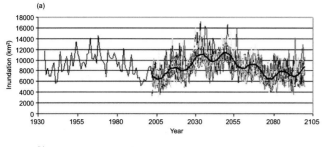

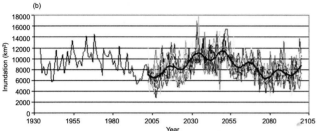

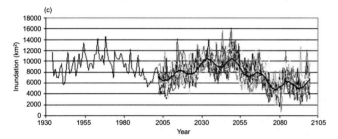

Figure 4.8 Model reconstructed flood inundation for 1934–2001 (thin black line) and model calculated flood inundation for ten realizations of rainfall and inflow under: (a) natural flow scenario; (b) abstraction scenario; and (c) climate change scenario, with average of all simulations superimposed in bold (after Wolski *et al.*, 2002).

comprised adjustment of regression coefficients for each explanatory variable using the minimum square method. The correlation coefficient between the observed and the simulated floods was 0.95, with a flood size estimation error of 600 km². The incorporation of potential evaporation (*E*) as a variable in the model did not improve model results. The model was used by Gumbricht *et al.* (2004) to estimate flood inundations back to 1934, a period for which rainfall and inflow records are available.

The series of model-calculated flood inundations for ten different realizations of inflow and rainfall for each of the scenarios are presented in Figure 4.8. The figure also shows the flood inundation since 1934, reconstructed using the model.

Step 3. Probability of given flood frequency associated with given inundation size. There are several factors that influence the way in which the flooding frequencies are calculated. First is the long-term variation observed in the generated inundation size; second, the nature of flood maps on which the procedure is based and finally, the continuous nature of the generated inundation size.

The Monte Carlo procedure for calculating flooding frequencies is as follows: (i) For each of the realizations, the inundation

size associated with each of the considered flooding frequencies is calculated. This is done by considering flooding frequency to be numerically equivalent to the probability of exceeding a given inundation size, as calculated from the 30 years of data. (ii) For each of the flooding frequencies *f* and for each of the flood size ranges *A*, the number of combinations of that frequency and flood size falling into that range is calculated. (iii) Equation (4.47) is used to obtain the probability of a given flooding frequency occurring within a given range.

$$p_{f/A} = \frac{n_{f/A}}{N} \qquad (4.47)$$

where $p_{f/A}$ is the probability that frequency *f* occurs within zone *A*; $n_{f/A}$ is the number of cases when inundation of size falling into zone *A* occurred with frequency *f*; and *N* is the number of cases in the sample. In this MCS the number of cases is 1,000.

As can be seen from Figure 4.8, the time series of generated flood inundation sizes is very strongly influenced by the 80-year cyclicity. One can determine flooding frequencies using the entire simulated period: 2003–2100. However, that would contain three relatively long periods of different, but relatively uniform, flooding conditions. It is likely that ecological change will occur in response to that long-term variation. As a result, average flooding frequencies obtained for the entire period would not be representative of the real conditions. The authors thus decided to divide the period of calculations into three parts, roughly corresponding to the three periods of different flooding conditions (2002–2031, 2032–2061, 2062–2091) and to determine flooding frequencies for each of them separately.

The aim of the entire procedure described in this case study is to determine the flooding frequency or, rather, the probability of flooding with a given frequency *for each point* in the delta. Information about floods is, however, obtained from a model in Equation (4.46) that represents the inundation as a whole. To represent the spatial distribution of the floods obtained from the model, maps of the spatial distribution of flood inundations of various sizes are used. The maps were obtained by classification of satellite images and their derivation presented by McCarthy *et al.* (2003). The area prone to flooding covers approximately 12,500 km². The flood-prone area is divided in zones (spatial unit of analysis) that represent areas characterized by a similar flooding pattern. The zones cover 500 km² each and represent areas progressively flooded with an increase in flood inundation.

For each of the realizations, the association of flood inundation and flooding frequency is determined by considering the flooding frequency equivalent to the probability of exceeding a given flood inundation. For the purpose of simpler calculation assumption of a discrete distribution of both inundation and flooding frequency is introduced. Only the following flooding frequencies are considered: 0.03, 0.1, 0.2, 0.3, 0.4, 0.5, 0.6, 0.7, 0.8, 0.9, and 0.97 in

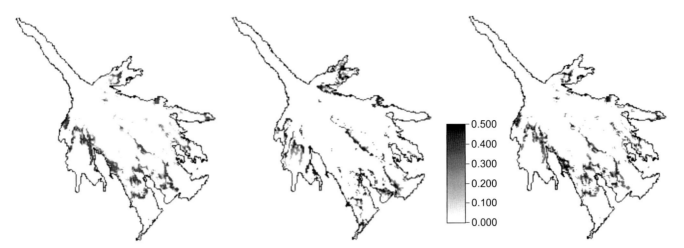

Figure 4.9 Probability of flooding with frequency 0.5 for climate change scenario: (left) 2002–2031, (center) 2032–2061, (right) 2062–2091. Areas covered by inundation smaller than 4,000 km² are not shown (after Wolski *et al.*, 2002).

discrete spatial zones. The flooding frequency probability is thus expressed by Equation (4.47).

Step 4. Mapping flood frequency probability. The probability of occurrence of all flooding frequencies is calculated for each of the flood inundation ranges. As the ranges have a specified representation in space, the probability of occurrence of a given flooding frequency is directly transferred to the spatial zones that represent the appropriate ranges. In this way, maps representing the probability of occurrence of a given frequency are obtained. Such maps are prepared for each of the frequencies considered and each scenario (Wolski *et al.*, 2002). Figure 4.9 represents selected maps for the climate change scenario and frequency of flooding equal to 0.5.

This study has much broader objectives than just to look into impacts of climate change on flooding in the Okavango Delta. The main conclusions are: (i) Long-term oscillations in rainfall/flood inundation dominate the variation in hydrologic variables influencing the delta. (ii) The effects of the simulated abstractions (100 MCM/year) on the flooding regime are very small and will mainly occur in the central and distal delta. (iii) The effects of global climate change may be significant. However, they will not be visible before the end of this century.

This case study also provides an interesting example of the application of the Monte Carlo probabilistic simulation approach to the management of flood risk.

4.5.2 Evolutionary optimization for flood risk management

Evolutionary optimization algorithms are becoming more prominent in the water management field (Simonovic, 2009). Significant advantages of evolutionary algorithms include: (i) no need for initial solution; (ii) easy application to non-linear problems and

to complex systems; (iii) production of acceptable results over longer time horizons; and (iv) generation of several solutions that are very close to the optimum.

Evolutionary programs are probabilistic optimization algorithms based on similarities with the biological evolutionary process. In this concept, a *population* of individuals, each representing a search point in the space of feasible solutions, is exposed to a collective learning process that proceeds from generation to generation. The population is arbitrarily initialized and subjected to the processes of *selection*, *recombination*, and *mutation* such that the new populations created in subsequent generations evolve towards more favorable regions of the search space. This is achieved by the combined use of a *fitness* evaluation of each individual and a selection process that favors individuals with higher fitness values, thus making the entire process resemble the Darwinian rule known as the *survival of the fittest*.

BASICS OF EVOLUTIONARY OPTIMIZATION

The evolution program is a probabilistic algorithm that maintains a population of individuals, $P(t) = \{x_1^t, \ldots, x_n^t\}$, for iteration t. Each individual represents a potential solution to the problem at hand, and in any evolution program is implemented as a (possibly complex) data structure S. Each solution x_i^t is evaluated to give some measure of its *fitness*. Then a new population (iteration $t + 1$) is formed by selecting the more fit individuals (selection step). Some members of the new population undergo transformations (transformation step) by means of *genetic operators* to form new solutions. There are low-order transformations m_i (mutation type), which create new individuals by a small change in a single individual $(m_i : S \to S)$, and higher-order transformations c_j (crossover type), which create new individuals by combining parts from several (two or more) individuals $(c_j : S \times \cdots \times S \to S)$. After a number of generations the program converges. It is hoped

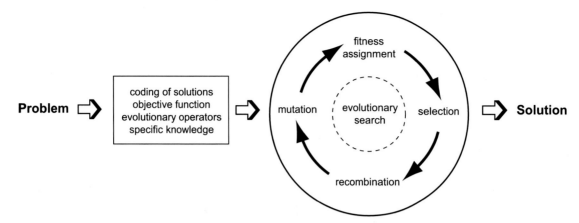

Figure 4.10 Schematic presentation of the evolutionary algorithm (after Simonovic, 2009).

that the best individual represents a near-optimum (reasonable) solution. Figure 4.10 is a schematic presentation of the evolutionary optimization process.

Let us consider the following example. We search for a network of levees that needs to satisfy some requirements (say, we search for the optimal topology of a network of levees according to criteria such as the cost of pumping and level of flood protection). Each individual in the evolution program represents a potential solution to the problem, i.e., each individual represents a levee network. The initial population of networks $P(0)$ (either generated randomly or created as a result of some heuristic process) is a starting point ($t = 0$) for the evolution program. The evaluation function is usually given – it incorporates the problem requirements. The evaluation function returns the fitness of each network, distinguishing between better and worse individuals. Several mutation operators can be designed that would transform a single network. A few crossover operators can be considered that combine the structure of two (or more) networks into one. Very often such operators incorporate problem-specific knowledge. For example, if the network we search for is connected, a possible mutation operator would delete a levee section from the network and add a new section to connect two disjoint subnetworks. The other possibility is to design a problem-independent mutation and incorporate this requirement into the evaluation function, penalizing networks that are not connected.

Clearly, many evolution programs can be formulated for a given problem. Such programs may differ in many ways. They can use different data structures for implementing single individual, genetic operators for transforming individuals, methods for creating an initial population, methods for handling constraints of the problem, and parameters (population size, probabilities of applying different operators, etc.). However, they all share a common principle: a population of individuals undergoes some transformations, and during the evolution process the individuals do their best to survive. The following processes constitute the main part of any evolutionary optimization algorithm.

Selection Selection determines which individuals are chosen for mating (recombination) and how many offspring each selected individual produces. The first step is fitness assignment by proportional fitness assignment, or rank-based fitness assignment. The actual selection is performed in the next step. Parents are selected according to their fitness by means of one of the following algorithms: roulette-wheel selection, stochastic universal sampling, local selection, truncation selection or tournament selection.

In selection the offspring-producing individuals are chosen. The first step is fitness assignment. Each individual in the selection pool receives a reproduction probability depending on its own objective value and the objective value of all other individuals in the selection pool. This fitness is used for the actual selection step that follows.

In rank-based fitness assignment, the population is sorted according to objective values. The fitness assigned to each individual depends only on its ranking and not on an actual objective value. Rank-based fitness assignment overcomes the scaling problems of proportional fitness assignment. The reproductive range is limited, so that no individuals generate an excessive number of offspring. Ranking introduces a uniform scaling across the population, and provides a simple and effective way of controlling selective pressure. Rank-based fitness assignment behaves in a more robust manner than proportional fitness assignment, and as a result it is the method of choice in most evolutionary optimization algorithms (including the one in the EVOLPRO algorithm, featured here).

Let us use *Nind* for the number of individuals in the population, *Pos* for the position of an individual in this population (the least-fit individual has *Pos* = 1, the fittest individual *Pos* = *Nind*), and *SP* for the selective pressure. The fitness value for an individual is calculated using linear ranking:

$$Fitness(Pos) = 2 - SP + 2 \times (SP - 1) \times \frac{(Pos - 1)_i}{Nind - 1}$$

$$(4.48)$$

Table 4.10 *Example data*

Number of individuals	1	2	3	4	5	6	7	8	9	10	11
Fitness value	2.0	1.8	1.6	1.4	1.2	1.0	0.8	0.6	0.4	0.2	0
Selection probability	0.18	0.16	0.15	0.13	0.11	0.09	0.07	0.06	0.03	0.02	0

Figure 4.11 Roulette-wheel selection.

Linear ranking allows values of selective pressure in [1.0, 2.0]. The other option is the use of non-linear ranking:

$$Fitness(Pos) = \frac{Nind \times X_i^{Pos-1}}{\sum_{i=1}^{Nind} X^{i-1}} \qquad (4.49)$$

where X is computed as the root of the polynomial:

$$0 = (SP - Nind) \times X^{Nind-1} + SP \times X^{Nind-2} + \cdots$$
$$+ SP \times X + SP \qquad (4.50)$$

The use of non-linear ranking permits higher selective pressures than the linear ranking method. Non-linear ranking allows values of selective pressure in [1, *Nind* – 2].

The simplest selection scheme is *roulette-wheel selection*, also called stochastic sampling with replacement. This is a stochastic algorithm and involves the following technique. The individuals are mapped to adjacent segments of a line, such that each individual's segment is equal in size to its fitness. A random number is generated and the individual whose segment spans the random number is selected. The process is repeated until the desired number of individuals is obtained (which is called the *mating population*). This technique is analogous to a roulette wheel with each slice proportional in size to the fitness (see Figure 4.11).

Example 8
Let us consider the optimization problem with Table 4.10, which shows the selection probability for 11 individuals, the linear ranking and selective pressure of two together with the fitness value.

Solution
Individual 1 is the most fit individual and occupies the largest interval, whereas individual 10 as the second least fit individual has the smallest interval on the line (Figure 4.11). Individual 11, the least fit individual, has a fitness value of 0 and gets no chance for reproduction.

For selecting the mating population, an appropriate number of uniformly distributed random numbers (between 0.0 and 1.0) is

independently generated. A sample of six random numbers is: 0.81, 0.32, 0.96, 0.01, 0.65, 0.42. Figure 4.11 shows the selection process of the individuals for the example in Table 4.10 with these sample random numbers. After selection, the mating population consists of the individuals 1, 2, 3, 5, 6, 9. The roulette-wheel selection algorithm provides a zero bias but does not guarantee minimum spread.

For more complex selection algorithms see, for example, Simonovic (2009).

Recombination-crossover Recombination produces new individuals by combining the information contained in two or more parents. This is done by combining the variable values of the parents. Depending on the representation of the variables, different methods can be used. Here we look at the discrete recombination method, which can be applied to all variable representations. The intermediate, line, and extended line recombination methods for real-valued variables as well as methods for binary-valued variables are not presented here, but can be found in the literature.

Discrete recombination performs an exchange of variable values between the individuals. For each position the parent that contributes its variable to the offspring is chosen randomly with equal probability according to the rule:

$$Var_i^O = Var_i^{P_1} \times a_i + Var_i^{P_2} \times (1 - a_i) \quad i \in (1, 2, \ldots, Nvar)$$
$$a_i \in \{0, 1\} \quad \text{uniform at random, } a_i \text{ for each } i \text{ new defined.}$$
$$(4.51)$$

Discrete recombination generates corners of the hypercube defined by the parents. Figure 4.12 shows the geometric effect of discrete recombination.

Example 9
Let us consider the following two individuals with three variables each (that is, three dimensions). Create new individuals using discrete recombination.

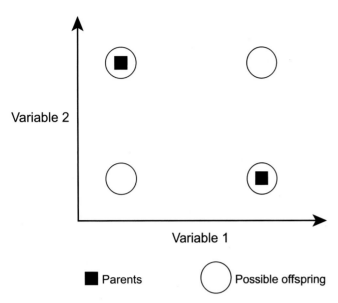

Figure 4.12 Possible positions of the offspring after discrete recombination.

| individual 1 | 12 | 25 | 5 |
| individual 2 | 123 | 4 | 34 |

Solution
For each variable the parent who contributes its variable to the offspring is chosen randomly with equal probability:

| sample 1 | 2 | 2 | 1 |
| sample 2 | 1 | 2 | 1 |

After recombination according to Equation (4.51) the new individuals are created:

| offspring 1 | 123 | 4 | 5 |
| offspring 2 | 12 | 4 | 5 |

Mutation Individuals are randomly altered by mutation. The variations (mutation steps) are mostly small. They will be applied to the variables of the individuals with a low probability (mutation probability or mutation rate). Normally, offspring are mutated after being created by recombination. Two approaches exist for the definition of the mutation steps and the mutation rate: either both parameters are constant during a whole evolutionary run, or one or both parameters are adapted according to previous mutations. We focus here only on real-valued mutation. (See the literature for binary mutation.)

Mutation of real variables means that randomly created values are added to the variables with a low probability. Thus, the probability of mutating a variable (*mutation rate*) and the size of the

change for each mutated variable (*mutation step*) must be defined. The probability of mutating a variable is inversely proportional to the number of variables (dimensions). The more dimensions one individual has, the smaller is the mutation probability. As long as nothing else is known, a mutation rate of $1/n$ is suggested.

The size of the mutation step is usually difficult to choose. The optimal step size depends on the problem considered, and may even vary during the optimization process. It is known that small steps are often successful, especially when the individual is already well adapted. However, large mutation steps can, when successful, produce good results much quicker. Thus, a good mutation operator should often produce small step sizes with a high probability and large step sizes with a low probability.

This operator is recommended:

$$Var_i^{Mut} = Var_i + s_i \times r_i \times a_i$$
$$i \in (1, 2, \ldots, n) \text{ uniform at random} \quad (4.52)$$

where $a_i = 2^{-k \times u}$, k is mutation precision, $u \in [0, 1]$ is uniform at random; $r_i = r \times domain$ r is mutation range (standard 10%); and $s_i \in \{-1, +1\}$ is uniform at random.

This mutation algorithm is able to generate most points in the hypercube defined by the variables of the individual and range of the mutation. (The range of mutation is given by the value of the parameter r and the domain of the variables.) Most mutated individuals will be generated near the individual before mutation. Only a few individuals will mutate farther away. That means the probability of small step sizes is greater than that of bigger steps.

Typical values for the parameters of the mutation operator from Equation (4.52) are: mutation precision k: $k \in \{4, 5, \ldots, 20\}$; and mutation range r: $r \in [0.1, 10^{-6}]$.

Reinsertion Once the offspring have been produced by selection, recombination, and mutation of individuals from the old population, the fitness of the offspring may be determined. If fewer offspring are produced than the size of the original population, then to maintain the size of the population the offspring have to be reinserted into the old population. Similarly, if not all offspring are to be used at each generation or if more offspring are generated than the size of the old population, a reinsertion scheme must be used to determine which individuals are to exist in the new population. Only local reinsertion that meets the used selection method (local selection – see above) is presented here.

In local selection, individuals are selected in a bounded neighborhood. The reinsertion of offspring takes place in exactly the same neighborhood. Thus, the locality of the information is preserved. The neighborhood structures used are the same as in local selection. The parent of an individual is the first selected parent in this neighborhood. The following schemes are possible for the selection of parents to be replaced and of offspring to reinsert:

- Insert every offspring and replace individuals in the neighborhood uniformly at random.
- Insert every offspring and replace the weakest individuals in the neighborhood.
- Insert offspring fitter than the weakest individuals in the neighborhood and replace the weakest individuals in the neighborhood.
- Insert offspring fitter than the weakest individuals in the neighborhood and replace their parents.
- Insert offspring fitter than the weakest individuals in the neighborhood and replace individuals in the neighborhood uniformly at random.
- Insert offspring fitter than their parents and replace the parents.

EVOLUTIONARY OPTIMIZATION, CLIMATE CHANGE, AND FLOOD RISK MANAGEMENT

Evolutionary optimization has been applied to flood control, operation of reservoirs, and other water resources management problems. Direct applications in flood risk management under climate change are not available yet in the literature. The following review provides some examples of the most recent broader application of this approach to water resources management.

Chang and Chen (1998) present genetic algorithms (GAs), one type of evolutionary optimization tool, as fairly successful in a diverse range of optimization problems, providing an efficient and robust way for guiding a search (i) in a complex system and (ii) in the absence of domain knowledge. Due to the temporal and spatial variability in rainfall and high mountains and steep channels upstream of all watersheds on Taiwan Island, water reservoirs are the most effective means for mitigating natural disasters such as floods. Rules for reservoir operation are intended to guide and manage such systems so that the releases made are in the best interests of the system's objectives consistent with inflows and existing storage levels. For optimal control of reservoirs, prediction of the inflow hydrograph and optimization of the release must be accurate. Due to the natural uncertainties of the predicted inflow hydrograph and complexity of reservoir operating rules, reservoir operation highly depends on the experience-based knowledge of the operator(s), such as during an extreme hydrologic situation like flooding. In this paper, two types of GA, real-coded and binary-coded, are examined for function optimization and applied to the optimization of a flood control reservoir model. The results show that both GAs are more efficient and robust than the random search method, with the real-coded GA performing better in terms of efficiency and precision than the binary-coded GA.

One of the most significant papers on the topic of evolutionary optimization of reservoir system operations is by Wardlaw and Sharif (1999). Several alternative formulations of a GA for reservoir systems are evaluated using the four-reservoir, deterministic, finite-horizon problem. This is done with a view to presenting fundamental guidelines for implementation of the approach to practical problems. Alternative representation, selection, crossover, and mutation schemes are considered. It is concluded that the most promising GA approach for the four-reservoir problem comprises real-value coding, tournament selection, uniform crossover, and modified uniform mutation. The real-value coding operates significantly faster than binary coding and produces better results. The known global optimum for the four-reservoir problem can be achieved with real-value coding. A nonlinear four-reservoir problem is considered also, along with one with extended time horizons. The results demonstrate that a GA could be satisfactorily used in real-time operations with stochastically generated inflows. A more complex ten-reservoir problem is also considered, and results produced by a GA are compared with previously published results. The GA approach is robust and is easily applied to complex systems. It has potential as an alternative to more traditional stochastic optimization approaches. This contribution is of more theoretical value since the flood risk management was not dealt with explicitly in the reservoir systems formulations.

Shafiei and Haddad (2005) implement evolutionary optimization (GA) in determining optimal flood levee setback. It is been demonstrated that GAs provide robust and acceptable solutions to the levee setback optimization problem. Past practice in flood levee design usually used occasional observations of flood stages and empirical judgments on required project scales. More recently, several studies have addressed the economic aspects of flood levee design, usually with benefit–cost analysis and optimization. In this work a static model is formulated to maximize the benefit of flood levee construction, considering levee construction cost and resultant protected land value benefit due to levees. This simple model allows preliminary quantitative examination of the tradeoff between optimal setback and optimal height in designing a new levee.

Kumar and Reddy (2006) revisit the general reservoir optimization problem. In this paper a metaheuristic technique called ant colony optimization (ACO) is introduced to derive operating policies for a multi-purpose reservoir system. Most of the real-world reservoir problems often involve non-linear optimization in their solution with high dimensionality and a large number of equality and inequality constraints. Often the conventional techniques fail to yield global optimal solutions. In this study, it is intended to test the usefulness of ACO in solving such types of problems and compare them to evolutionary optimization algorithms. The reservoir operation problem is approached by considering a finite time series of inflows, classifying the reservoir volume into several class intervals, and determining the reservoir release for each period with respect to a predefined optimality criterion. The proposed techniques are applied to a case study of Hirakud reservoir, which is a multi-purpose reservoir system located in India. The multiple objectives comprise minimizing flood risks, minimizing irrigation deficits, and maximizing hydropower production, in that order of priority. The developed model is applied for monthly operation, and consists of two models, namely, for short-time horizon operation and for long-time horizon operation. To evaluate the performance of ACO, the developed models are also solved using real-coded GA. The results of the two models indicate that the ACO model performs well and provides higher annual power production, while satisfying irrigation demands and flood control restrictions.

Makkeasorn *et al.* (2008) address flood forecasting in the context of climate change using evolutionary optimization and artificial neural networks. To more efficiently use the limited amount of water in the changing world or to resourcefully provide adequate time for flood warning, the issues led the authors to seek advanced techniques for improving streamflow forecasting on a short-term basis. This study emphasizes the inclusion of sea surface temperature (SST) in addition to the spatio-temporal rainfall distribution via the Next Generation Radar (NEXRAD), meteorological data via local weather stations, and historical stream data via USGS gage stations to collectively forecast discharges in a semi-arid watershed in south Texas, USA. Two types of artificial intelligence models, including genetic programming (GP) and neural network (NN) models, are employed comparatively. Four numerical evaluators are used to evaluate the validity of a suite of forecasting models. Research findings indicate that GP-derived streamflow forecasting models are generally favored in the assessment in which both SST and meteorological data significantly improve the accuracy of forecasting. Among several scenarios, NEXRAD rainfall data are proven most effective for a 3-day forecast, and the SST Gulf-to-Atlantic index shows larger impacts than the SST Gulf-to-Pacific index on the streamflow forecasts. The most forward-looking GP-derived models can even perform a 30-day streamflow forecast ahead of time with an r-square of 0.84 and RMS error of 5.4 in this study.

Karamouz *et al.* (2009) present two optimization models. The first model is developed to determine an economical combination of permanent and emergency flood control options and the second one is used to determine the optimal crop pattern along a river based on the assigned flood control options by the first optimization model. The optimal combination of flood protection options is determined to minimize flood damage and construction cost of flood control options along the river using the GA optimization model. In order to consider the effects of flood control options on the hydraulic characteristics of flow, two hydrologic routing models for the reservoir and the river are coupled with the optimization model. The discharge–elevation and elevation–damage curves obtained, based on separate hydraulic simulations of the river, are used for flood damage calculations in the optimization model. The parameters of a hydrologic river routing model are also calibrated using the developed hydraulic model results. The

Table 4.11 *Rating curve data*

Flow (m³/s)	Stage (m)
0.000	1660.000
2.350	1661.225
3.678	1662.450
4.954	1663.675
6.029	1664.900
6.977	1666.125
7.834	1667.350
8.622	1668.575
9.355	1669.800
10.044	1671.025

proposed model is applied to the Kajoo River in the southeastern part of Iran. The results demonstrate an economical integration of permanent and emergency flood control options along the river, which include the minimum expected value of damage related to floods with different return periods and construction cost of flood control options. Finally, the resulting protection scheme is used for land use planning through identifying the optimal crop mix along the river. In this approach, the objective function of the optimization model is an economic function with a probabilistic framework to maximize the net benefit of agricultural activities.

The following includes a detailed presentation of an application of evolutionary optimization for learning purposes.

Example 10

Consider, for example, the problem of finding the best-fit analytical equation of the rating curve (flow vs. stage) given with ten pairs of (x, y) points in Table 4.11. A typical empirical equation for this curve is:

$$Q = AH^b \tag{4.53}$$

where Q is flow (m³/s), H is net stage (m) above datum (1660 m), and A and b are parameters. So, in the case of the rating curve given in Table 4.11 the net stage is the stage value from the table minus 1660 m. Determine parameters A and b in such a way that the difference of the sum of squares between the analytic and tabulated values of flow for all ten points is minimized.

Solution

This optimization problem can be formulated as: find the values of parameters A and b such that the value of the following objective function is minimized:

$$\min \sum_{i=1}^{10} (Q_i - A(H_i - 1660)^b)^2 \tag{4.54}$$

Flow–stage values are provided in Table 4.11 for each of the ten points. In addition, from other empirical studies related to similar curve fits it can be assumed that the most likely range for the values of parameter b is $(0,1)$ and for parameter A is $(0,10)$. To be on the safe side, in this example the values of parameter A are inspected in the range of $(0,20)$. The value of parameter b must be less than 1 since it is never a straight line, and it must be greater than 0 since values below 0 would not result in an increasing function, while it is known that the flow does increase with the increase in stage. Taking into account this simple knowledge about the problem reduces the search space to a value for parameter A in the interval $(0,20)$ and the value of b in the interval $(0,1)$, which has a significant impact on the solution efficiency.

We use the EVOLPRO computer program provided on the website to solve the problem. Input data for this optimization problem is in the directory Probabilistic tools, sub-directory Optimization, sub-sub-directory EVOLPRO, sub-sub-sub-directory Examples under the name *Example1.evolpro*.

The algorithm first goes through a process of initialization, where 100 solutions are generated in a purely random manner. The optimal values are $A = 2.1079$ and $b = 0.6542$ with the objective function value $OF = 0.039$. Observe that the problem has been solved with the maximum number of iterations set to 1,000 and tolerance level set to 0.00001.

EVOLPRO COMPUTER PROGRAM

EVOLPRO facilitates solving an optimization problem using an evolutionary algorithm. The software is capable of handling nonlinear objective functions and constraints with multiple decision variables. It uses the methodology introduced in the section above. The EVOLPRO algorithm includes:

(1) Initialization: assignment of a set of random values (genes) between the lower and upper bound for each decision variable (chromosome). The size of population is an input variable (must be greater than 20).

(2) Identification of feasible search region: each chromosome (one set of values for all decision variables) is checked against the set of constraints. If one of the constraints is not satisfied, the chromosome is discarded. This process is repeated until the required number of chromosomes for the population is obtained.

(3) Evaluation of objective function: for each chromosome the value of the objective function is calculated and ranked.

(4) Selection: the best-fit 30% of the population is taken to select parents. Parents are picked randomly.

(5) Recombination: using relationship (4.51) new offspring genes are generated.

(6) Mutation: new genes are disturbed using a factor $1 + 0.005 * (0.5 - Rnd())$.

(7) Feasibility check: the new chromosome produced from offspring genes is verified against the constraints and bounds. If the constraints are satisfied, the new offspring is ready for migration.

(8) Reinsertion: offspring totalling 30% of the population size are inserted to replace the least-fit 30% of the previous population.

(9) Step 3 is repeated with the new population.

Steps 4 to 9 are repeated until the desired accuracy is obtained or the maximum number of iterations is reached. Both the desired accuracy (tolerance level) and maximum number of iterations are program input variables. The Readme file within the EVOLPRO sub-sub-directory provides simple instructions for the program installation. The Help menus available within EVOLPRO give guidance on EVOLPRO use.

Example 11

Experiment with the Example 10 data to find out the sensitivity of final results to, for example, maximum number of iterations. The problem from Example 10 has been solved with maximum number of iterations set to 1,000 and tolerance level set to 0.00001. Run EVOLPRO using 500 iterations and compare the final results.

Solution

In the EVOLPRO computer program under the Computation menu, the Population size option defines the size of the population. After each new solution is created, its objective function is evaluated and compared with the worst objective function of the initial five solutions. We can review the progress of optimization by invoking the Computation>View Iterations Summary menu sequence and invoking the Computation>View Convergence Graph menu sequence. Example 10 has been solved with the maximum number of iterations set to 1,000 and tolerance level set to 0.00001. Table 4.12 shows the iteration summary for the Example 10 solution. Figure 4.13 shows the corresponding convergence graph. A second run of the program has been made with maximum number of iterations set to 500 and tolerance level set to 0.00001. Table 4.13 shows the iteration summary for this solution. Figure 4.14 shows the corresponding convergence graph.

Comparison of the results from these two experiments provides some insight into the functioning of the EVOLPRO optimization algorithm. In the first case (Table 4.12 and Figure 4.13) the optimal solution of $A = 2.107934$, $b = 0654187$, and OF (objective function) $= 0.039454$ has been reached in 25 iterations with the convergence as illustrated in Figure 4.13. In the second case (Table 4.13 and Figure 4.14) the optimal solution of $A = 2.10966$, $b = 0653802$, and $OF = 0.039457$ has been reached in 45 iterations with the convergence as illustrated in Figure 4.14.

In addition to number of iterations or tolerance level, the EVOLPRO algorithm is sensitive to change in size of the population. In our two experiments this value has been kept constant and equal to 100.

Table 4.12 *Iteration summary for 1,000 iterations*

Iteration	OF	A	b
0	1.602867	2.672821	0.554258
1	1.056181	1.664027	0.770402
2	1.056181	1.664027	0.770402
3	1.056181	1.664027	0.770402
4	0.680785	1.793612	0.739629
5	0.052154	2.163057	0.642186
6	0.052154	2.163057	0.642186
7	0.052154	2.163057	0.642186
8	0.052154	2.163057	0.642186
9	0.042618	2.118814	0.652908
10	0.042618	2.118814	0.652908
11	0.042618	2.118814	0.652908
12	0.042618	2.118814	0.652908
13	0.042618	2.118814	0.652908
14	0.039665	2.104588	0.655186
15	0.039665	2.104588	0.655186
16	0.039665	2.104588	0.655186
17	0.039665	2.104588	0.655186
18	0.039665	2.104588	0.655186
19	0.039616	2.112518	0.653361
20	0.039564	2.113073	0.65313
21	0.039481	2.107628	0.654115
22	0.039481	2.107628	0.654115
23	0.039481	2.107628	0.654115
24	0.039472	2.110862	0.653521
25	0.039454	2.107934	0.654187

Table 4.13 *Iteration summary for 500 iterations*

Iteration	OF	A	b
0	6.259741	3.257781	0.412605
1	2.037459	2.84014	0.515957
2	2.037459	2.84014	0.515957
3	0.806644	2.05506	0.646004
4	0.355062	2.314799	0.600989
5	0.040704	2.107269	0.653507
6	0.040704	2.107269	0.653507
7	0.040704	2.107269	0.653507
8	0.040704	2.107269	0.653507
9	0.040704	2.107269	0.653507
10	0.040704	2.107269	0.653507
11	0.040704	2.107269	0.653507
12	0.040704	2.107269	0.653507
⋮	⋮	⋮	⋮
34	0.03949	2.108365	0.653922
35	0.03949	2.108365	0.653922
36	0.03949	2.108365	0.653922
37	0.03949	2.108365	0.653922
38	0.03949	2.108365	0.653922
39	0.03949	2.108365	0.653922
40	0.03949	2.108365	0.653922
41	0.039473	2.110346	0.653549
42	0.039472	2.109593	0.653693
43	0.039472	2.109593	0.653693
44	0.03946	2.109321	0.653913
45	0.039457	2.10966	0.653802

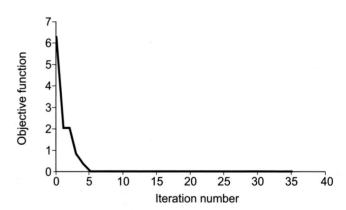

Figure 4.13 Convergence graph for 1,000 iterations.

Figure 4.14 Convergence graph for 500 iterations.

EVOLUTIONARY OPTIMIZATION FOR FLOOD FLOW FORECASTING WITH CLIMATE CHANGE IMPLICATIONS IN CHOKE CANYON RESERVOIR WATERSHED, SOUTH TEXAS

This section presents elements of the work conducted by Makkeasorn *et al.* (2008) that addressed flood forecasting in the context of climate change using evolutionary optimization and artificial NNs. This example is taken with the permission of the *Journal of Hydrology*. The presentation here is devoted to only one aspect of their work – evolutionary optimization. It is used as an example of real-world application of evolutionary optimization in management of flood risk under changing climatic conditions.

Introduction The importance of understanding and improving predictive capacity regarding all aspects of the global and regional water cycle is certain to continue to increase. One fundamental component of the water cycle is streamflow. Streamflow is related to the incidence of natural hazards, such as floods, that occur abruptly and may result in loss of life, destruction of the environment, and economic damage. Flood warning systems hold the highest possibility of reducing damage from the floods. Streamflow prediction therefore provides crucial information for adaptive water resources management.

Evolutionary optimization (especially the GP model) is selected as an approach for the development of a suite of streamflow forecasting models to meet various demands in this study. The strengths of the evolutionary approach include the natural selection process, and the white box characteristic. Since multiple input variables may be used in the prediction of flood flows, there is a need to identify the importance of each variable. The evolutionary process and natural selection techniques embedded in this modeling approach allow the screening of the multiple input variables to be executed explicitly for achieving the best result. The white box character of the modeling approach provides insight into internal structures of all the created models.

This example takes advantage of lead–lag regression in creating forecasting models. The lead–lag regression model is a statistical model that identifies differences of timing in fluctuations through a system. Three-step-behind inputs are paired with a current discharge, four-step-behind inputs are paired with a one-step-behind discharge, and so on. The multiple input variables of interest include historic streamflows, NEXRAD precipitation data, SSTs in the Pacific Ocean, the Atlantic Ocean, and the Gulf of Mexico, and local meteorological data collected from three weather stations in the watershed. Sea surface temperatures have been a primary expression of global climate anomalies for several decades. The El Niño–Southern Oscillation (ENSO) is the SST oscillation in the Pacific Ocean. A vast number of studies show influences of ENSO on climate changes in North, Central, and South America. The Pacific Decadal Oscillation (PDO) is another signature of climate change that largely affects decadal-scaled climate variations in the USA. Similar oscillations of SSTs also occur in the Atlantic Ocean and are known as the Atlantic Multi-decadal Oscillation (AMO) and North Atlantic Oscillation (NAO). All these studies of SST oscillations prove that climate variations in America and Europe are influenced by the connection of Pacific oscillations and NAO. Since river flow rates are largely driven by precipitation and climate variation is influenced by SST, there must be a connection between SST and river flow. This study incorporates SSTs of the Atlantic Ocean, the Gulf of Mexico, and the Pacific Ocean to aid in streamflow forecasting in south Texas. Specifically, the use of the sea surface temperature anomalies (SSTA) may capture some global climate change impacts and help improve flow volume predictions. This study investigates NEXRAD precipitation data in developing flood flow forecasting models to reduce the inaccuracy of interpolation and randomness of point rainfall data.

Hence, one of the goals of this study is to test the robustness of the forecasting models based on a short-term database. It is assumed that inter-annual variation may be driven by the SSTs that address the long-term impact due to the global climate changes. Effective lead-time streamflow forecast is one of the key aspects of successful flood risk management based on enlarged hydrometeorological datasets. Hence, it is the aim of this study to test the hypothesis that the inclusion of SSTs would significantly impact the accuracy of streamflow forecasting and the evolutionary optimization model can capture the underlying non-linear characteristics in a river system basin-wide.

Study area The Choke Canyon Reservoir watershed (CCRW) is a portion of the Nueces River basin, south Texas. It is composed of several land use and land cover patterns covering an area of approximately 15,000 km². The major uses of the land are agriculture and livestock. Intensive uses of groundwater for irrigation are highly concentrated in the middle and lower areas of the basin. The geography of the area strongly influences the hydrologic cycle of the watershed. In the upper portion of the watershed the steep slopes and arid terrain of the Balcones Escarpment rise into the Edwards Plateau. This area is the location of the unique formation of karst aquifer called the Edwards Aquifer. A karst formation refers to a carbonate rock formation of limestone and includes fractures and sinkholes (openings) that form a subsurface drainage system. A well-developed karst aquifer could transport water as fast as several miles per day.

The Edwards Aquifer is between 100 and 230 m thick and composed of porous limestone. The San Antonio segment is approximately 256 km long and between 8 and 64 km wide at the surface. The aquifer is characterized into three zones: the contributing zone, the recharge zone, and the artesian zone. The contributing zone of the CCRW is the area above the fault zone. The elevation changes between 740 m and 42 m from top to bottom. Faults exist in the upper area where the water recharges into the groundwater aquifer. USGS stream gages are located above and below the fault lines to capture the amount of water that is lost in the fault zone. Two USGS stream gages are located at the exit of the watershed to measure the amount of water released into the CCRW.

Although there are many conduits and large caverns in the Edwards limestone, the aquifer is not a good storage because of the high transmissivity rate. Most of the water is traveling in small pore spaces within the rock. Water enters the aquifer easily in the recharge zone, but the underground drainage is generally inadequate to hold all the water during large rain events. Recharge conduits and sinkholes quickly become filled up with water. This is the main reason why the region floods so easily. Downstream of the Balcones fault zone the landscape tends to flatten out as the streams continue southward and eastward into the South Texas brush country where slopes range from 0% to 10%. The stream-flow into the Choke Canyon Reservoir turns out to be critical in many flash flood events.

Description of a method and data This example follows the basic methodology of evolutionary optimization as discussed at the start of Section 4.5.2. However, its implementation is different. The evolutionary approach is used here to develop the model's structure (mathematical form of the objective function) and select the optimal set of model parameters based on the input data provided to the model. The EVOLPRO computer program presented above cannot be used in this way. The authors used a commercial product known as Discipulus (RML Technologies Inc., www.rmltech.com/, last accessed July, 2011). Regression models developed using evolutionary optimization are free from any particular model structure. The transparency of the evolutionary approach reveals structures of the regression models, which is the significant advantage of the approach. The evolutionary optimization models are the best for searching highly non-linear feasible spaces for global optima using adaptive strategies.

Given that the evolutionary model exhibits great potential to screen and prioritize the input variables, two sets of input data are defined, including all input variables and preselected input variables. While the former include all input variables fed into the model, the latter are those being picked up by the model. All input variables are used with the evolutionary model. While the training and the calibration data are used as the basis genotype to build models, another independent dataset is used to validate the generated models. Two groups of data used in this study include the existing national data from USGS Water Data for Nation, National Weather Service (NWS), National Data Buoy Center (NDBC), and the data collected by Makkeasorn *et al.* (2008) from three weather stations deployed in the study area. The National Water Information System (NWIS) (http://waterdata.usgs.gov, last accessed July, 2011) provides surface stream discharges. The NWS provides precipitation data obtained from NEXRAD. The NDBC provides SSTs of the seas and oceans around the USA. The research team has three meteorological stations located precisely at strategic locations. All of the datasets were collected or obtained on a daily basis between December 16, 2004, and May 5, 2005. The validation data are used only to test the fitness of the surviving models. In these, study data collected from December 15, 2004, to April 24, 2005, were used as the training and calibration datasets. Data collected from April 25 to June 4, 2005, were unseen data and used to validate models. The locations of USGS stream gages and the three weather stations deployed in the watershed are shown in Figure 4.15.

Gridded rainfall data are the measurement of precipitation with NEXRAD. Each grid square covers an area of 16 km^2 (4 km by 4 km). The daily data are published in inch per day per 16 km^2. Figure 4.16 shows the spatial distribution of daily rainfall intensity in mm per day. In order to match the precipitation rate to the discharge rate, unit conversion (inch per day to cubic meter per second) is performed for each grid of NEXRAD data. The sum of rainfall of all grids inside the watershed represents the average daily rainfall in cubic meters per second (m^3/s), which was used as the input data for evolutionary optimization models. The NEXRAD rainfall data are published (http://water.weather.gov/precip/, last accessed July, 2011). The total NEXRAD precipitation upstream of the watershed area is summed and converted into the runoff in cubic meters per second for the purpose of comparison against the discharge.

Sea surface temperature data are published at the website of the NDBC (www.ndbc.noaa.gov, last accessed July, 2011). Three buoy datasets were used in this study including buoy 42038 in the Gulf of Mexico, 46047 in the Pacific of Southern California, and 41002 in the Atlantic of South Carolina. The three locations of SST are converted to indices as follows: $SSTI_{sst-1} = SST_{Gulf}/SST_{Atlantic}$ and $SSTI_{sst-2} = SST_{Gulf}/SST_{Pacific}$ where SST is a temperature in °C. The indices are created under the assumption that hot air would rise to a high altitude, and cold air would move in to replace the hot air. Thus, if the SST in the Gulf of Mexico is higher than the SST in the Atlantic, the cold air over the Atlantic would transport moisture to Texas. Similarly, if the Pacific cold air moves toward Texas, moisture from the Pacific would be transported to Texas. Overall, SSTs from the three buoys are included as the inputs to aid in flood flow forecasting.

In addition to the stream gages, three meteorological stations were used to monitor changes at three locations in the CCRW from December 2004 to May 2005 (see Figure 4.15). The meteorological measurements collected at those stations include the air temperature, relative humidity, soil moisture, soil temperature, and precipitation. The main purpose of the stations is to fill in the gap in the information between the inflow gages, the discharge gages, the inland temperature, and the SST. Preliminary data analyses show: (i) that the peak of discharge occurred after each precipitation; (ii) that the soil moisture at three locations seems to follow the same pattern; and (iii) that the humidity changes do not seem to have any correlation with the discharge. However, humidity might have some kind of underlying correlation with discharge that the evolutionary programming (EP) models would be able to capture, which might improve the accuracy of the prediction. Hence, all of them were included in the EP model.

Scenario analyses Studies of the impacts of climate change on flood flows have been on the rise with great potential to improve adaptive water resources management. Models in this study were created under the assumption that SSTs of the eastern Pacific Ocean, the western Atlantic Ocean, and the Gulf of Mexico influence climate in the study area, which, in turn, characterizes the streamflow of rivers in the watershed. The USGS stream

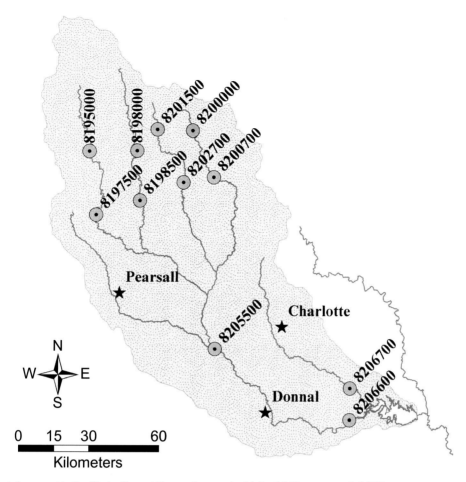

Figure 4.15 Weather stations used in the Choke Canyon Reservoir watershed (after Makkeasorn *et al.*, 2008).

gages therefore imply the capacity of streams to hold and transport the surface water. NEXRAD rainfall represents the water input. Soil moisture content implies the water storage capacity in the soil at three prescribed locations in the center and the south of the watershed. Creating flow forecasting models by incorporating the driving forces of precipitation using NEXRAD, the inherent watershed characteristics, the plausible climate impact with respect to SST indices, and the local micrometeorological factors, such as soil moisture, soil temperature, air temperature, precipitation, and relative humidity should result in a more transparent process of flood flow prediction. Such an effort has also sparked a marked interest in finding out the non-linear structure of streamflow variations among those valid driving forces on a short-term basis due to global climate change.

For the purpose of comparison, a base scenario is prepared using only USGS historical streamflow data (see dataset S0 in Table 4.14). Four datasets, S0, S1, S2, and S3, are then organized to form a set of scenarios of interest. S0 is a control dataset, which only contains seven sets of USGS streamflow data. The S1 dataset includes the USGS streamflow data (S0) and the NEXRAD rainfall data. The S2 dataset includes USGS streamflows, NEXRAD,

and the SST indices. The S3 dataset includes USGS streamflows, NEXRAD, SST indices, and 15 meteorological parameters collected from three weather stations. The S0 dataset is used to develop the base models. The S1 dataset is considered to identify how models respond to the NEXRAD rainfall data. The S2 dataset is purposely designed to investigate the impact of SST indices on

Table 4.14 *Datasets used in different scenarios*

	S0 dataset	S1 dataset	S2 dataset	S3 dataset
USGS streamflow data	✓	✓	✓	✓
NEXRAD		✓	✓	✓
Two SST indices derived			✓	✓
Fifteen local micrometeorological parameters at three stations				✓
Total discharge	✓	✓	✓	✓

After Makkeasorn *et al.*, 2008

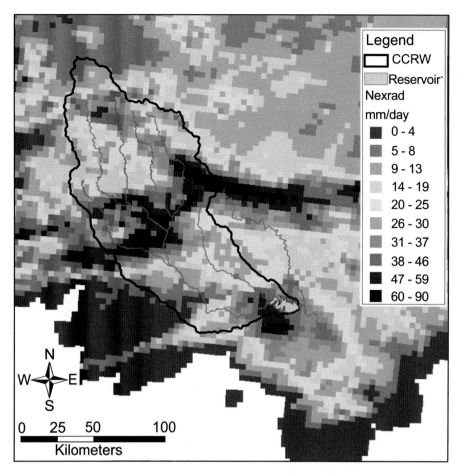

Figure 4.16 The south Texas NEXRAD rainfall imagery acquired on February 24, 2005 (after Makkeasorn *et al.*, 2008).

flood flow forecasting. The S3 dataset is organized to study the importance of local meteorological data on the performance of flood flow forecasting.

In addition to including four different datasets in forecasting models, three forecasting time spans were also used to test the robustness of models in terms of the relative significance of parameters associated with NEXRAD, SSTs, and the local meteorological data. In this study, the three forecasting time spans are 3, 7, and 30 days, respectively. Overall, a total of 12 evolutionary optimization scenarios were formed for the ultimate modeling analysis, thereby presenting a selective coverage of scenarios, denoted as EP-S0–3 day, EP-S1–3 day, EP-S2–3 day, EP-S3–3 day, EP-S0–7 day, EP-S1–7 day, EP-S2–7 day, EP-S3–7 day, EP-S0–30 day, EP-S1–30 day, EP-S2–30 day, and EP-S3–30 day.

Lead–lag regression is suitable for developing models with multiple input variables. Errors can be minimized because only historical and current input data are used instead of projecting multiple input data as drivers. One of the objectives of this study is to describe the characteristics of the lead–lag relation for flow forecasting. During the forecasting, the dependent variable (the

discharge) is offset from the independent variables (three USGS variables, NEXRAD rainfall, three buoys variables, and 15 meteorological variables) by a specific lead time step. For instance, input variables are paired, with three-lead-time-step discharges for 3-day forecasting models, with seven-lead-time-step discharges for 7-day forecasting models, and with 30-lead-time-step discharges for 30-day forecasting models.

Four statistical measures are applied in this case study to check the forecasting model performance: (i) ratio of standard deviation of predicted to observed discharges, CO; (ii) root-mean-square error, RMSE; (iii) square of the Pearson product moment correlation coefficient, r-square; and (iv) mean of percentage error, PE (see Makkeasorn *et al.*, 2008; page 346).

Forecasting analyses Available computer power allows for derivation of highly complex non-linear functions using EP for forecasting the streamflow. A very large number (millions) of EP-based models are created and measured during the evolutionary process. The better the fitness value, the better the model. Only the top 20 models with the highest level of fitness are selected for further evaluation. Amongst the top 20 models, the best model based on the fitness of the training data may not perform well in cases based on the unseen data. Therefore, the EP model that performed well on both the unseen dataset and the calibration dataset is chosen in this study.

The best 30-day flow forecasting model obtained by EP is:

$$30\_FF = 2 \times (2^{A1} - 1) \times V4^2$$
$$A1 = |\sin(A2)| + V2 - A3$$
$$A2 = \sin(2 \times V10 + 3.075786) + V2$$
$$A3 = [-\sin((-V1) \times (V8) \times (V16) - V15) \quad (4.55)$$
$$\times (2^{A4} - 1 + V2)]$$
$$A4 = \sqrt{|\sin(A5) + V9|}$$
$$A5 = \left(\frac{\sin((-V1) \times (V8) \times (V16) - V15)}{1.29105}\right.$$
$$\left. - 1.0309 + V2 - V4\right)^2 + V2$$

where $V1$ is the stream data at gage id# 8197500, $V2$ is the stream data at gage id# 8198500, $V4$ is $SST_{Gulf}/SST_{Atlantic}$, $V8$ is the soil temperature at Donnal, $V9$ is the precipitation at Donnal, $V10$ is the volumetric water content at Donnal, $V15$ is the volumetric water content at Charlotte, and $V16$ is the air temperature at Pearsall. See Figure 4.15 for the location of the listed sites.

The best 7-day flow forecasting model obtained by EP is:

$$7\_FF = [|2^{B2} - 1| - 2 \times V4] \times (-V15)$$
$$B2 = -0.6205 \times \{2^{B3} - 1 - [\sin(-B4) - V12]^2\}$$
$$B3 = [\sin(-B4) - V12]^2$$
$$B4 = (2^{0.69816 \times (B5)} - 1) \times 2^{-V15}$$
$$B5 = -1.9659 \times (B6) + V12 - 0.9 - 2 \times V15 \times \sin(V15)$$
$$B6 = [\sin(V15) \times (-V15)] - V4 \quad (4.56)$$

where $V4$ is $SST_{Gulf}/SST_{Atlantic}$, $V12$ is the relative humidity at Charlotte, and $V15$ is volumetric water content at Charlotte.

The best 3-day forecasting model obtained by EP is:

$$3\_FF = 2^{C1} - 1.3729$$
$$C1 = 2^{C2} - 1$$
$$C2 = 2^{C3} - 1$$
$$C3 = 2^{C4} - 1.3729 C3 = 2^{C4} - 1.3729$$
$$C4 = 2^{C5} - 1$$
$$C5 = |||C6| - C7 + V4 - V19| - C7|$$
$$C6 = (-1.2029) \times [-(C8^4) - V15]^2 \times (V6)$$
$$C7 = [(C12 - C11) \times (C11)] \times [-(C8^4) - V15]^2$$
$$C8 = \cos^2(C9) - \sin^2(|C10 - V15|)$$
$$C9 = \frac{-\sin(C20) \times \sin(C20)}{\cos[\sin(C20)]}$$
$$C20 = 2^{(-V12) \times (V20 + 1.4684)} - 1 \quad (4.57)$$

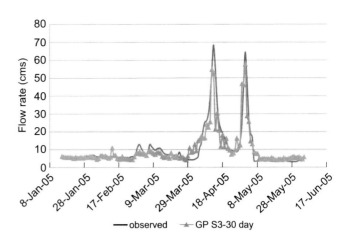

Figure 4.17 The performance of EP-derived flow forecasting model with a 30-day forecasting scheme.

where $V4$ is $SST_{Gulf}/SST_{Atlantic}$, $V5$ is $SST_{Gulf}/SST_{Pacific}$, $V6$ is the air temperature at Donnal, $V12$ is the relative humidity at Charlotte, $V15$ is the volumetric water content at Charlotte, $V18$ is the soil temperature at Pearsall, $V19$ is the rain at Pearsall, and $V20$ is the volumetric water content at Pearsall.

Results Figure 4.17 presents the flow forecasting based on the EP-derived model with the 30-day forecasting scheme. The forecasting time period started from April 21, 2005, to cover the first peak discharge during the calibration process. Such a choice allows the model to gain the opportunity to estimate the peak flow for the purpose of demonstration. A general agreement between the predicted and observed values can be confirmed within this diagram, indicative of the model accuracy.

The performance of the optimal forecasting models has been assessed using statistical measures and the results are presented in Table 4.15.

The r-square value increases as the input dataset moves from S1 to S3 in all EP-derived models. The r-square and RMSE show an agreement that training the models with the S3 dataset improves the forecasting accuracy. The RMSE, r-square, PE, and CO values indicate that EP-derived models are generally very accurate in terms of the prediction accuracy. Comparison of both RMSE and PE values shows general agreement in such a way that the increase of dimension of the inputs may significantly decrease the forecasting errors. The near-term forecasting models with a 3-day or 7-day scheme may produce relatively higher accuracy and smaller errors than the longer lead-time forecasting models with a 30-day scheme. Nonetheless, the 30-day EP-derived forecasting model may still exhibit predictions with very high accuracy ($r^2 = 0.84$) and considerably small error (RMSE = 5.4). This numerical potential reinforces the value of using such models as an early flood warning system at various time scales. Finally, the findings

Table 4.15 *Summary of statistical performance measures of optimal models*

Model	r-square	RMSE	PE	CO
EP-S0–3d	0.49	12.4	77.5	0.83
EP-S1–3d	0.75	8.6	54.0	0.81
EP-S2–3d	0.83	9.2	16.4	0.76
EP-S3–3d	0.90	6.1	45.3	0.90
EP-S0–7d	0.83	11.2	24.0	0.42
EP-S1–7d	0.34	14.5	13.9	0.52
EP-S2–7d	0.76	8.5	20.7	0.86
EP-S3–7d	0.79	7.9	65.0	0.77
EP-S0–30d	0.72	7.4	61.9	0.87
EP-S1–30d	0.54	9.5	51.2	0.90
EP-S2–30d	0.84	5.7	46.1	0.94
EP-S3–30d	0.84	5.4	39.2	0.81

After Makkeasorn *et al.*, 2008

also support the idea that micrometeorological parameters collected within the watershed can obviously improve the credibility of flow forecasting models.

4.5.3 Probabilistic multi-objective goal programming for flood risk management

Multi-objective decisions, in contrast to single-objective, do not have an optimal solution. As a result, the focus of multi-objective analysis is on the assessment of trade-offs between alternatives based on using more than one objective function.

BASIC CONCEPTS OF MULTI-OBJECTIVE ANALYSIS

Section 4.5.2 showed that a single-objective programming problem consists of optimizing one objective subject to a constraint set. On the other hand, a multi-objective programming problem is characterized by an *r*-dimensional vector of objective functions:

$$Z(x) = [Z_1(x), Z_2(x), \ldots, Z_r(x)]$$
$$\text{subject to}$$
$$x \in X \quad (4.58)$$
$$\text{where } X \text{ is the feasible region:}$$
$$X = \{x : x \in R^n, g_i(x) \leq 0, x_j \geq 0 \quad \forall i, j\}$$

where R is a set of real numbers, $g_i(x)$ is a set of constraints, and x is a set of decision variables.

The word *optimization* has been deliberately kept out of the definition of a multi-objective programming problem since we cannot, in general, optimize a priori a vector of objective functions. The first step of the multi-objective analysis consists of identifying the set of non-dominated solutions within the feasible region X. So, instead of seeking a single optimal solution, a set of non-inferior solutions is sought.

The essential difficulty with multi-objective analysis is that the meaning of the optimum is *not defined* as long as we deal with multiple objectives that are truly different. For example, suppose we are trying to determine the best design of a system of dams on a river, with the objectives of promoting national income, reducing deaths by flooding, and increasing employment. Some designs will be more profitable, but less effective at reducing deaths. How can we state which is better when the objectives are so different, and measured in such different terms? How can we state with any accuracy what the relative value of a life is in terms of national income? If we resolved that question, then how would we determine the relative value of new jobs and other objectives? The answer is, with extreme difficulty. The attempts to set values on these objectives are, in fact, most controversial.

In practice it is frequently awkward if not impossible to give every objective a relative value. Therefore, the focus of multi-objective analysis in practice is to sort out the mass of clearly dominated solutions, rather than determine the single best design. The result is the identification of a small subset of feasible solutions that are worthy of further consideration. Formally, this result is known as the set of *non-dominated solutions*. Mathematically, given a set of feasible solutions X, the set of non-dominated solutions is defined as follows:

$$S = \{x : x \in X, \text{ there exist no other } x' \in X$$
$$\text{such that } z_q(x') > z_q(x) \text{ for some } q \in \{1, 2, \ldots, r\} \quad (4.59)$$
$$\text{and } z_k(x') \leq z_k(x) \text{ for all } k \neq q$$

It is obvious from the definition of S that as we move from one non-dominated solution to another non-dominated solution and one objective function improves, then one or more of the other objective functions must decrease in value. For detailed discussion of multi-objective analysis see Simonovic (2009) and Goicoechea *et al.* (1982).

STOCHASTIC MULTI-OBJECTIVE ANALYSIS

A large number of problems in flood risk management under climate change require that decisions be made in the presence of uncertainty. A key difficulty in multi-objective analysis under uncertainty is in dealing with an uncertainty space that is huge and frequently leads to very large-scale optimization models. The main approaches to stochastic multi-objective analysis that are recognized in the literature are: (i) stochastic sensitivity analysis, (ii) decision-theoretic programming models, and (iii) risk programming in linear programming, which includes:

- Two-stage programming;
- Stochastic linear programming;
- Transition probability programming; and
- Chance-constrained programming.

A typical problem in stochastic sensitivity analysis arises when the random variations in the parameters of a linear programming (LP) problem allow the optimal solution to retain the same basis as the parameters vary. When this is the case, it has been shown that the objective function can be expanded into a Taylor's series around the expected-value solution and, furthermore, that this series has an asymptotic normal distribution. A level of complexity is added when parametric errors are significant and the same optimal basis cannot be retained throughout the entire range of those parametric errors. One approach to this situation is to attempt to build suitable approximations of the LP problem subject to parametric errors. These approximations are termed *deterministic equivalents*, in the sense that a probability statement is transformed into an algebraic expression (i.e., an equality or inequality) that no longer contains any random variables. As a result, these deterministic equivalents are often non-linear, and the dimensionality of the original LP problem increases.

For the implementation of stochastic multi-objective analysis in flood risk management I present a special form of stochastic goal programming that considers constraints as stochastic and expands one-sided probabilistic goal constraints into the case of two-sided probabilistic goal constraints with proper formulation, which preserves the original characteristics of goal programming. This two-sided probabilistic goal-constrained method is more efficient than the existing methods in that it has fewer constraints and decision variables.

The resulting chance-constrained goal programming (CCGP) method provides an effective way of adapting the LP simplex method, which takes into account the non-linear behavior of the parameters of a model. Problem formulations obtained by the application of this approach can be solved using LP computer tools that are widely available. A later section presents the LINPRO program that is provided on the accompanying website, which can be used quite effectively for solving flood risk management problems.

STOCHASTIC GOAL PROGRAMMING METHOD

Goal programming, originally introduced by Charnes and Cooper (1961), deals with the problem of achieving a set of conflicting goals. The objective function seeks to minimize deviations from the set of pre-assigned goals. Two types of goals are commonly used in goal programming models. The first type is a two-sided goal applicable to goals that must be achieved exactly; any deviation, either upward or downward, would result in penalty. The second type is a one-sided goal with which only upward or downward deviation would be penalized. When uncertainties exist in the goal programming problem, the chance-constrained formulation is often used.

For educational purposes I start with the traditional goal programming formulation with a one-sided goal for which over-achievement would result in an undesirable penalty. The objective

is to determine decision variables, x, that make, if possible, the objective function value not exceed the pre-specified goal, G, as

$$\sum_{j=1}^{n} c_j x_j \leq G \tag{4.60}$$

where c is a vector of the objective function coefficient. The deterministic goal programming model for such a goal is expressed as:

$$\sum_{j=1}^{n} c_j x_j + d^- - d^+ = G \tag{4.61}$$

where d^- and d^+, respectively, represent the under- and over-achievement with respect to the goal in which a positive penalty is associated with d^+ and a negative penalty with d^- in the objective function. When the elements in c are random variables, two approaches are found in the literature that convert the deterministic equation into the chance-constrained one (a good basic text on stochastic programming is that by Birge and Louveaux, 1997). One approach is to utilize Equation (4.61) by simply converting it into a probabilistic form as:

$$P\left[\sum_{j=1}^{n} c_j x_j + d^- - d^+ = G\right] \geq \alpha \tag{4.62}$$

where α is the goal compliance reliability. A very common problem with this formulation is that the modelers fail to recognize that the probability of a continuous random variable equaling a fixed value is zero.

The next step in the implementation of goal programming is the derivation of the chance constraint from Equation (4.60).

$$P\left[\sum_{j=1}^{n} c_j x_j \leq G\right] \geq \alpha \tag{4.63}$$

The corresponding deterministic equivalent of Equation (4.63) is derived as:

$$\sum_{j=1}^{n} E(c_j) x_j + Z_\alpha \left[x' \sum x\right]^{0.5} \leq G \tag{4.64}$$

where $E(c_j)$ is the mean value of the jth objective coefficient and Σ is the variance-covariance matrix of the random objective function coefficients. The CCGP equation is then obtained by adding deviational variables in Equation (4.64) as:

$$\sum_{j=1}^{n} E(c_j) x_j + d^- - d^+ = G - Z_\alpha \left[x' \sum x\right]^{0.5} \tag{4.65}$$

Note that the original goal G from Equation (4.61) has been changed in Equation (4.65).

The concept of the proposed chance-constrained goal formulation is considered for the three possible goal types of stochastic goal programming. The first type represents a two-sided goal for which the upward and downward deviations are to be penalized. In this case, it is desired to obtain the shortest interval that attains the goal, a pre-assigned reliability of α_r. Second (upward deviation) and third (downward deviation) types represent one-sided goals. In these cases, only upward or downward deviations are to be penalized. The reliability of attaining each goal is denoted by α_s and α_t, respectively.

A general form of the CCGP model, based on the goal programming format, may be formulated as:

$$\text{Minimize } D_0 = \sum_{r=1}^{R} \left(d_r^+ + d_r^-\right) + \sum_{s=1}^{S} d_s^+ + \sum_{t=1}^{T} d_t^+ \quad (4.66)$$

subject to

$$P\left[G_r - d_r^- \leq c_r' x \leq G_r + d_r^+\right] > \alpha_r, \quad r = 1, 2, \ldots, R \quad (4.67)$$

$$P\left[c_s' x \leq G_s + d_s^+\right] > \alpha_s, \quad s = 1, 2, \ldots, S \quad (4.68)$$

$$P\left[c_t' x \geq G_t - d_t^-\right] > \alpha_t, \quad t = 1, 2, \ldots, T \quad (4.69)$$

$$P\left[Ax \leq b\right] > \beta \quad (4.70)$$

$$x \geq 0 \quad (4.71)$$

$$d_r^-, d_r^+, d_s^+, d_t^- \geq 0, \quad \forall r, s, \text{ and } t \quad (4.72)$$

where R is the number of two-sided goals, S is the number of one-sided goals for upside deviation, and T is the number of one-sided goals for downside deviation. In addition, α_r, α_s, and α_t are pre-assigned reliabilities of attaining the corresponding goals. Constraint (4.70) represents the regular constraint in a probabilistic form that takes into account the random character of A and b. It can be represented as a deterministic equivalent when the elements of A and b are all constants.

Constraints (4.67) through (4.70) in a general formulation are all probabilistic. In order to solve a stochastic goal programming problem, these constraints must be transformed to their respective deterministic equivalents. The transformation procedure can be found elsewhere (for example in Taha (1976) or Birge and Louveaux (1997)). The final form of the deterministic equivalent formulation of the problem given by Equations (4.66) to (4.72) is as follows:

$$\text{Minimize } D_0 = \sum_{r=1}^{R} \left(d_r^+ + d_r^-\right) + \sum_{s=1}^{S} d_s^+ + \sum_{t=1}^{T} d_t^+ \quad (4.73)$$

subject to

$$E(c_r')x - 0.5d_r^+ + 0.5d_r^- = G_r, \quad \forall r \quad (4.74)$$

$$E(c_r')x - d_r^+ + F_z^{-1}\left[\frac{1+\alpha_r}{2}\right] \times \left[x' \sum_r x\right]^{0.5} \leq G_r, \quad \forall r$$
$$(4.75)$$

$$E(c_r')x - d_s^+ + F_z^{-1}[\alpha_s] \times \left[x' \sum_s x\right]^{0.5} \leq G_s, \quad \forall s \quad (4.76)$$

$$E(c_r')x + d_t^- + F_z^{-1}[1 - \alpha_t] \times \left[x' \sum_t x\right]^{0.5} \geq G_t, \quad \forall t \quad (4.77)$$

$$x \geq 0 \quad (4.78)$$

$$d_r^-, d_r^+, d_s^+, d_t^- \geq 0, \quad \forall r, s, \text{ and } t \quad (4.79)$$

where F_z^{-1} is the inverse cumulative distribution function of the standard normal random variable. In this transformation an assumption has been used that all random variables are normally distributed. The above model can be generalized even further by including other deterministic constraints, probabilistic constraints, or a mixture of both.

The deterministic goal programming model represented follows an LP format, which can be easily solved by the LP simplex algorithm. However, the deterministic equivalent transformation of CCGP constraints represented by Equations (4.73) to (4.79) is non-linear, and cannot be solved directly by the LP technique. Therefore, the problem becomes one of non-linear optimization, which can be solved by various non-linear programming techniques. Alternatively, a procedure can be adopted to linearize the non-linear terms of the chance constraints and solve the linearized model iteratively. The "linearized" constraints in the CCGP model are obtained by moving the non-linear terms in constraints (4.75), (4.76), and (4.77) to the right hand side of the constraints and can be written as:

$$E(c_r')x - d_r^+ \leq G_r - F_z^{-1}\left[\frac{1+\alpha_r}{2}\right] \times \left[x'^0 \sum_r x\right]^{0.5}, \quad \forall r$$
$$(4.80)$$

$$E(c_r')x - d_s^+ \leq G_s - F_z^{-1}[\alpha_s] \times \left[x'^0 \sum_s x\right]^{0.5}, \quad \forall s$$
$$(4.81)$$

$$E(c_r')x + d_t^- \geq G_t - F_z^{-1}[1 - \alpha_t] \times \left[x'^0 \sum_t x\right]^{0.5}, \quad \forall t$$
$$(4.82)$$

here x'^0 is an assumed solution vector to the optimal CCGP model. Consequently, the linearized CCGP model can be solved by the LP technique iteratively, each time comparing the values of the current solutions with those obtained in the previous iteration, then updating the assumed solution values, and using them to compute the variance term on the right hand sides, until convergence criteria are met. Of course, alternative stopping rules, such as specifying the maximum number of iterations, can also be used in order to prevent excessive iteration during the computation. However, prior to the application of these procedures, an assumption for the distribution of the standardized random variable Z must be made so that the terms $F_z^{-1}\left[\frac{1+\alpha_r}{2}\right]$, $F_z^{-1}[\alpha_s]$, and $F_z^{-1}[1 - \alpha_t]$ in constraints (4.80), (4.81), and (4.82) can be evaluated.

Due to the non-linear nature of the CCGP model, the optimum solution obtained, in general, cannot be guaranteed to be the global optimum. Thus, it is recommended that a few runs of these procedures with different initial solutions should be carried out to ensure that the model solution converges to an overall optimum.

STOCHASTIC GOAL PROGRAMMING, CLIMATE CHANGE, AND FLOOD RISK MANAGEMENT

Application of stochastic multi-objective analysis in flood risk management under climate change is not available in the published literature. However, a few examples presented below, related to flood risk management in general, illustrate the initial benefits from considering stochastic multi-objective analysis in the field.

Raju and Kumar (1998) used stochastic and fuzzy linear programming (FLP) for the evaluation of management strategies for a case study of the Sri Ram Sagar Project, Andhra Pradesh, India. Three conflicting objectives – net benefits, crop production, and labor employment – are considered in the irrigation planning scenario. The flood control function of the multi-purpose project was treated as one of the constraints. Uncertainty in the inflows is considered by stochastic programming. The monthly inflows into the Sri Ram Sagar Reservoir are assumed to follow a log-normal distribution. Twenty-three years of historical inflow data are used to obtain the various dependability levels of inflows. Optimization of each individual objective (labor, production, and benefits) is performed with an LP algorithm that gives the upper and lower bounds for the multi-objective analysis. Much stronger emphasis in this paper is given to fuzzy multi-objective program formulation.

Foued and Sameh (2001) present a stochastic optimization solution for the operation and control of multi-purpose reservoir systems, with an example from Tunisia. The problem involves finding appropriate releases from various reservoirs in the system in order to satisfy multiple conflicting objectives, such as the satisfaction of demands with the required salinity, the minimization of pumping cost, the minimization of flood damage, and the minimization of drought impacts. To solve this problem, the authors consider two problem characteristics: multiple objectives and stochastic variables.

A five reservoir system in northern Tunisia is used as a case study. The system is simplified by aggregation of three reservoirs of the extreme north into one corresponding storage. Reservoir inflows and demands are treated as random variables. Flood control, as well as drought control, purposes of the reservoir system are considered of importance only during a certain period of the year. The overall formulation includes three probabilistic goals: (i) minimization of the difference between the reservoir system

release and demand for water, (ii) minimization of salinity at Beyeoua reservoir, and (iii) minimization of pumping costs at Beyeoua and Echkel reservoirs. The first goal was constructed from four stochastic components: (i) drinking water supply, (ii) irrigation water supply, (iii) minimum flood control storage, and (iv) minimum storage for drought control. The stochastic goal program has been converted into the deterministic equivalent, linearized and then solved using the traditional LP algorithm. In this formulation, the impacts of climate change on the random variables are not considered. However, the same formulation may be applied after addressing the impacts of climate change on the random input variables.

Lund (2002) looks into economical integration of permanent and emergency flood control options as a long-standing problem in water resources planning and management. A two-stage LP formulation of this problem is proposed and demonstrated, which provides an explicit economic basis for developing integrated floodplain management plans. The approach minimizes the expected value of flood damage and costs, given a flow or stage frequency distribution. In this work, again, climate change is not taken into consideration. However, the formulation can be very easily applied to investigate the climate-caused change in flood conditions, after the appropriate modification of flow and stage frequency distributions. A variety of permanent and emergency floodplain management options can be examined in the method, and interactive effects of options on flood damage reduction can be represented. This work suffers from the major problem of combining, using economic quantification, some emergency flood control measures that in practice will never be considered using economic criteria. This limitation reduces the value of the paper, but does not detract from the presented methodology, which can be implemented correctly for the solution of many flood risk management problems. The general formulation in this paper includes: (i) the damage in a floodplain resulting from water stage, given a set of permanent floodplain management actions, and emergency flood response actions; and (ii) an implementation cost to each of these management actions. The overall economic objective of managing the floodplain is then the expected value of the sum of these costs and damage, with the average taken using the stage-probability distribution. The complete formulation is a two-stage linear or integer-linear program. This form of problem can often be solved by readily available linear program solvers. The paper examples (very simple and hypothetical) are solved using spreadsheet linear program solvers. Limitations of the method in terms of forecast uncertainty and concave additive damage function forms are discussed also, along with extensions for addressing these more difficult situations.

Chang *et al.* (2007) developed a decision-making tool that can be used by government agencies in planning for flood emergency logistics. In this article, the flood emergency logistics problem with uncertainty is formulated as two stochastic programming models that allow for the determination of a rescue resource distribution system for urban flood disasters. The decision variables include the structure of rescue organizations, locations of rescue resource storehouses, allocations of rescue resources under capacity restrictions, and distributions of rescue resources. By applying the data processing and network analysis functions of the GIS, flooding potential maps can estimate the possible locations of rescue demand points and the required amount of rescue equipment. The decision problem does not consider the impacts of climate change on the flood risk. However, the methodology can be easily adopted to address the impact of climate change on flood emergency logistics and therefore reduction of flood risk. Due to the uncertainty of flood disasters and the quick response requirement, authorities must create a flood disaster response plan or standard operation procedure and reserve enough rescue equipment, resources, and rescue teams at proper places in advance. Once the disaster has occurred, the rescue operation will be launched into schedule without chaos. Basic flood rescue operations follow guidelines that divide the disaster area among several groups and emphasize intra-group distribution and inter-group backup, despite the varying structures of rescue organizations among different counties. Under such a disaster rescue system, the flood emergency logistics network has a multi-group, multi-echelon, and multi-level structure. First, the grouping and classifying model is used to group the disaster rescue areas and classify their level of emergency by minimizing the expected shipping distance. Then, based on the results of the first model, the two-stage stochastic programming model determines the selected local rescue bases that need to be set up after the disaster as well as the quantity of rescue equipment in the storehouses of all echelons and the transportation plans for rescue equipment. The former is the first-stage decision and the two latter cases are the second-stage decisions based upon the first-stage decision and the realized uncertainty scenarios. The corresponding costs for distributing the rescue equipment in all rescue demand scenarios are considered in the first-stage decision. Thus, the objective is to minimize the current facility setup cost and the equipment average cost as well as the expected future transportation cost, the supply-shortage cost, and the demand-shortage penalty during the period of rescue operation. The proposed models are solved using a sample average approximation scheme. The proposed model is applied to Taipei City, located in Northern Taiwan. Taipei City is 272 square kilometers in area and has a population density of over 10,000 people per square kilometer. Many places in Taipei City often face the risk of flood during typhoon seasons. To reduce the damage of disasters, the government requires that the entire disaster rescue operation should be completed within 36 hours. Furthermore, the Taipei City Government is designated as the head disaster rescue center; 12 administrative district offices are set up as regional rescue centers; and 84 offices of the subdivision of the district are chosen as candidate points for the local rescue bases.

LINPRO COMPUTER PROGRAM

Stochastic multi-objective goal programming often ends in the solution obtained by the implementation of the traditional LP simplex algorithm. The website accompanying this book includes the LINPRO software for the solution of an LP problem. The Readme file within the LINPRO sub-sub-directory provides simple instructions for the program installation. The Help menus available within LINPRO give guidance on LINPRO use. The enclosed computer program implements the LP solution using the simplex algorithm. A detailed presentation of the simplex method is available elsewhere. The recommended source is Simonovic (2009, Chapter 9). The following example illustrates the implementation of multi-objective stochastic goal programming (presented earlier) and use of LP software in its solution.

Example 12

The Water Authority considers investment in nine mutually exclusive flood project alternatives to mitigate potential increases in flood damage due to climate change. The net present values for each project are given together with a certain configuration of construction costs over a 2-year period. The objective is to maximize the present values of these investments in the context of a fraction of the adopted flood project investment, given a budget constraint of $50 M for the first period and $20 M for the second period.

The Water Authority expects that each investment in these projects should yield a certain amount of flood damage reduction benefits in each period and utilize a specified number of man-hours per day. Table 4.16 lists the project proposals along with their net present values and the present values of construction costs, flood damage benefits, and man-hours for each period.

The Water Authority would like to achieve the following goals: (i) The flood protection as a whole must yield a net present value of at least $32.40 M. Hence it is a one-sided goal because there is no penalty if it is more than this amount. (ii) The flood damage reduction benefits must be at least $70 M for the first period. Again it is a one-sided goal because if the value is more than this amount, it is desirable, but if it is less, it causes losses for the Water Authority that will have an impact on other water management programs. (iii) The flood damage reduction benefits must be at least $84 M for the second period. This is a one-sided goal similar to goal (ii). (iv) The man-hours of labor per day must be exactly 40 for the first period. This is a two-sided goal because any deviations (whether upward such as payment for overtime or downward such as idle time of employees) are not desirable. (v) The man-hours

Table 4.16 *Data for the flood project alternative selection example (millions of $)*

Flood project	PV of project costs	PV of construction costs		Flood damage reduction benefits		Man hours	
		Period 1	Period 2	Period 1	Period 2	Period 1	Period 2
1	14	12	3	14	15	10	12
2	17	54	7	30	42	16	16
3	17	6	6	13	16	13	13
4	15	6	2	11	12	9	13
5	40	30	35	53	52	19	16
6	12	6	6	10	14	14	14
7	14	48	4	32	34	7	9
8	10	36	3	21	28	15	22
9	12	18	3	12	21	8	13

of labor per day are to be exactly 40 for the second period. Again, it is a two-sided goal similar to goal (iv).

Because of the uncertainty associated with the objective function coefficients in each goal, the Water Authority wishes to achieve the goals with certain pre-assigned reliabilities α_1, α_2, α_3, α_4, and α_5. In addition, it is assumed that the objective function coefficients are independent normal random variables, with the means given in Table 4.16 and standard deviations equal to some percentage of the means.

Solution

The mathematical formulation of the problem is as follows:

$$Min\ d_0 = d_1^- + d_2^- + d_3^- + d_4^- + d_5^- + d_5^+ \qquad (4.83)$$

subject to

$$P\left[c_1'x \geq G_1 - d_1^-\right] > \alpha_1 \qquad (4.84)$$

$$P\left[c_2'x \geq G_2 - d_2^-\right] > \alpha_2 \qquad (4.85)$$

$$P\left[c_3'x \geq G_3 - d_3^-\right] > \alpha_3 \qquad (4.86)$$

$$P\left[G_4 - d_4^- \leq c_4'x \leq G_4 + d_4^+\right] \geq \alpha_4 \qquad (4.87)$$

$$P\left[G_5 - d_5^- \leq c_5'x \leq G_5 + 5\right] \geq \alpha_5 \qquad (4.88)$$

$$a_1'x \leq b_1 \qquad (4.89)$$

$$a_2'x \leq b_2 \qquad (4.90)$$

$$1 \leq x \leq 0 \qquad (4.91)$$

$$d_1^-, d_2^-, d_3^-, d_4^-, d_4^+, d_5^-, d_5^+ \geq 0, \qquad (4.92)$$

where vectors c_1 through c_5 are the coefficients of goal constraints, representing the present values of alternative project investment, the flood damage reductions in period 1 and period 2, and man-hours in period 1 and period 2, respectively. Vectors a_1 and a_2 in constraints (4.89) and (4.90) are the coefficients of regular

deterministic constraints representing the present values of construction costs in period 1 and period 2, respectively.

The problem represented by Equations (4.83) to (4.92) is first converted into the deterministic equivalent by rewriting probabilistic constraints (4.84) to (4.88) in their deterministic form as shown in Equations (4.73) to (4.79). The corresponding LP problem has been solved using the LINPRO program. A comprehensive sensitivity analysis is performed for:

(1) reliabilities α_1, α_2, α_3, α_4, and α_5 equal to 85%, 90%, and 95%, and

(2) various values of standard deviation of 5%, 10%, 25%, and 50%.

A summary of the results is shown in Table 4.17. A quick examination of the results presented in Table 4.17 indicates that the total amount of optimal deviations from goals is sensitive to various standard deviations and reliabilities. For a given uncertainty level of model parameters, the total amount of optimal deviations is increased as the reliability requirement of the goal constraints increases.

The optimal results show that in order to achieve the goals with high reliability, more total deviations should be anticipated in the objective functions. By increasing the standard deviations for the same level of reliability, as expected, the value of the objective function is increased.

STOCHASTIC MULTI-OBJECTIVE ANALYSIS FOR FLOOD EMERGENCY LOGISTICS PREPARATION IN TAIPEI CITY, TAIWAN

This section is based on the work conducted by Chang *et al.* (2007) on the introduction of a scenario planning approach for the flood emergency logistics preparation problem under

Table 4.17 *Optimal results for various standard deviations and reliabilities (millions of $)*

Standard deviation	5%			10%			25%			50%		
Reliability	0.85	0.90	0.95	0.85	0.90	0.95	0.85	0.90	0.95	0.85	0.90	0.95
x_1	0.332	0.305	0.266	0.220	0.168	0.090	0.000	0.000	0.000	0.000	0.000	1.000
x_2	0.000	0.000	0.000	0.000	0.000	0.000	0.000	0.000	0.000	0.000	0.000	0.000
x_3	1.000	1.000	1.000	1.000	1.000	1.000	0.918	0.816	0.682	0.536	0.642	0.327
x_4	1.000	1.000	1.000	1.000	1.000	1.000	1.000	1.000	1.000	1.000	1.000	1.000
x_5	0.251	0.253	0.255	0.258	0.261	0.255	0.285	0.302	0.325	0.350	0.332	0.324
x_6	0.000	0.000	0.000	0.000	0.000	0.000	0.000	0.000	0.000	0.000	0.000	0.000
x_7	0.551	0.557	0.564	0.575	0.586	0.602	0.623	0.625	0.627	0.630	0.628	0.423
x_8	0.000	0.000	0.000	0.000	0.000	0.000	0.000	0.000	0.000	0.000	0.000	0.000
x_9	0.000	0.000	0.000	0.000	0.000	0.000	0.000	0.000	0.000	0.000	0.000	0.000
d_1^-	0.000	0.000	0.000	0.000	0.000	0.000	0.000	0.000	0.000	0.000	0.000	0.000
d_2^-	11.8	12.3	13.0	13.8	14.7	16.2	19.7	21.9	25.1	28.9	32.2	33.4
d_3^-	20.8	21.3	22.0	22.9	24.0	25.5	29.5	32.1	35.8	40.2	43.3	45.0
d_4^-	0.000	0.000	0.000	0.000	0.000	0.000	0.000	0.000	0.000	0.000	0.000	0.000
d_4^+	0.000	0.000	0.000	0.000	0.000	0.000	0.000	0.000	0.000	0.000	0.000	5.27
d_5^-	0.000	0.000	0.000	0.000	0.000	0.000	0.000	0.000	0.000	0.000	0.000	0.000
d_5^+	0.000	0.000	0.000	0.000	0.000	0.000	0.000	0.000	0.000	0.000	3.32	14.1
d_0	32.6	33.6	35.0	36.7	38.7	41.7	49.2	54.0	60.9	69.1	78.8	97.7

uncertainty, and its real implementation for Taipei City in Taiwan. The occurrence of flood disasters under changing climatic conditions in an area and the demand for emergency management are both uncertain. Therefore, there is a need for emergency planning that will provide rescue agencies with the best options for uncertain future flood disaster conditions. Comprehensive flood preparation planning tools currently do not exist (Simonovic, 2011). Chang *et al.* (2007) have developed a decision-making tool that can be used in practice by disaster prevention and rescue agencies for planning flood emergency logistics preparation. This tool builds on the practical issues of emergency logistics and the current organizational structure of disaster prevention agencies is taken into consideration. In addition, the disaster rescue demand point and the amount of demand are treated as stochastic. A suitable rescue resource preparation plan must be determined prudently by considering all possible rescue demand scenarios. In this example, the authors explore how to optimize the expected performance over all scenarios and incorporate scenario planning into flood emergency logistics preparation modeling.

Introduction Flood emergency logistics planning involves dispatching commodities (medical materials, specialized rescue equipment, rescue teams, etc.) to distribution centers in affected areas as soon as possible to accelerate flood relief operations. In the past this problem has been addressed using a single-commodity, multi-modal network flow model on a capacitated network over a multi-period planning horizon. The sum of incurred costs is minimized for the transportation and storage of dispatched commodities (food for example). Most of the literature on emergency logistics emphasizes how to dynamically distribute limited rescue resources to an area after a disaster has occurred. Planning for flood emergency logistics preparation has received limited attention. None of the studies used stochastic programming to deal with this issue. The proposed methodology will apply stochastic programming theory to determine flood emergency logistics preparation. Although some published supply chain network design models and the described emergency logistics preparation planning problem have some common characteristics, they are not the same. Therefore, the chain network design models cannot be adapted to specific flood conditions for two reasons: the rescue demand points in a flood disaster occur randomly, and the distribution of rescue resources requires mutual assistance among multi-group, multi-echelon, and multi-level suppliers. The random character of emergency assistance demand is further emphasized by climatic change. The following model is developed to provide a more adequate solution to the stochastic multi-backup location problem.

Model description Due to the uncertainty of flood disasters under climate change and the quick response requirement, authorities must create a flood disaster response plan or standard operation procedure and plan for enough rescue equipment, resources, and rescue teams at proper places in advance. Basic flood

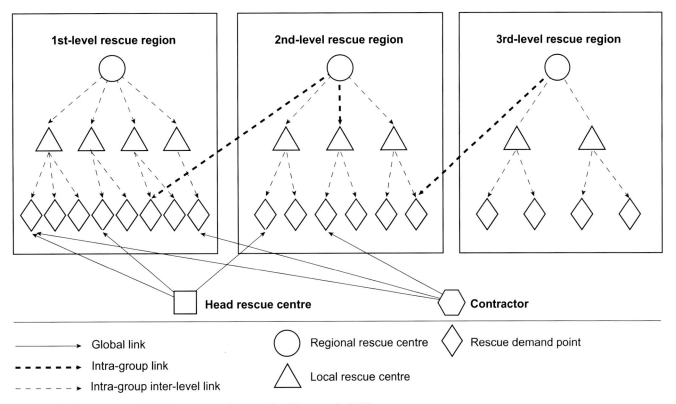

Figure 4.18 The structure of a flood rescue organization (after Chang *et al.*, 2007).

rescue emergency operations divide the disaster area among several groups and emphasize intra-group distribution and inter-group backup (some variations are possible among different counties). Therefore, the flood emergency logistics network has a multi-group, multi-echelon, and multi-level structure. The structure of rescue organizations and the relation of mutual backup are illustrated in Figure 4.18.

As shown in the figure, nodes are classified into five types: a head rescue center, regional rescue centers, local rescue bases, contractors, and demand points. Regional rescue centers, local rescue bases, and demand points can be further classified according to the district in which they are located. In addition, there are four types of flows (activities), which are shown as global links, intra-group inter-level links, intra-group intra-level links, and inter-group links in Figure 4.18. The structure of rescue organizations and the flood rescue division is explained in detail in Chang *et al.* (2007). It is evident that flood emergency logistics preparation planning is rigid and complex due to uncertain demands. In this study (Chang *et al.*, 2007), the flood emergency logistics preparation problem with uncertainty is formulated as two stochastic programming models (decomposition of the multi-objective analysis problem) that determine the rescue resource distribution plan for urban flood disasters, including the structure of the rescue organization, the location of rescue resource storehouses, the allocation of rescue resources within capacity restrictions, and the distribution of rescue resources. The basic

assumptions of this work are: (i) the rescue resource may be supplied either by the government or by private enterprises with a rental contract; (ii) the stochastic rescue demands are predictable under all possible flooding scenarios (including climate change); and (iii) transportation costs are deterministic.

Input variables used in the models are:

$L(\omega)$ set of demand points under rainfall event ω;

$L^k(\omega)$ set of demand points under rainfall event ω that lie within the service area of regional rescue center k;

$L_n(\omega)$ set of demand points that belong to the level-n areas under rainfall event ω;

$L_n^m(\omega)$ set of demand points that belong to group-m and level-n areas under rainfall event ω;

$J_n^m(\omega)$ set of local rescue bases that belong to group-m and level-n area under rainfall event ω;

$K_n^m(\omega)$ set of regional rescue centers that belong to group-m and level-n area under rainfall event ω;

$K_2(\omega)$ set of regional rescue centers that belong to level-2 areas under rainfall event ω;

$M(l)$ group of demand point l;

$N(j)$ level of local rescue base j;

C^i unit penalty for shortage of equipment type i during the period of rescue operation;

$d_l^i(\omega)$ amount of equipment type i required at demand point l under rainfall event ω;

e^i average cost of equipment type i amortized to the period of rescue operation (purchase and maintenance cost);

f_j average operation cost of local rescue base j during the period of rescue operation;

O^i unit pump capacity of equipment type i;

$P(\omega)$ probability of rainfall event ω;

$\overline{Q_j}$ storage capacity of local rescue base j;

$\overline{Q_k}$ storage capacity of regional rescue center k;

$\overline{Q_o}$ storage capacity of head rescue center o;

r^i unit rent cost of equipment type i;

t^i_{jl} unit transportation cost of equipment type i shipped to demand point l from local rescue base j;

t^i_{kl} unit transportation cost of equipment type i shipped to demand point l from regional rescue center k;

t^i_{ol} unit transportation cost of equipment type i shipped to demand point l from the head rescue center o;

v^i amount of equipment type i; and

δ_{kl} shortest distance from regional rescue center k to demand point l.

The decision variables include:

$q^i_{jl}(\omega)$ amount of equipment type i shipped to demand point l from local rescue base j under rainfall event ω;

$q^i_{kl}(\omega)$ amount of equipment type i shipped to demand point l from regional rescue center k under rainfall event ω;

$q^i_{ol}(\omega)$ amount of equipment type i shipped to demand point l from head rescue center o under rainfall event ω;

S^j_i amount of equipment type i stored in local rescue base j;

S^j_k amount of equipment type i stored in regional rescue center k;

S^j_o amount of equipment type i stored in head rescue center o;

$X^{mn}_k = 1$ if regional rescue center k belongs to group m and level n, otherwise $= 0$;

$y_j = 1$ if local rescue bases j are selected for setup, otherwise $= 0$;

$Y^m_{kk'} = 1$ if regional rescue centers k' and k both belong to group m, otherwise $= 0$;

$\alpha^i_l(\omega)$ amount of shortage of equipment type i at demand point l under rainfall event ω;

$\beta^i_l(\omega)$ amount of surplus of equipment type i at local rescue base j under rainfall event ω;

$\beta^i_k(\omega)$ amount of surplus of equipment type i at regional center k under rainfall event ω; and

$\beta^i_o(\omega)$ amount of surplus of equipment type i at head center o under rainfall event ω.

Model 1: Grouping and classifying. The first model groups the disaster rescue areas and classifies their level of emergency by minimizing shipping distance. The optimal solution of this model provides association of a rescue supplier with a specific disaster rescue group and a level of emergency. The optimization problem in mathematical form is:

$$Minimize \sum_m \sum_k \sum_{k' \neq k} \left[Y^m_{kk'} \sum_\omega P(\omega) \sum_i \sum_{l \in L^k(\omega)} \delta_{k'l} d^i_l(\omega) \right]$$

(4.93)

subject to

$$2Y^m_{kk'} \leq \sum_n X^{mn}_{k'} + \sum_n X^{mn}_k \leq Y^m_{kk'} + 1 \quad \forall m, k', k \quad (4.94)$$

$$\sum_\omega P(\omega) \sum_{l \in L^k(\omega)} \sum_i O^i d^i_l(\omega) X^{m1}_k$$
$$\geq \sum_\omega P(\omega) \sum_{l \in L^{k'}(\omega)} \sum_i O^i d^i_l(\omega) X^{m2}_{k'} \quad \forall m, k' \neq k, k$$

(4.95)

$$\sum_\omega P(\omega) \sum_{l \in L^k(\omega)} \sum_i O^i d^i_l(\omega) X^{m2}_k$$
$$\geq \sum_\omega P(\omega) \sum_{l \in L^{k'}(\omega)} \sum_i O^i d^i_l(\omega) X^{m3}_{k'} \quad \forall m, k' \neq k, k$$

(4.96)

$$\sum_k X^{mn}_k \geq 1 \quad \forall m, n \quad (4.97)$$

$$\sum_m \sum_n X^{mn}_k = 1 \quad \forall k \quad (4.98)$$

$$Y^m_{kk'} \in \{0, 1\} \quad \forall m, k', k \quad (4.99)$$

$$X^{mn}_k \in \{0, 1\} \quad \forall m, n, k \quad (4.100)$$

Model 2: Location allocation. Based on the results of the first model, the following two-stage stochastic programming model determines the selected local rescue bases that need to be set up after the disaster as well as the quantity of rescue equipment in the storehouses of all echelons and the transportation plans for rescue equipment. The former is the first-stage decision and the two latter are the second-stage decisions based upon the first-stage decision and the realized uncertainty scenarios. The corresponding costs for distributing the rescue equipment in all rescue demand scenarios must be considered in the first-stage decision. Thus, the objective is to minimize the current facility setup cost and the equipment average cost as well as the expected future transportation cost, the supply-shortage cost, and the demand-shortage penalty during the period of rescue operation. The mathematical formulation of the second model is as follows:

$$Min \sum_j f_j y_j + \sum_i e^i \left[\sum_j S^i_j + \sum_k S^i_k + S^i_o \right]$$
$$+ \sum_\omega P(\omega)TC(\omega) + \sum_\omega P(\omega)RC(\omega) + \sum_\omega P(\omega)SC(\omega)$$

(4.101)

subject to

$$TC(\omega) = \sum_{l \in L_1(\omega)} \sum_i \left[\sum_{j \in J_1^{M(l)}} t_{jl}^i q_{jl}^i(\omega) + \sum_{k \in K_1^{M(l)}} t_{kl}^i q_{kl}^i(\omega) \right.$$
$$\left. + t_{ol}^i q_{ol}^i(\omega) + \sum_{k \in K_2^{M(l)}} t_{kl}^i q_{kl}^i(\omega) \right]$$
$$+ \sum_{l \in L_2(\omega)} \sum_i \left[\sum_{j \in J_2^{M(l)}} t_{jl}^i q_{jl}^i(\omega) + \sum_{k \in K_2^{M(l)}} t_{kl}^i q_{kl}^i(\omega) \right.$$
$$\left. + t_{ol}^i q_{ol}^i(\omega) + \sum_{k \in K_3^{M(l)}} t_{kl}^i q_{kl}^i(\omega) \right]$$
$$+ \sum_{l \in L_3(\omega)} \sum_i \left[\sum_{j \in J_3^{M(l)}} t_{jl}^i q_{jl}^i(\omega) + \sum_{k \in K_3^{M(l)}} t_{kl}^i q_{kl}^i(\omega) \right.$$
$$\left. + t_{ol}^i q_{ol}^i(\omega) \right] \tag{4.102}$$

$$RC(\omega) = \sum_i r^i + \sum_l \alpha_l^i(\omega) \tag{4.103}$$

$$SC(\omega) = \sum_i C^i \left[\sum_j \beta_j^i(\omega) + \beta_o^i + \sum_k \beta_l^i(\omega) \right] \tag{4.104}$$

$$\alpha_l^i(\omega) = d_l^i(\omega) - \sum_{j \in J_1^{M(l)}} q_{jl}^i(\omega) - \sum_{k \in K_1^{M(l)}} q_{kl}^i(\omega) - q_{ol}^i(\omega)$$
$$- \sum_{j \in K_2^{M(l)}} q_{kl}^i(\omega) \quad \forall i, l \in L_1(\omega), \omega \tag{4.105}$$

$$\alpha_l^i(\omega) = d_l^i(\omega) - \sum_{j \in J_2^{M(l)}} q_{jl}^i(\omega) - \sum_{k \in K_2^{M(l)}} q_{kl}^i(\omega) - q_{ol}^i(\omega)$$
$$- \sum_{j \in K_3^{M(l)}} q_{kl}^i(\omega) \quad \forall i, l \in L_2(\omega), \omega \tag{4.106}$$

$$\alpha_l^i(\omega) = d_l^i(\omega) - \sum_{j \in J_3^{M(l)}} q_{jl}^i(\omega) - \sum_{k \in K_3^{M(l)}} q_{kl}^i(\omega)$$
$$- q_{ol}^i(\omega) \quad \forall i, l \in L_3(\omega), \omega \tag{4.107}$$

$$\beta_j^i(\omega) = S_j^i - \sum_{l \in L_{N(j)}^{M(j)}(\omega)} q_{jl}^i(\omega) \quad \forall i, j, \omega \tag{4.108}$$

$$\beta_k^i(\omega) = S_k^i - \sum_{l \in L_1^{M(j)}(\omega)} q_{kl}^i(\omega) \quad \forall i, k \in K_1, \omega \tag{4.109}$$

$$\beta_k^i(\omega) = S_k^i - \sum_{l \in L_1^{M(j)}(\omega)} q_{kl}^i(\omega) - \sum_{l \in L_2^{M(j)}(\omega)} q_{kl}^i(\omega) \quad \forall i, k \in K_2, \omega \tag{4.110}$$

$$\beta_k^i(\omega) = S_k^i - \sum_{l \in L_3^{M(j)}(\omega)} q_{kl}^i(\omega) - \sum_{l \in L_2^{M(j)}(\omega)} q_{kl}^i(\omega) \quad \forall i, k \in K_3, \omega \tag{4.111}$$

$$\beta_o^i(\omega) = S_o^i - \sum_{l \in L(\omega)} q_{ol}^i(\omega) \quad \forall i, \omega \tag{4.112}$$

$$\sum_i v^i S_j^i \le \overline{Q_j} y_j \quad \forall j \tag{4.113}$$

$$\sum_i v^i S_k^i \le \overline{Q_k} \quad \forall k \tag{4.114}$$

$$\sum_i v^i S_o^i \le \overline{Q_o} \tag{4.115}$$

$$y_j \in \{0, 1\} \quad \forall j \tag{4.116}$$

and all decision variables ≥ 0.

Solution procedure Two problems representing the flood emergency logistics preparation, formulated above using Equations (4.93) to (4.116), are approached using the following solution procedure developed by Chang *et al.* (2007). The key difficulty in solving this stochastic program is in the evaluation of the expectation in the objectives – it is difficult to estimate the rescue equipment demand for different rainfall events (especially if we look far into the future in order to be prepared for the potential impacts of climate change on flood risk). There are far too many future flood scenarios to consider. The goal of scenario planning is to specify a set of scenarios that represent the possible realizations of unknown problem parameters and to consider the range of scenarios in determining a robust solution. The authors selected a heuristic approach based on the sample average approximation scheme. The detailed steps are presented below. The flood risk maps for different rainfall events are used for developing scenarios that present all possible locations of rescue demand points and the required amount of rescue equipment. The GIS technology and its network analysis functions are utilized for this task.

Scenario development using GIS. The locations of demand points and the quantity of emergency rescue equipment required under different rainfall scenarios can be estimated using the GIS analysis function. Then, the estimated corresponding quantity of emergency rescue equipment can be used to determine the quantity of rescue equipment required for demand points under different flooding scenarios. The procedure is as follows:

Step 1. Generate map layers for: (i) point map with locations of rescue centers; (ii) polyline map of the road network; and (iii) polygon map of administrative districts.

Step 2. Overlap analysis. Overlap the flood potential map for various precipitation events with map layers from Step 1. Identify centroid of each flooded area as a rescue demand point.

Step 3. Data analysis. Calculate the flood volume at each location. Estimate the required quantities of each type of rescue equipment for each demand point and each rainfall event. Create a database.

Sample average approximation solution procedure. The expected value of objectives is obtained through external sampling (known as the sample average approximation). Two stochastic models in Equations (4.93) to (4.116) are solved using the sampling procedure as follows:

Step 1. Generate a number of independent samples of flood scenarios. For each sample, solve the two models sequentially.

Step 2. Compute the mean and variance of all solutions to model 1 and model 2.

Step 3. Choose two feasible solutions from Step 1 and estimate the true objective function values. Then, generate another independent sample of flood scenarios. Estimate the variances of these solutions. This new sample is much larger and can reflect the time horizon of the climate change scenarios.

Step 4. At the end, compute the optimality gap of the solutions from Step 1 and Step 3. For the details of the computational algorithm consult Chang *et al.* (2007, page 746).

Implementation of the modeling procedure for Taipei City

The proposed modeling procedure is applied to Taipei City, located in Northern Taiwan. Taipei City covers an area of 272 square kilometers and has a population density of over 10,000 people per square kilometer. During the typhoon season many places in Taipei City face the risk of flooding. To reduce the damage caused by disasters, the government requires that the entire disaster rescue operation should be completed within 36 hours. The local Taipei City Government is designated as the head disaster rescue center; 12 administrative district offices are set up as regional rescue centers; and 84 offices on the sub-district level are chosen as candidate points for the local rescue bases. Small-size auto-prime engine pumps are among the most important rescue equipment used in areas with disrupted water drainage. For the sake of demonstration, the focus of the following example is on these pumps. There are two kinds of small-size

Table 4.18 *Rescue equipment data*

Equipment type	Unit rent (NT$)	Average operating cost (NT$)	Shortage unit penalty (NT$)	Pumping capacity (m³/h)
Type A	35,640	67	67	72
Type B	36,396	94	94	108

After Chang *et al.*, 2007

Table 4.19 *Rescue facility data*

Facility type	Storage capacity (pumps)	Facility operating cost (NT$)
Head rescue center	281	–
Regional rescue center	35	–
Local rescue base	22	382

After Chang *et al.*, 2007

auto-prime engine pumps, i.e., Type A and Type B. They differ in pumping capacity, have no significant difference in size and have the same unit transportation cost. All the necessary input data are provided in Tables 4.18 and 4.19.

The rainfall is classified into three types of event by intensity: 300 mm/day, 450 mm/day, and 600 mm/day, with probabilities 0.208, 0.025, and 0.003, respectively. The flood risk maps are created in GIS and overlaid with the administrative district map as shown in Figure 4.19.

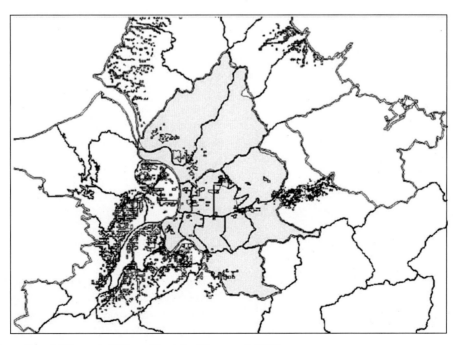

Figure 4.19 Flood risk map for rainfall event of 600 mm/day (after Chang *et al.*, 2007).

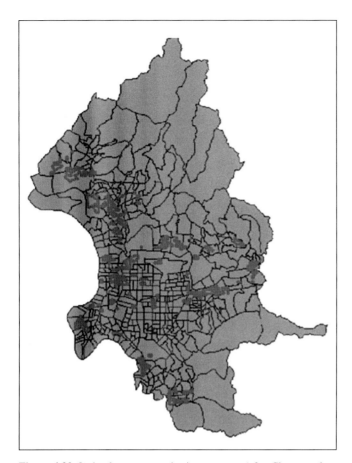

Figure 4.20 Optimal rescue organization structure (after Chang *et al.*, 2007).

Table 4.20 *Optimal rescue organization structure*

Rescue group	Rescue level		
	Level 1	Level 2	Level 3
Group 1	10	9	1
Group 2	4	12	11
Group 3	8	3	2
Group 4	7	5	6

After Chang *et al.*, 2007

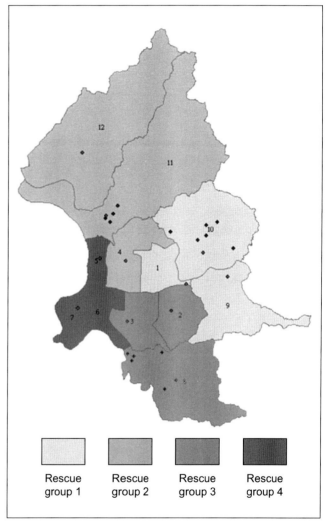

Figure 4.21 Optimal planning results (after Chang *et al.*, 2007).

The GIS spatial data analysis of the example problem for three rainfall events identified the following number of rescue demand points: 35, 91, and 159. The flood inundation example map is shown in Figure 4.20. After the spatial data analysis, the LP package was used to solve the proposed models. For the implementation of the sample average approximation method the number of flood scenarios (initial, small sample, large sample) is set to 3, 30, and 2000, respectively. The 12 administrative districts in Taipei City are divided into four disaster rescue groups, each classified into three levels. The results of the rescue organization structure are shown in Table 4.20. The distribution of all disaster rescue centers/bases is illustrated in Figure 4.21, and the storage at each disaster rescue supply point is given in Table 4.21. The total storage numbers of the Type A pump and the Type B pump are 642 and 232, respectively. The solution of the proposed models indicates that only one head rescue center, 10 regional rescue centers, and 15 local rescue bases are sufficient for Taipei City to perform flood rescue operations.

The authors performed the sensitivity analysis and concluded that as the storage capacity at each rescue echelon increases, the average total rescue cost goes down. These results confirm that the allocation of the optimal amount of rescue resources to the proper location leads to the best outcome. In order to observe the effect of the variability of the uncertain demand, three different levels of the variability of the uncertain demand were considered. It is observed that with the increase in the variability of the uncertain demand, the range of total rescue cost of the stochastic solution increases at a slower rate than that of the mean-value problem solution.

Table 4.21 *Optimal rescue equipment distribution*

Equipment type	Facility type												
	Head rescue center	Regional rescue center											
	1	1	2	3	4	5	6	7	8	9	10	11	12
Type A	264	9	0	33	26	30	0	25	26	13	13	2	28
Type B	17	1	0	2	9	5	0	10	9	22	22	7	7

	Local rescue base												
	22	23	24	25	26	45	47	48	50	51	58	59	60	61	62
Type A	20	21	21	22	21	0	4	14	1	18	10	1	1	0	19
Type B	2	1	1	0	1	22	18	5	21	4	12	11	1	20	2

After Chang *et al.*, 2007.

4.6 CONCLUSIONS

In flood risk management, probability is used based on historical records and called objective because it is determined by long-run observations of the occurrence of an event. However, many flood risk management decisions are made under the influence of subjective risk that cannot be described using probabilistic tools. In this chapter we review the probabilistic approach to flood risk management from the practical application point of view. Some very traditional tools are presented in the context of climate change. The focus of the presentation is on probabilistic simulation (Monte Carlo simulation), optimization (evolutionary programming) and multi-objective analysis (stochastic goal programming). All three techniques are illustrated with simple examples for educational purposes. Two computer programs are provided on the accompanying website for the implementation of the evolutionary optimization and LP that can be used for the solution of stochastic goal programming problems.

Three interesting real-world examples from Africa, the USA, and Asia are used to demonstrate the utility of the probabilistic approach to flood risk management under climate change. However, these examples (taken from the literature) do not explore the broad range of issues that climate change brings to flood risk management. There is ample opportunity for wider use of the presented techniques, as well as other probabilistic techniques that are not included in the book but are available in the literature, in assisting flood risk management.

4.7 EXERCISES

4.1. Extreme precipitation is one of the causes of flooding. What is the risk of the largest precipitation on record occurring this summer in London, Ontario?

4.2. Consider a reservoir for flood control of an urban area. What could be the risk from the failure of the dam?

4.3. Find the variance of the return period for different values of non-exceedance probabilities.

4.4. What is the probability that a 500-year flood will occur at least once in 100 years?

4.5. Suppose a levee is designed with a projected life of 100 years. The responsible water authority wants to take only a 5% chance that the levee will be overtopped within this period. What return period flood should be used for the levee design?

4.6. Consider the levee design problem from Example 7 in the text. Due to climate change there is a major shift in flow frequency. The discharge frequency curve under climate change is shown in the table below:

T_r (years)	Exceedance probability	Discharge (m^3/s)
200	0.005	898.8
67	0.015	676.1
18	0.055	538.5
14	0.070	423.0
7	0.145	298.8
5	0.200	222.5
3	0.333	158.4

a. Find the difference in the expected value of the levee height that will be required in order to address the impacts of climate change on the watershed.

b. Find the difference in the 90th percentile (the height of a levee that will account for the flood stage value at or below 90%) value of the levee height that will be required in order to address the impacts of climate change on the watershed. (Tip: Use MCS and MATLAB code provided on the website.)

4.7. A rectangular, underground reservoir for urban flood drainage control is to be proportioned. The flow rate, and thus the potential flood damage reduction obtained by the reservoir, is directly proportional to the storage volume provided. Typical figures are as follows:

Volume (m^3)	Flow (m^3/s)	Damage reduction ($)
100	0.35	500
500	0.40	2,500
1,000	0.50	5,000
2,000	0.65	10,000

The cost of constructing the reservoir is based on the following rates: base and top \$2/m^2; sides \$4/m^2; and ends \$6/m^2.

a. Find the dimensions for the maximum net benefits (damage reduction – construction cost). The damage reduction may be approximated by the function $d_1 = 5 \times$ volume in m^3. For solving the problem use the EVOLPRO computer program from the website.

b. Vary the desired accuracy (tolerance level) and the maximum number of iterations. Discuss the solutions.

c. Change the size of the population and discuss its impact on the optimal solution and optimization process.

4.8. Find the optimal size of the reservoir flood control volume. The total benefit (FCB) derived from the reservoir flood control volume (measured as a reduction in the flood damage) has been estimated as:

$$FCB = 100x - 0.00005x^2$$

where FCB is expressed in dollars per year (\$/yr) and x represents the available reservoir flood control volume expressed in 10^6 m^3 per year (MCM/yr). Estimates of the annual total cost (FCC) of a reservoir volume result in the expression:

$$FCC = 44.42x^{0.9} + 0.098x^{1.45}$$

where FCC is in \$/yr.

a. The agency responsible for the reservoir construction wishes to maximize the benefit/cost ratio. Solve the problem using the EVOLPRO computer program from the website.

b. Vary the desired accuracy (tolerance level) and the maximum number of iterations. Discuss the solutions.

c. Change the size of population and discuss its impact on the optimal solution and optimization process.

4.9. Consider the stochastic goal programming problem from Example 12 in the text.

a. Assume that goals (b) and (c) can be considered as two-sided goals (that flood control may take money from other water management programs if the desired goals are not reached). Reformulate the stochastic goal program and write its deterministic equivalent.

b. Use LINPRO provided on the website to solve the problem assuming the following values for standard deviations: 5%, 10%, 20%, 30%, and 50%, and reliabilities $\alpha_1, \alpha_2, \alpha_3, \alpha_4,$ and α_5 equal to 75%, 85%, 90%, and 95%.

c. Discuss your solutions.

4.10. Check the most recent literature for other applications of probabilistic simulation, optimization, and multi-objective analysis in flood risk management under climate change.

a. Explain what climate change issues are being addressed.

b. Indicate what are the advantages and disadvantages of the proposed solutions.

Part III
Flood risk management: fuzzy set approach

5 Risk management: fuzzy set approach

Flood risk management systems include people, infrastructure, and environment. These elements are interconnected in complicated networks across broad geographical regions. Each element is vulnerable to natural hazards or human error, whether unintentional, as in the case of operational errors and mistakes, or intentional, such as a terrorist act.

Until recently the probabilistic approach was the only approach for flood risk analysis. However, it fails to address the problem of uncertainty that goes along with climate change, human input, subjectivity, and a lack of history and records (as discussed in Section 3.2). There is a real need to convert to new approaches that can compensate for the ambiguity or uncertainty of human perception.

5.1 PARADIGM CHANGE

Probability is a concept widely used in flood risk management. To perform operations associated with probability, it is necessary to use sets, collections of elements, each with some specific characteristics (Simonovic, 2009). In flood risk management practice we use three major conceptual interpretations of probability: (i) classical interpretation – equally likely concept, (ii) frequency interpretation, and (iii) subjective interpretation.

5.1.1 Problems with probability in flood risk management

One of the main goals of flood risk management is to ensure that a system performs satisfactorily under a wide range of possible future conditions. This premise is particularly true of large and complex flood risk management systems (see example of the Red River basin flood of 1997 in Simonovic (2011), page 9). Flood risk management systems usually include structural elements like conveyance facilities such as canals, pipes, and pumps, diversion facilities such as diversion canals and floodways, and storage facilities such as reservoirs that function in parallel with many non-structural measures. Structural elements are interconnected in complicated networks serving broad geographical regions. Each

element is vulnerable to temporary disruption in service because of flood hazards or human error, whether unintentional, as in the case of operational errors and mistakes, or from intentional causes, such as a terrorist act.

Floods can cause serious damage or the total failure of flood risk management systems. Human error can also affect functioning of the systems. With the consideration of climate change the sources of uncertainty are many and as a result they provide a great challenge to flood risk management. The goal to ensure failsafe system performance may be unattainable. Adopting high safety factors is one way to avoid the uncertainty of potential failures. However, making safety the first priority may render the system solution infeasible. Therefore known uncertainty sources must be quantified.

Probabilistic (stochastic) reliability analysis has been used extensively to deal with the problem of uncertainty in flood risk management (see Chapter 4). Prior knowledge of the PDFs of both resistance and load, and their joint probability distribution function, is a prerequisite of the probabilistic approach. In practice, data on previous failure experience is usually insufficient to provide such information. Even if data are available to estimate these distributions, approximations are almost always necessary to calculate system reliability (Ang and Tang, 1984). Subjective judgment by the water resources decision-maker in estimating the probability distribution of a random event – the subjective probability approach of Vick (2002) – is another approach to dealing with data insufficiency. The third approach is Bayes's theory, where engineering judgment is integrated with observed information. The choice of a Bayesian approach or any subjective probability distribution presents real challenges – the degree of accuracy is strongly dependent on a realistic estimation of the decision-maker's judgment.

Fuzzy set theory was intentionally developed to try to capture judgmental belief, or the uncertainty that is caused by the lack of knowledge. Relative to probability theory, it has some degree of freedom with respect to aggregation operators, types of fuzzy sets (membership functions), and so on, which enables it to be adapted to different contexts. During the past 50 years, fuzzy set theory and fuzzy logic have contributed successfully to technological development in different application areas (Zadeh, 1965; Zimmermann, 1996) and in the past 10 years has started to gain ground in water resources management (Simonovic, 2009). However, application of the fuzzy set approach to flood risk management under climate change is still new.

This chapter explores the utility of fuzzy set theory in addressing various uncertainties in the management of flood risk under climate change. Since there is limited literature that applies fuzzy set theory to water resources management, this chapter introduces in detail some of the basic concepts of fuzzy sets and fuzzy arithmetic. However, the main objective here is to discuss fuzzy theory in relation to the simulation, optimization, and multi-objective analysis of flood risk management systems under climate change.

5.1.2 Fuzziness and probability

Probability and fuzziness are related, but they are different concepts. Fuzziness is a type of deterministic uncertainty. It describes the *event class ambiguity*. Fuzziness measures the *degree to which* an event occurs, not whether it occurs, and therefore could be of value in addressing the climate change. At issue is whether the event class can be unambiguously distinguished from its opposite. Probability, in contrast, arises from the question *whether or not* an event occurs. Moreover, it assumes that the event class is crisply defined and that the law of non-contradiction holds. Fuzziness occurs when the law of non-contradiction is violated.

In essence, whenever there is an experiment for which we are not capable of "computing" the outcome, a probabilistic approach may be used to estimate the likelihood of a possible outcome belonging to an event class. A fuzzy theory extends the traditional notion of the probability when there are outcomes that belong to several event classes at the same time, but to different degrees. Fuzziness and probability are orthogonal concepts that characterize different aspects of human experience. Hence, it is important to note that neither fuzziness nor probability governs physical processes in nature. These concepts were introduced by humans to compensate for our own limitations.

Let us review the example from Simonovic (2009) that shows the difference between fuzziness and probability. That the laws of non-contradiction and excluded middle can be violated was pointed out by Bertrand Russell with the tale of the barber. Russell's barber is a bewhiskered man who lives in a town and shaves a man if and only if he does not shave himself. The question is, who shaves the barber? If he shaves himself, then by definition he does not. But if he does not shave himself, then by definition he does. So he does and he does not. This is a contradiction or *paradox*. It has been shown that this paradoxical situation can be numerically resolved as follows. Let S be the proposition that the barber shaves himself and not-S the proposition that he does not. Since S implies not-S and vice versa, the two propositions are

Figure 5.1 Ordinary set classification (after Pedrycz and Gomide, 1998).

logically equivalent, i.e., $S = $ not-S. Fuzzy set theory allows for an event class to coexist with its opposite at the same time, but to different degrees, or in the case of paradox to the same degree, which is different from zero or one.

There are many similarities between fuzziness and probability. The largest, but superficial and misleading, similarity is that both approaches quantify uncertainty using numbers in the unit interval [0,1]. This means that both approaches describe and quantify the uncertainty numerically. The structural similarity arising from lattice theory is that both approaches algebraically manipulate sets and propositions associatively, commutatively, and distributively. These similarities are misleading because a key distinction comes from what the two approaches are trying to model. Another distinction is in the idea of observation. Clearly, the two concepts possess different kinds of information: fuzzy memberships, which quantify similarities of objects to imprecisely defined properties; and probabilities, which provide information on expectations over a large number of experiments.

5.2 INTRODUCTION TO FUZZY SETS

Selected fuzzy set concepts are presented here for those not familiar with the approach. The material presented in this section of the text is based on Kaufmann and Gupta (1985), Zimmermann (1996), Pedrycz and Gomide (1998), and Simonovic (2009). First I will present basic definitions of fuzzy sets, followed by algebraic operations, which will then form the basis for further consideration in this chapter.

5.2.1 Basic definitions

The basic concept of set theory is a collection of objects that have similar properties or general features. Humans tend to organize objects into sets so as to generalize knowledge about objects through the classification of information. The ordinary set classification imposes a dual logic. An object either belongs to a set or does not belong to it, as set boundaries are well defined. For example, if we consider a set A in a universe X in Figure 5.1, it is obvious that object x_1 belongs to set A, while x_2 does not. We denote the acceptance of belonging to a set by 1 and rejection of belonging by 0. The classification is expressed through a characteristic membership function $\mu_{\tilde{A}}(x)$, for $x \in X$:

$$\mu_{\tilde{A}} = \begin{cases} 1, & \text{if } x \in A \\ 0, & \text{if } x \notin A \end{cases} \qquad (5.1)$$

where $A(x)$ is the characteristic function denoting the membership of x in set A.

The basic notion of fuzzy sets is to relax this definition, and admit intermediate membership classes to sets. Therefore, the characteristic function can accept values *between* 1 and 0, expressing the grade of membership of an object in a certain set. According to this notion, the fuzzy set will be represented as a set of ordered pairs of elements, each of which presents the element together with its membership value to the fuzzy set:

$$\tilde{A} = \{(x, \mu_{\tilde{A}}(x)) | x \in X\} \qquad (5.2)$$

The membership function is the crucial component of a fuzzy set, therefore all operations with fuzzy sets are defined through their membership functions. The basic definition of a fuzzy set is that it is characterized by a membership function mapping the elements of a domain, space, or universe of discourse X to the unit interval [0,1]:

$$\tilde{A} : X \rightarrow [0, 1] \qquad (5.3)$$

where $\tilde{A}$ is the fuzzy set in the universe of discourse X, and X is the domain, or the universe of discourse.

The function in Equation (5.3) describes the membership function associated with a fuzzy set $\tilde{A}$. A fuzzy set is said to be a *normal fuzzy set* if at least one of its elements has a membership value of 1.

The crisp set of elements that belong to the fuzzy set $\tilde{A}$ at least to the degree α is called the *α-level set*:

$$A_\alpha = \{x \in X | \mu_{\tilde{A}}(x) \geq \alpha\} \qquad (5.4)$$

and

$$A'_\alpha = \{x \in X | \mu_{\tilde{A}}(x) > \alpha\} \qquad (5.5)$$

is called a *strong α-level set* or *strong α-level cut*. α is also known as the *credibility level*. To illustrate the definition let us consider the following simple example.

Example 1

An engineer wants to present possible flood protection levee options to a client. One indicator of protection is the levee height in meters. Let $X = \{1, 1.1, 1.2, 1.3, 1.4, 1.5, 1.6, 1.7, 1.8\}$ be the set of possible levee heights. Create an example that will show the fuzzy set "safe levee."

Solution

One possible representation of the fuzzy set "safe levee" may be described as $\tilde{A} = \{(1.3, .2), (1.4, .4), (1.5, .6), (1.6, .8), (1.7, 1), (1.8, 1)\}$. Figure 5.2 shows the fuzzy set of the "safe levee."

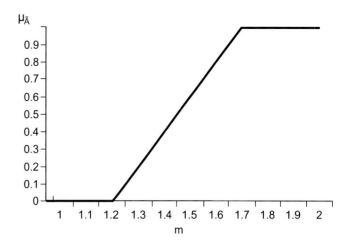

Figure 5.2 "Safe levee" fuzzy set.

A fuzzy set is convex if all α-level sets are convex, or:

$$\mu_{\tilde{A}}(\lambda x_1 + (1 - \lambda)x_2) \geq \min(\mu_{\tilde{A}}(x_{1}), \mu_{\tilde{A}}(x_2)) \qquad (5.6)$$

where $x_1, x_2 \in X$, and $\lambda \in [0, 1]$.

The membership function may have different shapes and may be continuous or discrete, depending on the context in which it is used. Figure 5.3 shows the four most common types of continuous membership function. Families of parameterized functions, such as the following triangular membership function, can represent most of the common membership functions explicitly:

$$\mu_{\tilde{A}} = \begin{cases} 0, & \text{if } x \leq a \\ \dfrac{x - a}{m - a}, & \text{if } x \in [a, m] \\ \dfrac{b - x}{b - m}, & \text{if } x \in [m, b] \\ 0, & \text{if } x \geq b \end{cases} \qquad (5.7)$$

where m is the modal value, and a, b are the lower and upper bounds of the non-zero values of membership.

A *fuzzy number* is a special case of a fuzzy set, having the following properties: (i) It is defined in the set of real numbers; (ii) its membership function reaches the maximum value, 1.0, i.e., it is a normal fuzzy set; and (iii) its membership function is unimodal (it is a convex fuzzy set).

A fuzzy number can be defined as follows:

$$\tilde{X} = \{(x, \mu_{\tilde{x}}(x)) : x \in R; \mu_{\tilde{x}}(x) \in [0, 1]\} \qquad (5.8)$$

where $\tilde{X}$ is the fuzzy number, $\mu_{\tilde{x}}(x)$ is the membership value of element x to the fuzzy number, and R is the set of real numbers.

A *support of a fuzzy number* is the ordinary set, which is defined as follows:

$$S(\tilde{X}) = \tilde{X}(0) = \{x : \mu_{\tilde{x}}(x) > 0\} \qquad (5.9)$$

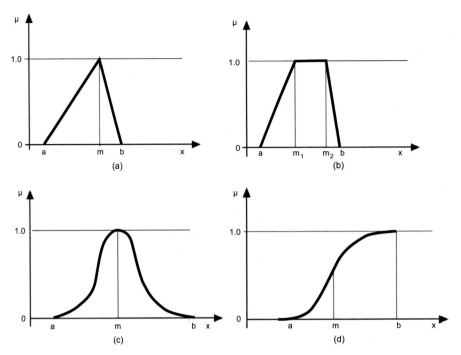

Figure 5.3 Triangular (a), trapezoid (b), Gaussian (c), and sigmoid (d) membership functions.

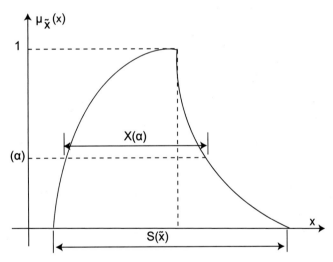

Figure 5.4 Credibility level (α-cut) and support of a fuzzy set.

The fuzzy number support is the 0-level set and includes all the elements with the credibility level α higher than 0. Figure 5.4 illustrates these definitions.

5.2.2   Set-theoretic operations for fuzzy sets

Operations with fuzzy sets are defined using their membership functions. I present the concepts here as originally suggested by Zadeh (1965), since this provides a consistent framework for the theory of fuzzy sets. However, this is not the only possible way to extend classical set theory to the fuzzy domain.

INTERSECTION

The membership function $\mu_{\tilde{C}}$ of the intersection $\tilde{C} = \tilde{A} \cap \tilde{B}$ is defined by:

$$\mu_{\tilde{C}}(x) = min\{\mu_{\tilde{A}}(x),\ \mu_{\tilde{B}}(x)\}, \quad x \in X \qquad (5.10)$$

where $\mu_{\tilde{C}}$ is the membership of the fuzzy intersection of $\tilde{A}$ and $\tilde{B}$; $min(\ )$ is the ordinary minimum operator; $\mu_{\tilde{A}}$ is the membership of fuzzy set $\tilde{A}$; and $\mu_{\tilde{B}}$ is the membership of fuzzy set $\tilde{B}$.

UNION

The membership function $\mu_{\tilde{C}}$ of the union $\tilde{C} = \tilde{A} \cup \tilde{B}$ is defined by:

$$\mu_{\tilde{C}}(x) = max\{\mu_{\tilde{A}}(x),\ \mu_{\tilde{B}}(x)\}, \quad x \in X \qquad (5.11)$$

where $\mu_{\tilde{C}}$ is the membership of the fuzzy union of $\tilde{A}$ and $\tilde{B}$; $max(\ )$ is the ordinary maximum operator; $\mu_{\tilde{A}}$ is the membership of fuzzy set $\tilde{A}$; and $\mu_{\tilde{B}}$ is the membership of fuzzy set $\tilde{B}$.

COMPLEMENT

The membership function $\mu_{\bar{\tilde{C}}}$ of the complement of fuzzy set $\tilde{C}$ is defined by:

$$\mu_{\bar{\tilde{C}}}(x) = 1 - \mu_{\tilde{C}}(x), \quad x \in X \qquad (5.12)$$

where $\mu_{\bar{\tilde{C}}}$ is the membership of the complement of fuzzy set $\tilde{C}$; and $\mu_{\tilde{C}}$ is the membership of fuzzy set $\tilde{C}$.

Note that *min* and *max* are not the only operators that can be used to model the intersection and union of fuzzy sets, respectively.

Zimmermann (1996) presents an interpretation of intersection as "logical and" and the union as "logical or."

AND–OR OPERATORS

Assuming that $\wedge$ denotes the fuzzy AND operation and $\vee$ denotes the fuzzy OR operation, the definitions for both operators are:

$$\mu_{\tilde{A} \wedge \tilde{B}}(x) = \min\{\mu_{\tilde{A}}(x),\ \mu_{\tilde{B}}(x)\}, \quad x \in X \qquad (5.13)$$

and

$$\mu_{\tilde{A} \vee \tilde{B}}(x) = \max\{\mu_{\tilde{A}}(x),\ \mu_{\tilde{B}}(x)\}, \quad x \in X \qquad (5.14)$$

Detailed examples that illustrate the theoretic operations of fuzzy sets are available in Simonovic (2009, Chapter 6).

5.2.3 Fuzzy arithmetic operations for fuzzy sets

One of the most basic concepts of fuzzy set theory, which can be used to transfer crisp mathematical concepts to fuzzy sets, is the *extension principle*. It is defined as follows. Let X be a cartesian product of domains $X = X_1, \ldots, X_r$. Function f is mapped from X to a domain Y, $y = f(x_1, \ldots, x_r)$. Then the extension principle allows us to define a fuzzy set in Y by:

$$\tilde{B} = \{(y,\ \mu_{\tilde{B}}(y)) | y = f(x_1, \ldots, x_r), (x_1, \ldots, x_r) \in X\} \qquad (5.15)$$

where:

$$\mu_{\tilde{B}}(y) = \begin{cases} \sup_{(x_1, \cdots, x_r) \in f^{-1}(y)} \min\{\mu_{\tilde{A}_r}(x_1), \ldots, \mu_{\tilde{A}_1}(x_r)\} & \text{if } f^{-1}(y) \neq 0 \\ 0 & otherwise \end{cases} \qquad (5.16)$$

where f^{-1} is the inverse of f.

At any α-level, the fuzzy number $\tilde{A}$ can be represented in the interval form as follows:

$$\tilde{A}(\alpha) = [a_1(\alpha),\ a_2(\alpha)] \qquad (5.17)$$

where $\tilde{A}(\alpha)$ is the fuzzy number at α-level; $a_1(\alpha)$ is the lower bound of the α-level interval; and $a_2(\alpha)$ is the upper bound of the α-level interval.

From this definition and the extension principle, the arithmetic operations on intervals of real numbers (crisp sets) can be extended to the four main arithmetic operations for fuzzy numbers, i.e., addition, subtraction, multiplication, and division. The fuzzy operations of two fuzzy numbers $\tilde{A}$ and $\tilde{B}$ are defined at any α-level cut as follows:

$$\tilde{A}(\alpha)(+)\tilde{B}(\alpha) = [a_1(\alpha) + b_1(\alpha),\ a_2(\alpha) + b_2(\alpha)] \qquad (5.18)$$
$$\tilde{A}(\alpha)(-)\tilde{B}(\alpha) = [a_1(\alpha) - b_2(\alpha),\ a_2(\alpha) - b_1(\alpha)] \qquad (5.19)$$
$$\tilde{A}(\alpha)(\times)\tilde{B}(\alpha) = [a_1(\alpha) \times b_1(\alpha),\ a_2(\alpha) \times b_2(\alpha)] \qquad (5.20)$$
$$\tilde{A}(\alpha)(/)\tilde{B}(\alpha) = [a_1(\alpha)/b_2(\alpha),\ a_2(\alpha)/b_1(\alpha)] \qquad (5.21)$$

Note that for multiplication and division $[\tilde{A}(\alpha)(/)\tilde{B}(\alpha)](\times)\tilde{B} \neq \tilde{A}$ and for addition and subtraction $[\tilde{A}(\alpha)(-)\tilde{B}(\alpha)](+)\tilde{A} \neq \tilde{A}$.

Detailed examples that illustrate the arithmetic operations of triangular fuzzy sets are available in Simonovic (2009, Chapter 6).

5.2.4 Comparison operations on fuzzy sets

An important application of fuzzy set theory to flood risk management involves the ranking of fuzzy sets. The relative performance of different flood protection alternatives may be visually intuitive when looking at its fuzzy representation. However, in cases where many alternatives display similar characteristics, it may be impractical or even undesirable to make a visual selection. A method for ranking alternatives can automate many interpretations, and create reproducible results. A ranking measure may also be useful in supplying additional insight into decision-maker preferences, such as distinguishing relative risk tolerance levels.

The selection of a ranking method is subjective, and specific to the form of problem and the fuzzy set characteristics that are desirable. There exists an assortment of methods, ranging from horizontal and vertical evaluation of fuzzy sets to comparative methods (Zimmermann, 1996). Some of these methods may independently evaluate fuzzy sets, while others use competition to choose from a selection list. Horizontal methods are related to the practice of defuzzifying a fuzzy set by testing for a range of validity at a threshold membership value, and are not dealt with in this text. Vertical methods tend to use the area under a membership function as the basis for evaluation, such as the center of gravity. The comparative methods introduce other artificial criteria for judging the performance of a fuzzy set, such as the compatibility of fuzzy sets. The center of gravity method and the fuzzy acceptability method are introduced as common representatives of vertical and comparative methods, respectively. The reader is directed to Simonovic (2009, Chapter 6) for the presentation of other methods.

WEIGHTED CENTER OF GRAVITY METHOD (WCOG)

Given the desirable properties of a ranking method for many fuzzy applications to water resources systems management, one technique that may qualify as a candidate for ranking fuzzy sets in the range [0,1] is the centroid method. The centroid method appears to be consistent in its ability to distinguish between most fuzzy sets. One weakness, however, is that the centroid method is unable to distinguish between fuzzy sets that may have the same centroid, but differ greatly in their degree of fuzziness. This weakness can be somewhat alleviated by the use of weighting. If high membership values are weighted higher than low membership values, there is some indication of degree of fuzziness when comparing rankings from different weighting schemes. However, in the case of symmetrical fuzzy sets, weighting schemes will not distinguish relative fuzziness.

A weighted centroid ranking measure can be defined as follows:

$$WCoG = \frac{\int g(x)\mu(x)^q \, dx}{\int \mu(x)^q \, dx} \qquad (5.22)$$

where $g(x)$ is the horizontal component of the area under scrutiny and $\mu(x)$ the membership function values.

In practice, WCoG can be calculated in discrete intervals across the valid domain. It allows parametric control in the form of the exponent q. This control mechanism allows ranking for cases ranging from the modal value ($q = \infty$) – which is analogous to an expected case or most likely scenario – to the center of gravity ($q = 1$) – which signifies some concern over extreme cases. In this way, there exists a family of valid ranking values (which may or may not change too significantly). The final selection of appropriate rankings is dependent on the level of risk tolerance from the decision-maker.

The ranking of fuzzy sets with WCoG is by ordering from the smallest to the largest value. The smaller the WCoG measure, the closer the center of gravity of the fuzzy set to the origin. As a vertical method of ranking, WCoG values act on the set of positive real numbers.

FUZZY ACCEPTABILITY MEASURE

Another ranking method that shows promise is a fuzzy acceptability measure, *Acc*, based on Kim and Park (1990). Kim and Park derive a comparative ranking measure, which builds on the possibility to signify an optimistic perspective and supplements it with a pessimistic view similar to necessity. Therefore their measure relies on the concept of *compatibility*.

The compliance of two fuzzy membership functions can be quantified using the fuzzy compatibility measure. The basic concepts of *possibility* and *necessity* lead to the quantification of the compatibility of two fuzzy sets. The possibility measure quantifies the overlap between two fuzzy sets, while the necessity measure describes the degree of inclusion of one fuzzy set in another fuzzy set.

The *possibility* is then defined as:

$$Poss(\tilde{A}, \tilde{B}) = \sup[\min\{\mu_{\tilde{A}}(x), \mu_{\tilde{B}}(x)\}], \quad x \in X \quad (5.23)$$

where $Poss(\tilde{A}, \tilde{B})$ is the possibility measure of fuzzy numbers $\tilde{A}$ and $\tilde{B}$; $\sup[\ ]$ is the least upper bound value, i.e., *supremum*; and $\mu_{\tilde{A}}(x)$, $\mu_{\tilde{B}}(x)$ are the membership functions of the fuzzy numbers $\tilde{A}$ and $\tilde{B}$ respectively.

The possibility measure is a symmetrical measure, that is:

$$Poss(\tilde{A}, \tilde{B}) = Poss(\tilde{B}, \tilde{A}) \qquad (5.24)$$

The *necessity* measure is defined as:

$$Nec(\tilde{A}, \tilde{B}) = \inf[\max\{\mu_{\tilde{A}}(x), \mu_{\tilde{B}}(x)\}], \quad x \in X \quad (5.25)$$

where $Nec(\tilde{A}, \tilde{B})$ is the necessity measure of fuzzy numbers $\tilde{A}$ and $\tilde{B}$; $\inf[\ ]$ is the greatest lower bound value, i.e., *infimum*; and $\mu_{\tilde{A}}(x)$, $\mu_{\tilde{B}}(x)$ are the membership functions of the fuzzy numbers $\tilde{A}$ and $\tilde{B}$ respectively.

The necessity measure is an asymmetrical measure, that is:

$$Nec(\tilde{A}, \tilde{B}) \neq Nec(\tilde{B}, \tilde{A}) \qquad (5.26)$$

Both measures hold the following relation:

$$Nec(\tilde{A}, \tilde{B}) + Poss(\bar{\tilde{A}}, \tilde{B}) = 1 \qquad (5.27)$$

where $\bar{\tilde{A}}$ is the fuzzy complement of fuzzy number $\tilde{A}$.

These two measures, *Poss* and *Nec*, can be combined to form an acceptability measure (*Acc*) as follows:

$$Acc = \omega Poss(G, L) + (1 - \omega) Nec(G, L) \qquad (5.28)$$

Parametric control with the acceptability measure (*Acc*) is accomplished with the ω weight and the choice of a fuzzy desirable state such as a goal, G. The ω weight controls the degree of optimism and degree of pessimism, and indicates (an overall) level of risk tolerance. The choice of a fuzzy goal is not so intuitive. It should normally include the entire range of alternative options L, but it can be adjusted to a smaller range either for the purpose of exploring the shape characteristics of L, or to provide an indication of necessary stringency. By decreasing the range of G, the decision-maker becomes more stringent in that the method rewards higher membership values closer to the desired value. At the extreme degree of stringency, G becomes a non-fuzzy number that demands the alternatives be ideal. As a function, G may be linear, but can also be adapted to place more emphasis or less emphasis near the best value.

The ranking of fuzzy sets using *Acc* is accomplished by ordering values from the largest to the smallest. That is, the fuzzy set with the greatest *Acc* is most acceptable. *Acc* values are restricted in the range [0,1] since both the *Poss* and *Nec* measures act on [0,1].

5.2.5 Development of fuzzy membership functions

A fuzzy set is characterized by a membership (characteristic) function that associates each member of the fuzzy set with a real number in the interval [0,1]. The membership function essentially embodies all fuzziness for a particular fuzzy set; its description is the essence of a fuzzy property or operation.

REVIEW OF METHODS FOR DEVELOPMENT OF MEMBERSHIP FUNCTIONS

There are numerous ways to assign membership values or functions to fuzzy variables – more ways than there are to assign PDFs to random variables. The following is a brief review of six

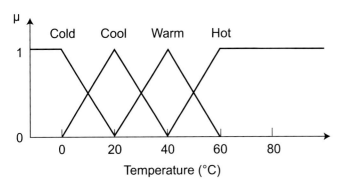

Figure 5.5 Membership functions for the fuzzy variable *temperature* (after Ross, 2010).

straightforward methods described in the literature. For further details the reader is directed to Ross (2010).

Intuition This method is simply derived from the capacity of humans to develop membership functions through their own natural intelligence and understanding. In order to utilize intuition, contextual and semantic knowledge about an issue is required. Thus, membership function development is dependent on the subjectivity of the individual or individuals consulted in its development. A single fuzzy variable may have more than one membership function, that is, there may be many partitions. As an example, consider the membership function for the fuzzy variable temperature. Figure 5.5 shows various shapes on the universe of temperature as measured in degrees C. Each curve corresponds to various fuzzy variables: cold, cool, warm, and hot. These curves are a function of context and the person involved with their development.

An important characteristic of these curves for the purpose of use in fuzzy operations is that different partitions overlap. The precise shapes of these curves are not so important in their utility. Rather, it is the approximate placement of the curves on the universe of discourse, the number of curves (partitions) used, and the overlapping character that are the most important.

Inference The inference method is based on our ability to perform deductive reasoning. When given a body of facts or knowledge we are able to deduce or infer a conclusion. The inference method can take many forms. Consider the example of identifying a triangle when we possess formal knowledge of geometry and geometric shapes (Ross, 2010). In the identification of a triangle, let A, B, and C be the inner angles of a triangle in the order $A \geq B \geq C \geq 0$ and let U be the universe of triangles, such that:

$$U = \{(A, B, C) | A \geq B \geq C \geq 0; A + B + C = 180°\}$$

$$(5.29)$$

The membership of different triangle types can be inferred based on a knowledge of geometry. For an approximate isosceles triangle, an algorithm for the membership meeting the constraint of Equation (5.29) can be developed as:

$$\mu_{\bar{I}}(A, B, C) = 1 - \frac{1}{60°} \min(A - B, B - C) \quad (5.30)$$

So, for example, if $A = B$ or $B = C$ the membership value of the isosceles triangle is $\mu_{\bar{I}} = 1$; however, if $A = 120°$, $B = 60°$, $C = 0°$ then $\mu_{\bar{I}} = 0$.

Rank ordering The rank ordering approach is based on the assessment of preferences by a single individual, a committee, a poll and/or other opinion methods that can be used to assign membership values to a fuzzy variable. Preferences are determined by pairwise comparisons, and these determine the ordering of the membership. This method is similar to finding relative preferences through a questionnaire and developing membership functions as a result. The next section of this text illustrates the use of the rank ordering approach.

Neural networks Neural networks use models that try to recreate the working of neurons in the human brain. Neurons are believed to be responsible for humans' ability to learn, thus the goal is to implement NNs in the development of membership functions. The use of NNs in membership function development is centered on a training process (learning as a result of available data for input) and an unsupervised clustering process (Ross, 2010). After training, the degree of membership function for a given input value may be estimated through NN computation. That is, each input value has a certain estimated degree of belonging to a cluster that is equivalent to the degree of the membership function represented by the cluster. Ross (2010, Chapter 6) presents a few examples of the implementation of the NN approach to the development of fuzzy membership functions.

Genetic algorithms Genetic algorithms, introduced in Section 4.5.2, are also used in the development of membership functions. The process starts by assuming some functional mapping for a system (membership functions and their shapes for fuzzy variables). The membership functions are then converted to a code familiar to the GA, bit strings (zeros and ones), which can then be connected together to make a longer chain of code for manipulation in the GA (i.e., selection, recombination, and mutation). An evaluation function is used to evaluate the fitness of each set of membership functions (parameters that define the functional mapping). Based on the fitness value, unsatisfactory strings are eliminated and reproduction of satisfactory strings proceeds for the next generation. This process of generating and evaluating strings is continued until the membership functions with the best fitness value are obtained. Ross (2010, Chapter 6) can be consulted for examples of GA implementation in the development of fuzzy membership functions.

Inductive reasoning This approach utilizes inductive reasoning to generate membership functions by deriving a general consensus from the particular. Inductive reasoning assumes availability of no information other than a set of data. Therefore, this approach is quite useful for complex systems where the data are abundant and static (more dynamic data would require regular modifications of membership functions).

The approach is to partition a set of data into classes based on entropy minimization. The entropy, E, where only one outcome is true is the expected value of the information contained in the data set and is given by:

$$E = -k \sum_{i=1}^{N} [p_i \ln p_i + (1 - p_i) \ln (1 - p_i)] \quad (5.31)$$

where the probability of the ith sample to be true is p_i, N is the number of samples, and k is a normalizing parameter. The minus sign in front of the parameter k ensures that entropy will be a positive value greater than or equal to zero.

Through iterative partitioning, the segmented data calculation of an estimate for entropy is possible. The result is a solution of points in the region of data interval used to define the membership function. The choice of shape of the membership function is arbitrary as long as some overlap is present between membership functions, therefore simple shapes such as triangles, which exhibit some degree of overlap, are often sensible. Ross (2010, Chapter 6) also offers examples of inductive reasoning applications.

DEVELOPMENT OF A FUZZY MEMBERSHIP FUNCTION FOR FLOOD RISK MANAGEMENT

Simonovic (2009, Chapter 6) presents one practical example of development of the membership function for flood risk management. This example offers a few different lessons: It illustrates sophisticated aggregation operators that are used in the development of membership functions; and it presents a practical procedure for the decomposition of complex systems for the development of a membership function. This example is from the flooding case study for the Red River, Manitoba, Canada. Part of this example is presented here with much less emphasis on the theoretical aggregation operators.

We have defined earlier a set S as the set of all elements in X having the property p. If the property p is such that it clearly separates the elements of X into two classes (those that have p and those that do not have p) we say that p defines a *crisp* subset S of X. When there is no such clear separation, we say that the property p defines an *ill-defined* subset S of X. A fuzzy membership function is a proper mathematical definition of an ill-defined set.

Let us suppose that we have an ill-defined set A that is too complex for the straightforward construction of its corresponding

Table 5.1 *Hierarchical approach for development of a fuzzy set*

Step 1	Decomposition of the complex fuzzy information (an ill-defined set A) into less complex components (ill-defined sets B_i). If any of the ill-defined sets B_i is still complex, continue with decomposition until the ill-defined sets obtained at the lowest hierarchical level are such that either construction of the corresponding fuzzy sets is not complex any more, or they cannot be decomposed any further.
Step 2	Construction of the fuzzy sets corresponding to the ill-defined sets at the bottom of the concept hierarchy using the existing methods for membership development from Section 5.2.5.
Step 3	Development of an aggregation model that will be able to perform reliable aggregation of the fuzzy sets constructed in Step 2 into a fuzzy set $\tilde{A}$ corresponding to the initial ill-defined set A.

fuzzy set $\tilde{A}$. If we want to design an automated tool for the construction of such a fuzzy set using hierarchical analysis, the procedure in Table 5.1 may be applied.

Each of the three steps in Table 5.1 is important and must be worked out carefully. For Step 1, some input from experts in the subject area of the ill-defined set A is usually necessary. In Step 2, the choice of a suitable method for membership evaluation has to be made.

When dealing with complex environments, such as flood risk management, it is good practice to decompose the ill-defined set A into more levels with fewer components as opposed to fewer levels with many components. This will simplify the aggregation process in Step 3 of Table 5.1. The overall decomposition should be intuitively clear in order to provide an effective evaluation of membership values in Step 2. In summary, if enough attention is given to the decomposition process in Step 1, then the complexity of Step 2 and Step 3 can be reduced significantly.

The rules by which membership functions are aggregated can be very complex and four of them: (i) Zimmermann's γ-family of operators, (ii) ordered weighted averaging (OWA) operators, (iii) composition under pseudomeasures (CUP), and (iv) polynomial composition under pseudomeasures (P-CUP), are presented in Simonovic (2009, Chapter 6) and will not be repeated here. The following example will concentrate more on the practical side of membership development for flood risk management.

The main purpose of the following example is to illustrate the practical application of the three aggregation methods (OWA, CUP, and P-CUP). Three local experts for flood risk management participated in the evaluation process, but only one of them (Expert 1) participated throughout the whole experiment. The membership functions for the sets from the lowest hierarchical

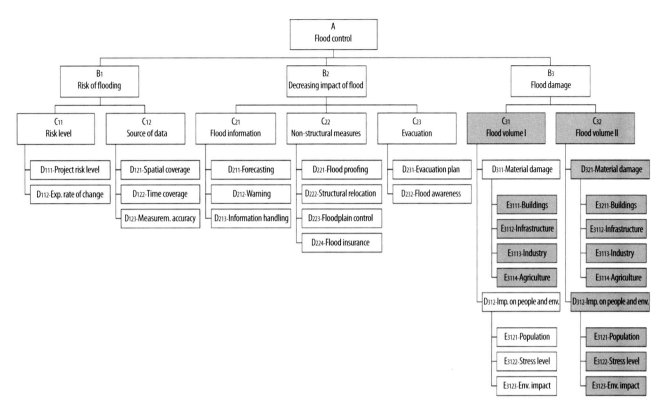

Figure 5.6 Outcome of the decomposition of flood control.

level in the flood risk management concept are considered a subject for a-priori evaluation.

The example defines *flood control*, which is difficult to represent via a fixed number of assessable parameters, and therefore it is defined here through its purpose. Good flood control should provide adequate protection against flooding and effective emergency management measures if flooding occurs. Both flood protection and emergency management measures are designed to prevent serious harm to people, and reduce material and environmental damage. Flood protection is highly associated with the flood return period, and subject to numerous uncertainties associated with eventual flooding.

The three major areas of concern can be inferred from this introduction. The first concern is about the safety level, below which flooding will not occur. The second one includes all the actions aimed at decreasing the negative impact of a flood should it occur. The third concern is with the total potential "bill" after the flood has occurred. At this point, it seems appropriate to start developing a concept hierarchy for flood control, since this will significantly reduce the complexity of further analytical thinking.

Example 2

Develop a concept hierarchy for *flood control* and develop its membership function for the Red River basin in Manitoba, Canada.

Solution

Step 1: decomposition. We start by dividing flood control into the three major components as described above (see Figure 5.6). Further decomposition can then be performed for each of these components. In the case of the risk of flooding (component B_1 in Figure 5.6), we can decompose it into the risk level – calculated from the available hydrologic data – and the reliability of the data. We also need to distinguish between the level of risk that will be steady over time and the level of risk that may significantly change in the near future, for example because of climate and other changes, such as land use upstream. The arrangement of these influential factors (one of several possible) that was adopted as our decomposition model for the component B_1 is shown in Figure 5.6.

In similar fashion, the other two components, B_2 and B_3, are analyzed and further decomposed. Components C_{31} and C_{32} are decomposed in the same way. The only difference is in the flood volume for which the total damage has to be determined. This classification can extend to as many different flood volumes as appropriate in a particular flooding region. The total material damage corresponding to each flood volume can be obtained from a stage–damage curve. The evaluation of overall flood damage is achieved by taking the weighted average of membership values for flood volume I, flood volume II and so on. The weights should correspond to the probability of each event.

Table 5.2 *Example query form: bottom level*

C_{22} Non-structural measures

D_{221} *Flood proofing*
1 = not existing
5 = ready, easy to do, reliable protection

D_{222} *Structural relocation*
1 = no plan
5 = relocation can be performed effectively

D_{223} *Floodplain control*
1 = no control
5 = effective control over the whole area

D_{224} *Flood insurance*
1 = no good flood insurance plan
5 = flood insurance covers most of the potential damage

NON-STRUCTURAL MEASURES?
1 = not included or poorly developed
5 = maximum possible, actions are well planned and organized

Table 5.3 *Example query form: top level*

A Flood control

B_1 *Risk of flooding*
(risk level, potential changes of the risk level in the future, spatial and time coverage of the data used for calculation and the accuracy of these data)

1 = high risk, no data at all, frequency of flooding is expected to increase in the future

5 = very low risk of flooding, reliable hydrologic data, no major increase expected

B_2 *Decreasing impact of flood*
(forecasting techniques, warning system, efficiency of distributing this information, what types of non-structural measures are considered for protection and how easy is it to implement them, whether there is an evacuation plan and the local population is aware and prepared for the eventual flooding)

1 = no forecasting or warning system, no extra measures for flood protection planned, no evacuation plan
5 = up-to-date forecasting and warning systems, an excellent communication system for information handling, plan for evacuation is very effective and people are flood educated

B_3 *Potential flood damage*
(the total material value in the flood zone, how many people live there, their age structure and their overall mobility, whether there is any possibility of major environmental impairment in the case of flooding)

1 = level of material damage is very high, flood zone is highly populated, flood can cause major problems to the whole environment

5 = flood practically does not endanger anything

FLOOD CONTROL?
1 = very unsatisfactory
5 = excellent; no need for improvement

Material damage, component D_{311}, can be expressed using a crisp number, and the appropriate membership values can easily be assigned. The components E_{3111}, E_{3112}, E_{3113}, and E_{3114} are included only to indicate that the importance of different categories of capital concentration may vary.

However, we should disconnect the issue of importance from the potential impact on people and the environment, since this impact is considered separately through the component D_{312}. Therefore, the aggregation of components C_{31} and C_{32} into B_3, as well as the aggregation of components E_{3111}, E_{3112}, E_{3113} and E_{3114} into D_{311}, was excluded from the case study.

Step 2: preparation of questionnaire. Every single component within the flood control concept hierarchy must be carefully formulated and prepared for evaluation by defining its extremes, i.e., the worst and the best conditions, which will correspond to the membership values 0 and 1. Three experts are selected based on their expertise in the area of flood risk management. Experts not only need to agree on these boundary conditions, they also need to agree on their meaning.

In this case study, these requirements were not met and the experts were exposed individually to these definitions through query forms. (Samples of two query forms, one corresponding to the bottom and the other one to the top level of the concept hierarchy, are shown in Tables 5.2 and 5.3.)

The preparation of a good questionnaire is an important step of the process, and some useful guidelines are available in Bates (1992). In this case study, the chosen scale for evaluation was from 1 to 5. This was later converted into a scale from 0 to 1, which is more appropriate for most fuzzy aggregation methods.

Step 3: evaluation. All the experts were provided with two sets of query forms and asked to use any number from the interval [1, 5] for the evaluation. The first set was prepared with input containing all the combinations of 5s and 1s for each decomposition. The evaluation of these combinations yielded the pseudomeasures necessary for the use of CUP and P-CUP aggregation methods. The second set of query forms had input randomly generated from the set {1, 2, 3, 4, 5}. Neither expert had strong objections on the way flood control was decomposed – disparities were mainly due to their different understanding of extremes.

Tables 5.4 and 5.5 provide sample results of the experts' evaluations. The results of evaluations for the first set of sample data

Table 5.4 *Example evaluation of the first set of data for component C_{23}: evacuation*

Sample data		Evaluation for C_{23}		
D_{231}	D_{231}	Expert 1	Expert 2	Expert 3
1	1	1	1	1
5	1	3	2	1
1	5	3	3	2
5	5	5	5	4

Table 5.5 *Example evaluation of the second set of data for component B_3: potential flood damage*

Sample data	D_{311}	3	3	5	2	4	3	3	1
	D_{312}	4	2	4	3	4	3	5	2
Expert 1	B_3	4	2.5	5	2.5	4.5	3.5	4	1.5

(combinations of 1s and 5s shown in the first two columns) are given in Table 5.4, while the evaluation for the second set of data, performed only by Expert 1, is given in Table 5.5.

A detailed set of evaluation results is provided in Simonovic (2009).

Step 4: implementation of aggregation operators. Three aggregation operators, OWA, CUP, and P-CUP, were implemented using the input obtained through Steps 1 to 3. Figure 5.7 shows a portion of the results of the exercise. Membership functions obtained using all three aggregation techniques are presented for the elements A, B_1, B_2, B_3, C_{11}, C_{12}, D_{111}, and D_{112} of the original concept hierarchy of flood control from Figure 5.6. Despic and Simonovic (2000) calculated the total error (as the sum of squared errors) and the maximum error for each aggregation method to show that P-CUP for this set of data gives a much better fit than the other two methods.

Different experts have different experiences, and these were the keystone for the judgments used in the case study. Every expert can be expected to be consistent in his or her own judgment, but not necessarily consistent with other experts. Lumping all the data together to find the best aggregation operator would result really in modeling no one's reasoning. It is not only that fuzzy relationships are context dependent: the meaning of a relationship depends on who is expressing it at the linguistic level. In practice, the best we can hope for is a reliable fuzzy set construction by one person in one specific setting.

5.3 FUZZY RISK DEFINITION

Fuzzy set theory is adopted for use in flood risk management under climate change through the introduction of the very innovative concept of fuzzy risk definition as a triplet: (i) a combined reliability–vulnerability index, (ii) a robustness index, and (iii) a resiliency index. Originally this idea was introduced to municipal water supply management (El-Baroudy and Simonovic, 2004) and further extended to flood risk management (Ahmad and Simonovic, 2007, 2011; and Simonovic and Ahmad, 2007).

The majority of flood risk analyses rely on the use of a probabilistic approach as presented in Chapter 4. Both resistance and load used in the conceptual definition of risk (Section 4.1) are considered to be random variables. However, the characteristics of resistance and/or load cannot always be measured precisely or treated as random variables. Therefore, the fuzzy representation of them is examined. The first use of both fuzzy resistance and fuzzy load can be found in Shrestha and Duckstein (1997).

The key concept involved in the implementation of the three fuzzy risk indices is the concept of partial failure. It is discussed below. The discussion of all three fuzzy risk indices follows.

5.3.1 Partial failure

The calculation of fuzzy risk indices depends on the exact definition of unsatisfactory system performance. It is difficult to arrive at a precise definition of failure because of the uncertainty in determining system resistance, load, and the accepted threshold below which performance is taken to be unsatisfactory. Figure 5.8 depicts a typical system performance (resistance time series), with a constant load during the operation horizon. According to the classical definition, the failure state is the state when resistance falls below the load, margin of safety (difference between the resistance and load) $M < 0.0$, or safety factor $\Theta < 1.0$, which is represented by the ratio between the system's resistance and load, shown in Figure 5.8 by the dashed horizontal line.

A simple example of a flood levee used for flood management in an urban area will be considered in the introduction of fuzzy risk definition. The failure state occurs when resistance falls below the load. In the case of the simple levee example, load can be represented by the floodwater level and the degree of resistance by the elevation of the levee embankment used to protect an area from flooding. The failure state will occur when the levee height is lower than the floodwater elevation. The margin of safety (the difference between load and resistance) $M(D) < 0.0$, or safety factor (the ratio between load and resistance) $\Theta(D) < 1.0$, shown in Figure 5.8 by the dashed horizontal line, are used to define the partial system failure. The area above the line is the area of failure. The area below the dashed line is the area of safety.

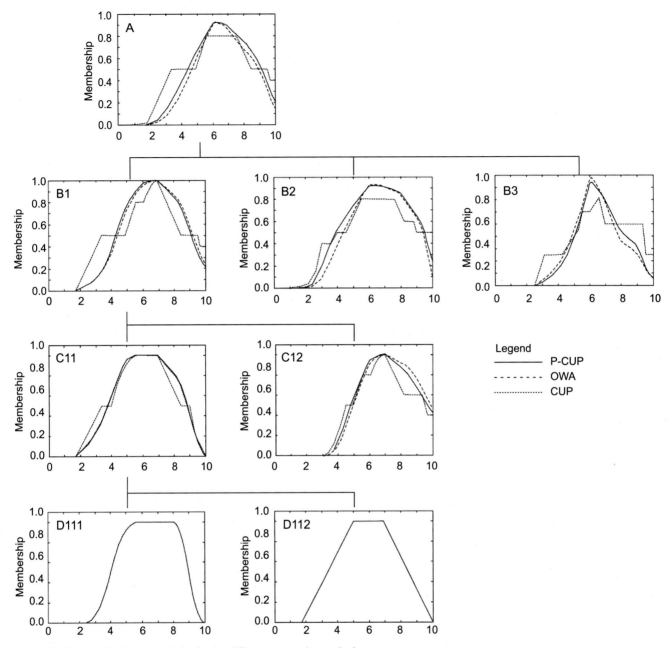

Figure 5.7 Membership functions obtained using different aggregation methods.

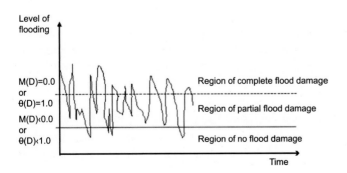

Figure 5.8 Variable flood levee performance.

It is known from practice that failure of the levee, for example, does not occur immediately when the floodwater levels exceed the levee height. We may observe different resistance of the levee structure from one location to another, or fluctuating load (floodwater levels). Consequently, a certain level of levee failure may be tolerated. The precise identification of failure is neither realistic nor practical. It is more realistic to build in the inevitability of partial failure. A degree of acceptable levee failure was introduced using the solid horizontal line, as shown in Figure 5.8. The area between the two lines represents the area of *partial failure* and the area above the solid line is the area of complete failure.

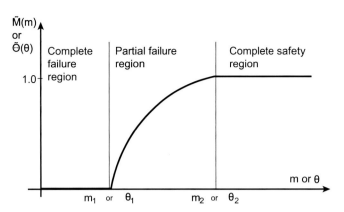

Figure 5.9 Fuzzy membership function of acceptable performance.

FUZZY REPRESENTATION OF THE ACCEPTABLE PERFORMANCE

The boundary of the partial failure region is ambiguous and varies from one decision-maker to another depending on the perception of risk. Boundaries cannot be determined precisely. Fuzzy sets are capable of representing the notion of imprecision better than ordinary sets. Therefore, the fuzzy membership of the levee performance can be expressed in mathematical form as:

$$\tilde{M}(m) = \begin{cases} 0, & if\ m \leq m_1 \\ \varphi(m), & if\ m \in [m_1, m_2] \\ 1, & if\ m \geq m_2 \end{cases}$$

or (5.32)

$$\tilde{\Theta}(\theta) = \begin{cases} 0, & if\ \theta \leq \theta_1 \\ \varphi(\theta), & if\ \theta \in [\theta_1, \theta_2] \\ 1, & if\ \theta \geq \theta_2 \end{cases}$$

where $\tilde{M}$ is the fuzzy membership function of margin of safety; $\varphi(m)$ and $\varphi(\theta)$ are functional relationships representing the subjective view of the partial failure (acceptable risk); m_1, m_2 are the lower and upper bounds, respectively, of the partial failure region; $\tilde{\Theta}$ is the fuzzy membership function of factor of safety; and θ_1, θ_2 are the lower and upper bounds of the partial failure region, respectively.

Figure 5.9 is a graphical representation of the definition presented in Equation (5.32). The lower and upper bounds of the partial failure region are introduced in Equation (5.32) by m_1 (or θ_1) and m_2 (or θ_2). The value of the margin of safety (or factor of safety) below m_1 (or θ_1) is definitely unacceptable. Therefore, the membership function value is zero. The value of the margin of safety (or factor of safety) above m_2 (or θ_2) is definitely acceptable and therefore belongs to the completely safe region. Consequently, the membership value is 1. The membership of the in-between values varies with the subjective assessment of partial failure by a decision-maker. Different functional forms may be

used for $\varphi(m)$ (or $\varphi(\theta)$) to reflect the subjective decision-maker's perception and acceptability of partial failure.

High system reliability is reflected through the use of high values for the margin of safety (or factor of safety), i.e., high values for both m_1 and m_2 (or θ_1 and θ_2). The difference between m_1 and m_2 (or θ_1 and θ_2) inversely affects the system reliability, i.e., the larger the difference, the lower the reliability. Therefore, the reliability reflected by the definition of an acceptable level of system performance can be quantified in the following way:

$$LR = \frac{m_1 \times m_2}{m_2 - m_1}$$
or (5.33)
$$LR = \frac{\theta_1 \times \theta_2}{\theta_2 - \theta_1}$$

where LR is the reliability measure of the acceptable level of performance.

The subjectivity of decision-makers will always result in a degree of ambiguity of risk perception. This alternative definition of failure allows for a choice between the lower bound, the upper bound, and the shape of function $\varphi(m)$ or $\varphi(\theta)$.

FUZZY REPRESENTATION OF THE SYSTEM-STATE

System resistance and load can be represented in a fuzzy form to capture the uncertainty inherent in system performance. The fuzzy form allows for the determination of the membership function of the resistance and load in a straightforward way even when there is limited available data. Fuzzy arithmetic, introduced in Section 5.2.3, can be used to calculate the resulting margin of safety (or factor of safety) membership function as a representation of the system-state at any time:

$$\tilde{M} = \tilde{X}(-)\tilde{Y}$$
and (5.34)
$$\tilde{\Theta} = \tilde{X}(/)\tilde{Y}$$

where $\tilde{M}$ is the fuzzy margin of safety; $\tilde{X}$ is the fuzzy resistance capacity; $\tilde{Y}$ is the fuzzy load; $(-)$ is the fuzzy subtraction operator; $(/)$ is the fuzzy division operator; and $\tilde{\Theta}$ is the fuzzy factor of safety.

FUZZY COMPATIBILITY

The purpose of comparing two fuzzy membership functions, as introduced in Section 5.2.4, is to illustrate the extent to which the two fuzzy sets match. The reliability assessment, discussed here, involves a comparative analysis of the system-state membership function and the acceptable performance membership function. Therefore, the compliance of two fuzzy membership functions can be quantified using the fuzzy compatibility measure.

Possibility (Equation 5.23) and necessity (Equation 5.25) lead to the quantification of the compatibility of two fuzzy sets. The

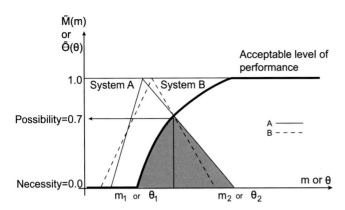

Figure 5.10 Compliance between the system-state and acceptable performance membership functions.

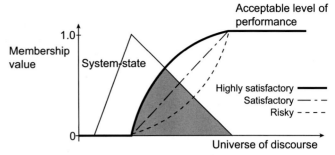

Figure 5.11 Compatibility between membership functions of the system-state and different levels of acceptable performance.

possibility measure quantifies the overlap between two fuzzy sets, while the necessity measure describes the degree of inclusion of one fuzzy set in another fuzzy set. However, in some cases (see Figure 5.10) high possibility and necessity values do not clearly reflect the compliance between the system-state membership function and the acceptable performance membership function. Figure 5.10 shows two system-state functions, A (solid line triangle) and B (dashed line triangle), which have the same possibility and necessity values. However, system-state A has a larger overlap with the performance membership function than system-state B (the shaded area in Figure 5.10).

The overlap area between the two membership functions, as a fraction of the total area of the system-state, illustrates compliance more clearly than the possibility and necessity measures in Equations (5.23) and (5.25) respectively:

$$C_{S,L} = \frac{OA_{S,L}}{A_S} \qquad (5.35)$$

where $C_{S,L}$ is the compliance between the system-state membership function (S) and the acceptable performance membership function (L); $OA_{S,L}$ is the overlap area between the system-state membership function (S) and the acceptable performance membership function (L); and A_S is the area of the system-state membership function (S).

An overlap in a high significance area (that is, an area with high membership values) is preferable to an overlap in a low significance area. Therefore, the compliance measure should take into account the weighted area approach by modifying Equation (5.35) into:

$$CM_{S,L} = \frac{WOA_{S,L}}{WA_S} \qquad (5.36)$$

where $CM_{S,L}$ is the compatibility measure between the system-state membership function (S) and the acceptable performance membership function (L); $WOA_{S,L}$ is the weighted overlap area

between the system-state membership function (S) and the acceptable performance membership function (L); and WA_S is the weighted area of the system-state membership function (S).

5.3.2   Fuzzy reliability–vulnerability

Reliability and vulnerability are used to provide a complete description of system performance in the case of failure, and to determine the magnitude of the failure event. Once an acceptable level of performance is determined in a fuzzy form, the anticipated performance in the event of failure as well as the expected severity of failure can be determined.

When certain values are specified for the lower and upper bounds (m_1 and m_2 (or θ_1 and θ_2) in Equation 5.32), thus establishing a predefined acceptable performance, the anticipated system failure is limited to a specified range. In order to calculate system reliability, several acceptable levels of performance must be defined to reflect the different perceptions of decision-makers.

A comparison between the fuzzy system-state membership function and the predefined fuzzy acceptable performance membership function provides information about both system reliability and system vulnerability at the same time (see Figure 5.11). The system reliability is based on the proximity of the system-state to the predefined acceptable performance. The measure of proximity is expressed by the compatibility measure suggested in Equation (5.36). The combined fuzzy reliability–vulnerability index is formulated as follows:

$$RE_f = \frac{\max_{i \in K}\{CM_1, CM_2, \ldots, CM_i\} \times LR_{max}}{\max_{i \in K}\{LR_1, LR_2, \ldots, LR_i\}} \qquad (5.37)$$

where RE_f is the combined fuzzy reliability–vulnerability index; LR_{max} is the reliability measure of acceptable level of performance corresponding to the system-state with maximum compatibility value; LR_i is the reliability measure of the ith acceptable level of performance; CM_i is the compatibility measure for the system-state with the ith acceptable level of performance; and K is the total number of defined acceptable levels of performance.

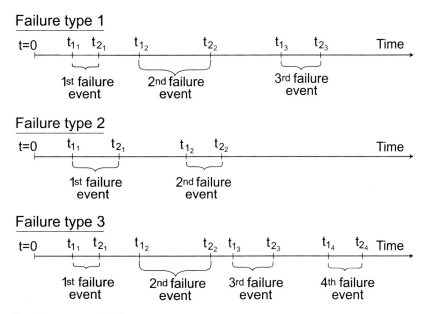

Figure 5.12 Recovery times for different types of failure.

The reliability–vulnerability index is normalized to attain a maximum value of 1.0 by the introduction of the $\max_{i \in K}\{LR_1, LR_2, \ldots, LR_i\}$ value as the maximum achievable reliability.

5.3.3 Fuzzy robustness

Robustness measures the system's ability to adapt to a wide range of possible future load conditions, at little additional cost. The fuzzy form of change in future conditions can be obtained through a redefinition of the acceptable performance and/or a change in the system-state membership function. As a result, the system's robustness is defined as the change in the compatibility measure:

$$RO_f = \frac{1}{CM_1 - CM_2} \tag{5.38}$$

where RO_f is the fuzzy robustness index; CM_1 is the compatibility measure before the change in conditions; and CM_2 is the compatibility after the change in conditions.

According to Equation (5.38), the higher the change in compatibility, the lower the value of fuzzy robustness. Therefore, high robustness values allow the system to adapt better to new conditions.

5.3.4 Fuzzy resiliency

Resilience measures how fast the system recovers from a failure state. The time required to recover from the failure state can be represented as a fuzzy set. Because the reasons for a failure may be different, system recovery times will vary depending on the type of failure, as shown in Figure 5.12.

A series of fuzzy membership functions can be developed to allow for various types of failure. The maximum recovery time is used to represent the system recovery time (Simonovic, 2009, after El-Baroudy and Simonovic, 2004):

$$\tilde{T}(\alpha) = \left[\max_{j \in J}[t_{1_1}(\alpha), t_{1_2}(\alpha), \ldots, t_{1_J}(\alpha)], \right.$$
$$\left. \max_{j \in J}[t_{2_1}(\alpha), t_{2_2}(\alpha), \ldots, t_{2_J}(\alpha)] \right] \tag{5.39}$$

where $\tilde{T}(\alpha)$ is the system fuzzy maximum recovery time at the α-level; $t_{1_j}(\alpha)$ is the lower bound of the jth recovery time at the α-level; $t_{2_j}(\alpha)$ is the upper bound of the jth recovery time at the α-level; and J is the total number of failure events.

The center of gravity of the maximum fuzzy recovery time can be used as a real number representation of the system recovery time. Therefore, system resilience is determined to be the inverse value of the center of gravity:

$$RS_f = \left[\frac{\int_{t_1}^{t_2} t\, \tilde{T}(t)\, dt}{\int_{t_1}^{t_2} \tilde{T}(t)\, dt} \right]^{-1} \tag{5.40}$$

where RS_f is the fuzzy resiliency index; $\tilde{T}(t)$ is the system fuzzy maximum recovery time; t_1 is the lower bound of the support of the system recovery time; and t_2 is the upper bound of the support of the system recovery time.

The inverse operation can be used to illustrate the relationship between the value of the recovery time and the resilience. The longer the recovery time, the lower the system's ability to recover from the failure, and the lower the resilience.

5.4 FUZZY TOOLS FOR FLOOD RISK MANAGEMENT UNDER CLIMATE CHANGE

Climate change brings to flood risk management the need for proper consideration of uncertainties that arise, for example, from the choice of the GCM and/or emissions scenario, from the choice of the downscaling tool, or from the subjective perception of climate change significance and its impact on extreme flows. The fuzzy set approach is gaining in importance for sustainable management of flood risk due to climate change. If we see risk management as a climate change adaptation mechanism, the set of tools to be discussed in this section of the book may have a very wide application domain. Three fuzzy-based management tools, fuzzy simulation, optimization, and multi-objective analysis, are presented in detail. Selected examples are also included in order to illustrate the utility of these tools. The main purpose of the following discussion is to introduce these tools to an audience that has not used these tools before for flood risk management under climate change.

5.4.1 Fuzzy simulation for flood risk management under climate change

The computer simulation model is a formal attempt to construct a model of a complex flood risk management system to make adequate predictions of its behavior under different initial and boundary conditions that may be caused by climate change. Fuzzy simulation can be an appropriate approach to include various inherent uncertainties of flood risk management systems in the simulation process. Among the several commonly used classes of fuzzy simulation models are relational equations, fuzzy neural networks, and fuzzy regression models.

The approach used for the introduction of fuzzy simulation is derived from utilizing the fuzzy inference method, based on the representation of human knowledge in IF-THEN rule-based form. Using rule-based simulation, the inference of a conclusion or fact (consequent) given an initial known fact (premise, hypothesis, antecedent) can be made (Ross, 2010).

BASICS OF FUZZY RULE-BASED SIMULATION
A typical way of presenting knowledge is in the form of a natural language expression of the form:

$$IF\ premise\ (antecedent),\ THEN\ conclusion\ (consequent)$$
$$(5.41)$$

The form is commonly referred to as the IF-THEN rule-based form or deductive form. This form of knowledge representation is appropriate in the context of linguistics where the linguistic variables can be naturally represented by fuzzy sets and logical connectives of these sets.

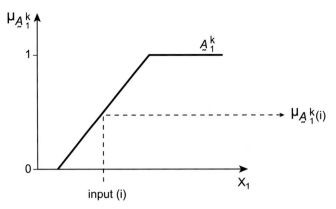

Figure 5.13 Fuzzification of scalar input from known membership function.

Mamdani's fuzzy inference method is the most commonly seen fuzzy simulation methodology in practice and in the literature, and is the methodology presented here (Mamdani and Assilian, 1975). The method was originally proposed as an attempt to control a steam engine and boiler combination by synthesizing a set of linguistic control rules obtained from experienced human operators. The Mamdani inference method is a graphical technique that follows five main steps: (i) development of fuzzy sets and linguistic rules, (ii) fuzzification of inputs, (iii) application of fuzzy operators, (iv) aggregation of all outputs, and (v) defuzzification of aggregated output.

Step 1: Development of fuzzy sets and linguistic rules. To begin, the Mamdani form rules are described by a collection of r linguistic IF-THEN expressions. The following is an expression for a fuzzy system with two non-interactive inputs x_1 and x_2 (antecedents) and a single output (consequent) y. The concept holds for any number of antecedents (inputs) and consequents (outputs).

$$IF\ x_1\ is\ \underset{\sim}{A}_1^k\ and\ (or)\ x_2\ is\ \underset{\sim}{A}_2^k\ THEN\ y^k\ is\ \underset{\sim}{B}^k$$
$$for\ k = 1, 2, \ldots, r \qquad (5.42)$$

where $\underset{\sim}{A}_1^k$ and $\underset{\sim}{A}_2^k$ are the fuzzy sets representing the kth antecedent pairs; and $\underset{\sim}{B}^k$ is the fuzzy set representing the kth consequent. The membership functions for the fuzzy sets may be generated with one of the methods discussed in Section 5.2.5.

Step 2: Fuzzification of inputs. The inputs to the system x_1 and x_2 are scalar values. In order to proceed with the inference method, the corresponding degree to which the inputs belong to the appropriate fuzzy sets via membership functions needs to be found. Fuzzification of the input thus requires the membership function of the fuzzy linguistic set to be known. The corresponding degree of membership for the scalar input belonging to the universe of discourse is found through the function evaluation. Figure 5.13 outlines the procedure in graphical form.

It should be noted that input to any fuzzy system may be a membership function, such as a water level reading. Either way,

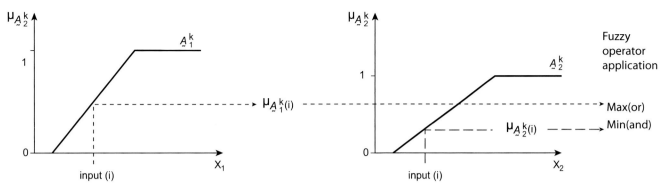

Figure 5.14 Fuzzy operator used for the generalized rule representation.

the simulation methodology remains the same as one that employs fuzzy singletons (scalar values) as input.

Step 3: Application of fuzzy operators. Once the inputs are in fuzzy form, the degree to which each condition of the antecedent is satisfied is known for each rule. If there are multiple antecedent conditions for each rule, as in the case of Equation (5.42), then a fuzzy operator is used to obtain one number that represents the antecedent for that rule. This number is applied to the output function producing a single truth value for the rule. The description of logical operators commonly used follows. Equation (5.42) has conjunctive antecedents and in brackets for illustration shows disjunctive antecedents.

For conjunctive antecedents, assuming a new fuzzy subset A_s as:

$$A_s^k = A_1^k \cap A_2^k \quad for\, k = 1, 2, \ldots, r \qquad (5.43)$$

expressed by means of a membership function, shown in Figure 5.14:

$$\mu_{A_s^k}(x) = \min[\mu_{A_1^k}, \mu_{A_2^k}] \quad for\, k = 1, 2, \ldots, r \quad (5.44)$$

For a disjunctive antecedent a similar procedure follows. This time fuzzy set A_s is defined as:

$$A_s^k = A_1^k \cup A_2^k \quad for\, k = 1, 2, \ldots, r \qquad (5.45)$$

expressed by means of membership function, shown in Figure 2.2:

$$\mu_{A_s^k}(x) = \max[\mu_{A_1^k}, \mu_{A_2^k}] \quad for\, k = 1, 2, \ldots, r \quad (5.46)$$

Given the above, the compound rule may be rewritten as

$$IF\ A_s^k\ THEN\ B_s^k \quad for\, k = 1, 2, \ldots, r \qquad (5.47)$$

Step 4: Aggregation of outputs. It is common for a rule-based system to include more than one rule. Therefore, in order to reach a decision, or overall conclusion, aggregation of individual consequents or outputs contributed by each rule is required.

In this way all the outputs are combined into a single fuzzy set, which may be defuzzified in the final step to obtain a scalar solution.

The aggregation of outputs may be achieved in two ways: (i) max-min truncation; or (ii) max-product scaling. Discussion in this book is limited to the first one. Ross (2010) is an excellent source of information about the second one. In the max-min case, the aggregation is achieved by the use of the minimum or maximum membership function value from the antecedents (depending on the logical operator used in the rule). This value is propagated through to the consequent and in so doing truncates the membership function for the consequent of each rule. This procedure is implemented for each rule. The truncated membership functions for each rule are then combined. This is achieved through the use of disjunctive or conjunctive rules, and the same fuzzy operators as in Step 3.

If the system of rules needs to be jointly satisfied, the truncated outputs should be aggregated as a conjunctive system – the rules are connected by "and" connectives. In the case where the objective is for at least one rule to be satisfied, the aggregation of outputs may be treated by the definition of a disjunctive system – the rules are connected by "or" connectives. Figure 5.15 illustrates the aggregation of outputs into a single fuzzy membership function. Each antecedent is treated as conjunctive and the aggregation of outputs of each rule is treated as a disjunctive system.

Step 5: Defuzzification of aggregated output. The final step of the rule-based system simulation is typically identification of a single value obtained from the defuzzification of the aggregated fuzzy set of all outputs. Many defuzzification methods are available in the literature: max membership principle, centroid method, weighted average method, and numerous other methods. There is no one most suitable defuzzification method. Selection of the best method for defuzzification is context or problem dependent. I recommend the use of the centroid method introduced in Section 5.2.4, Equation (5.22). It is simple, most widely used in practice, and physically appealing (Ross, 2010).

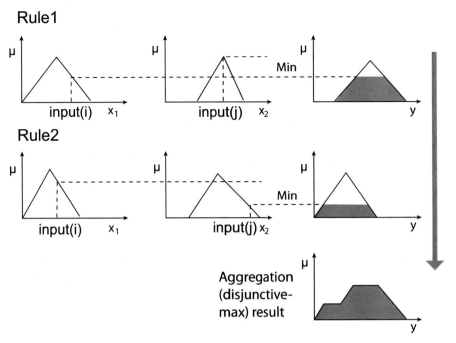

Figure 5.15 Aggregation of rule outputs into a single fuzzy membership function.

FUZZY RULE-BASED SIMULATION, CLIMATE CHANGE, AND FLOOD RISK MANAGEMENT

Very limited literature is available on the use of fuzzy simulation in climate change flood studies. The potential of the approach has not yet been fully investigated, creating plenty of opportunities to push the research and practical application boundaries ahead. The following are some applications that can be found in related research areas.

Bardossy *et al.* (1995) applied a fuzzy rule-based methodology to the problem of classifying daily atmospheric circulation patterns (CP). This example is not in the area of fuzzy rule-based simulation, but provides an interesting use of fuzzy rules in classification studies. The purpose of the approach is to produce a semi-automated classification that combines the expert knowledge of the meteorologist and the speed and objectivity of the computer. Rules are defined corresponding to the geographical location of pressure anomalies. A CP is described by the location of four different possible pressure anomalies. The rules are formulated with fuzzy sets, allowing certain flexibility because slightly different pressure maps may correspond to a given CP. Accordingly, the degree of fulfilment of a rule is defined in order to measure the extent to which a pressure map may indeed belong to a CP type. As an output of the analysis, the CP on any given day is assigned to one, and only one, CP type with a varying degree of credibility. The methodology is applied to a European case study. The subjective classification of European CPs provides a basis for constructing the rules. The classification obtained can be used, for example, to simulate local precipitation conditioned on the 700 hPa pressure field. The information content of the fuzzy classification as measured by precipitation-related indices is similar to that of existing subjective classifications. The fuzzy rule-based approach thus has potential to be applicable to the classification of GCM-produced daily CPs for the purpose of predicting the effect of climate change on space-time precipitation over areas where only a rudimentary classification exists or where none at all exists.

Cameron *et al.* (2000) explore the potential for assessing the impacts of climate change on flood frequency for one watershed in the UK (Plynlimon, Wales) by taking into account uncertainty in modeling rainfall–runoff processes under current conditions. The methodology is implemented within a generalized likelihood uncertainty estimation framework that allows for uncertainty in the model parameters and for the realization effect in reproducing the apparent statistics of potential flood events represented by the short series of observations. The same approach is used by Beven and Blazkova (2004) for the simulation of both high intensity and low intensity rainfall events, and snowmelt events, over subcatchments in contributing to the flood frequency distribution. Their case study is a dam site in a large catchment (1,186 km^2) in the Czech Republic. A fuzzy rules method is used to evaluate each model run, based on the available observations of peak discharges, flow duration curves, and snow water equivalents. This yields a combined likelihood measure that is used to weight the flood predictions for each behavioral parameter set. The cumulative distribution for flood peaks for any chosen probability of exceedance over all behavioral models can then be estimated. This can be used to assess the risk of a potential flood peak (or duration or volume) within a risk-based dam safety assessment.

Bardossy *et al.* (2005) used a fuzzy rule-based methodology for downscaling local hydrologic variables from large-scale atmospheric circulation models. The method is used to estimate the frequency distribution of daily precipitation conditioned on daily geopotential fields. This work does not directly consider flood risk management but provides essential input for the estimation of flood risk. Hydrologic model transformation of precipitation into runoff will be necessary to use the results of this work. The downscaling task is accomplished in two steps. First, the exceedance probabilities corresponding to selected precipitation thresholds are estimated by fuzzy rules defined between geopotential fields (premises) and exceedance events (conclusion). Then a continuous probability distribution is constructed from the discrete exceedance probabilities and the observed behavior of precipitation. The methodology is applied to precipitation measured at Essen, a location in the Ruhr catchment, Germany. Ten years of precipitation data (1970–1979) are used for the development of fuzzy rules and another 10 years (1980–1989) for validation. The 700 hPa geopotential fields are used to characterize large-scale circulation. The application example demonstrates that this direct downscaling method is able to capture the relationship between the premises and the conclusions; namely, both the estimated exceedance probabilities and the frequency distribution reproduce the empirical data observed in the validation period.

Hall *et al.* (2007) attempt to construct probability distributions over socio-economic scenarios, a task that is often resisted by the research community. Their work is also not directly related to flood risk management but has a very important link to it. Variation between published probability distributions of climate sensitivity attests to incomplete knowledge of the prior distributions of critical parameters and structural uncertainties in climate models. The authors address these concerns by adopting a fuzzy rule-based approach. The socio-economic scenarios are considered as fuzzy linguistic constructs. Any precise emissions trajectory (which is required as input for climate modeling) is assigned a degree of membership in a fuzzy scenario. Next, this work demonstrates how fuzzy scenarios can be propagated through a low-dimensional climate model, MAGICC. Fuzzy scenario uncertainties and imprecise probabilistic representation of climate model uncertainties are combined using random set theory to generate lower and upper cumulative probability distributions for the global mean temperature anomaly. Finally, they illustrate how non-additive measures provide a flexible framework for aggregation of scenarios, which can represent some of the semantics of socio-economic scenarios that defy conventional probabilistic representation. This interesting application of the fuzzy rule-based approach can be directly translated into flood risk modeling for the development of fuzzy socio-economic flood scenarios.

Mahabir *et al.* (2007) deal with spring ice jam floods (characteristic of northern communities) where lack of sufficient data for development of predictive warning models is very common. The ability to transfer a model between river basins is highly desirable but has not previously been achieved due to the statistical, site-specific nature of most river breakup models. Building on previous research, the ability to apply soft computing methods to model spring river ice breakup was further investigated with the goal of evaluating basin transferability and climate change scenarios using logic-based fuzzy and neuro-fuzzy modeling techniques. The Athabasca River (prototype development site) and the Hay River (model transfer site) in northern Canada were chosen as test sites due to their propensity for ice-jam-related floods. The fuzzy rule-based model was transferable between basins, in that extreme flood years could be distinguished from years when little flooding was reported; but the high quantitative accuracy of the neuro-fuzzy model was not reproduced at the second site. Climate change scenarios for the Athabasca River indicate a continuously decreasing risk of severe ice jams, while the frequency in the Hay River basin is expected to first increase for a period before declining.

Hanson and her collaborators (2010) deal with climate change effects on the world's coasts, a problem that at broad scales has typically proven difficult to analyze in a systematic manner. This research explores an outcome-driven deductive methodology for geomorphological analysis that recognizes the non-linearity of coastal morphology and organizes current knowledge and understanding using fuzzy set concepts. Building on recent large-scale coastal investigations and with reference to a case study of the East Anglian coast, UK, the methodology defines the active coastal system using a flexible generic classification and integrates expert opinion, using the notion of possibility, as a basis for the assessment of potential future geomorphological response to changes in sea level and sediment supply. Working over the medium term at a scale where data availability and detailed process understanding is variable, this work explores the integrated assessment of coastal units, such as sedimentary cells and subcells, to allow the investigation of coastal planning (including shoreline management) in response to climate change, including sea level rise. This includes the development of a capacity for the simulation of changes in coastal flood and erosion risk, biodiversity, and social and economic resources. The overall assessment effort is termed the Coastal Simulator. The fuzzy logic offers powerful concepts and methodology, including the incorporation of multiple outcomes, for Coastal Simulator. It allows for a level of abstraction and generalization commensurate with a regional scale of coastal analysis because it uses a collection of fuzzy sets (classes with inexact boundaries) and rules to simplify and reason about data. Preliminary results for the East Anglian coast suggest that the constraining of the active coastal system by sea defences is already having, and will continue to be, a significant influence on coastal evolution irrespective of the rate of sea level rise. Therefore, significant potential exists to guide future coastal evolution toward

Table 5.6 *Discharge frequency data*

T_r (years)	Exceedance probability	Discharge (m³/s)
500	0.002	898.8
200	0.005	676.1
100	0.010	538.5
50	0.020	423.0
20	0.050	298.8
10	0.100	222.5
5	0.200	158.4

Table 5.7 *Stage discharge data*

Discharge (m³/s)	Stage (m)
898.8	8.32
676.1	7.57
538.5	6.70
423.0	5.80
298.8	4.76
222.5	4.00
158.4	3.24

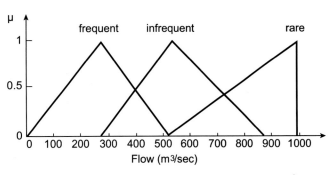

Figure 5.16 Triangular fuzzy membership functions for flow (m³/s).

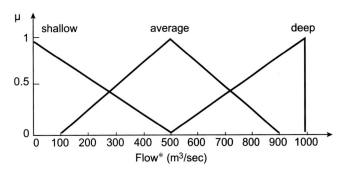

Figure 5.17 Triangular fuzzy membership functions for stage using flow* variable (m³/s).

preferred outcomes by using this approach as a component of climate change adaptive shoreline management. This methodology could be applied to a wide range of flood risk problems outside of the coastal domain.

Example 3

Let us revisit the flood protection levee design Example 7 from Section 4.5.1. The problem is to determine the height of a levee for 100-year return period flood protection with the discharge frequency curve as provided in Table 5.6 and the stage discharge curve in Table 5.7. The freeboard value is 1 m.

Assume that 580 m³/s design discharge is representative of a 100-year return period flow. Assume triangular membership functions for the linguistic variables. Use Tables 5.6 and 5.7 as an aid for the membership function development.

Solution

The fuzzy problem of flood protection levee design is solved using the fuzzy rule-based Mamdani inference method.

Step 1: Development of fuzzy membership functions. We start with partitioning the flow input space into three linguistic partitions within the interval of [0 m³/s, 1000 m³/s]: (i) *frequent*, (ii) *infrequent*, and (iii) *rare*. Similarly we partition the stage input space according to flow (using variable flow*) into three fuzzy membership functions described linguistically within the interval of [0 m³/s, 1000 m³/s] as: (i) *shallow*, (ii) *average*, and

(iii) *deep*. The output variable safety that describes the required safety level of the dyke is represented with a fuzzy set with three linguistic partitions: (i) *low*, (ii) *medium*, and (iii) *high* within the interval of [1 m, 10 m]. The triangular shape of fuzzy membership functions is assumed for illustrative simplicity. The range for each partition and maximum membership value of 1 govern the triangular membership function shape. These parameters were subjectively chosen in this example by the author. The fuzzy sets and their triangular membership functions are shown in Figures 5.16, 5.17, and 5.18.

In Figure 5.16 the flow input space membership functions are developed using subjective selection of flow of 280 m³/s to represent the frequent flow, 520 m³/s to represent the infrequent flow, and 1,000 m³/s to represent the rare flow. Each of the partitions has an ambiguous range surrounding the value representing the full degree of membership in the fuzzy set. In the case of rare flow the ambiguous range is one sided. Similarly, the parameters that govern the shapes of triangular membership functions in Figures 5.17 and Figures 5.18 are determined.

Step 2: Input fuzzification. The design flow input of 580 m³/s is fuzzified in order for the fuzzy inference procedure to proceed. Using the appropriate membership functions, the scalar inputs are fuzzified and their results (results of rule 2 and 3 firing) are shown in Figure 5.19.

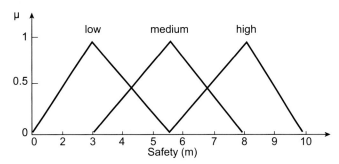

Figure 5.18 Triangular fuzzy membership functions for safety (M).

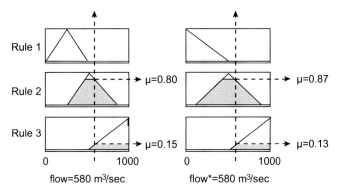

Figure 5.19 Fuzzification of the design flow input.

Step 3: Application of fuzzy operators. As the antecedents are disjunctive the max operator is used. The antecedents for each rule are represented by a single membership value (see Figure 5.19):

$$\mu_1 = max[0, 0] = 0$$
$$\mu_2 = max[0.8, 0.87] = 0.87 \qquad (5.48)$$
$$\mu_3 = max[0.15, 0.13] = 0.15$$

where μ_1, μ_2, and μ_3 are fuzzy membership values corresponding to rule 1, 2, and 3, respectively.

Step 4: Aggregation of outputs. The fuzzy membership functions corresponding to the output for each rule are truncated with respect to the membership values found in the previous step. These membership values are further aggregated using a disjunctive rule (max) system definition. The aggregation of outputs for flood levee height is illustrated in Figure 5.20.

Step 5: Defuzzification of the aggregated output. Finally, the aggregated output is defuzzified using the centroid method given by Equation (5.22). The defuzzified value location is shown in Figure 5.20 as $Y*$:

$$H = Y^* = 5.88 \, \text{m} \qquad (5.49)$$

This is the final value of the levee height designed to account for the floodwater elevation. The final elevation is:

$$H_t = H + H_f = 5.88 + 1 = 6.88 \, \text{m} \qquad (5.50)$$

The flood levee height of 6.88 m is determined using the fuzzy simulation approach.

FUZZY SIMULATION FOR THE DESIGN OF MUNICIPAL WATER INFRASTRUCTURE UNDER CLIMATE CHANGE

This section illustrates the implementation of the fuzzy simulation approach for the design of municipal water infrastructure under climate change. The example presented here is generic in nature. The use of the procedure is very much motivated by the current need of municipal engineers for a design procedure that will accommodate impacts of climate change. Most of the municipalities in Canada (and many other countries over the world) are addressing climate change through the modification of intensity–duration–frequency (IDF) curves (Prodanovic and Simonovic, 2008; Simonovic and Peck, 2009) that are used for the design of municipal water infrastructure. The Canadian Standards Association (2010) initiated the process of IDF modification in order to address the current water resources management needs. A few

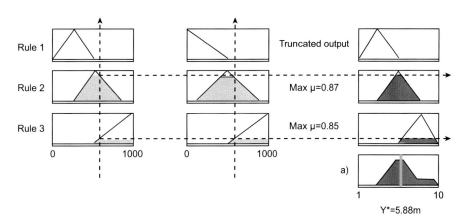

Figure 5.20 Outline of aggregation procedure for flood levee height design: (a) disjunctive aggregation of rules.

important observations from the Draft Technical Guide (Canadian Standards Association, 2010) may be of interest at this point:

> Under climate change, scientists project a warming climate with increases in the intensity and frequency of extreme precipitation with the result that infrastructure designed with historical IDF values may be at greater risk from damage or failure. An increase in the number of damaging flood events exceeding design conditions would result in increased stress to the infrastructure and potentially a reduction in service life (e.g., increased high flow events in storm sewers could result in increased strain on pipe joints, thereby increasing leakage and resulting in failure or possible sink hole formation). It is thus important to understand how extreme precipitation and IDF values are changing in the current climate and recognize our state of knowledge, uncertainties and assumptions in projecting how they could change in the future. New analytical approaches may need to be developed and existing IDF methodologies investigated (e.g., analyses of annual maximum series (AMS) and partial duration series (PDS)) to determine how they can be applied to future IDF analyses.

> Short-duration rainfall intensities of less than 24 hours are important considerations in planning, design, and management of many drainage and stormwater systems. Almost all such infrastructure has, to date, been designed and built using design information calculated from historical climate data, with a typical life cycle design of 50 years or longer. With a projected increase in the frequency and severity of intense rainfall events in the future climate, design criteria are expected to change. Under climate change, it will be important to review and update climatic design values and incorporate changes into building codes and other infrastructure standards. "No-regrets" adaptation actions such as these will be needed in order to maintain an acceptable service level or risk level throughout the expected lifecycle of the infrastructure.

The update of IDFs to include climate change suffers from the multiple sources of uncertainty, discussed in Chapter 2, such as choice of GCM or downscaling tool among others. Therefore, the fuzzy simulation design procedure is presented here to be used with IDF curves updated for climate change. To ensure adequate stormwater conveyance, a municipality could replace existing sewers susceptible to surcharging with pipes of a greater capacity. In general, this method of coping with an increase in future rainfall is very expensive and inflexible. However, there may be cases when sewer replacement may be a viable alternative: (i) sewer pipes are scheduled to be replaced due to lack of capacity and resulting damage, or structural and/or foundation

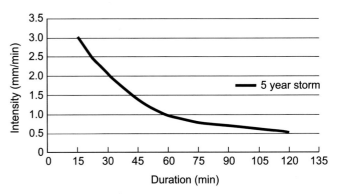

Figure 5.21 Intensity–duration–frequency curve.

problems; (ii) sewer pipes must be replaced because of redevelopment. The following example is developed to illustrate the application of fuzzy simulation for sewer pipe design and can be easily expanded to more complex municipal water infrastructure design and operations problems.

Example 4

The design problem considers a basin with an area of 2 hectares and a runoff coefficient of 0.6 where a concrete ($n = 0.013$) sewer pipe is to be installed at a slope of 0.5%. The preliminary basin investigations determined the longest flow path time to the proposed pipe location to be 15 minutes. Determine the appropriate pipe size using the updated IDF curve for climate change shown in Figure 5.21.

Solution

The traditional pipe design procedure involves the use of time of concentration, the time required for stormwater to flow from the most remote point in the basin to the pipe inlet structure. It is sometimes referred to as the hydraulic length. The peak discharge under a constant rate of effective rainfall will be reached if the effective rain duration is equal to the time of concentration:

$$T_c = t_0 + t_f \tag{5.51}$$

where t_0 (inlet time) is the time required for stormwater to reach an inlet, t_f is the flow time in the pipes upstream of the design point and T_c is the time of concentration.

The flow time in the pipes upstream of the design point can be determined using:

$$t_f = \sum_{j=1}^{N} \frac{L_j}{V_j} \tag{5.52}$$

where L_j is the length of the jth pipe, V_j is the average velocity in the jth pipe and N is the number of pipe segments upstream along the flow path considered. The inlet time can be calculated using Table 5.8. For the selected return period, the intensity of the design rainfall is obtained from the IDF curve, assuming the storm

Table 5.8 *Inlet time common values*

Densely developed impervious surfaces directly connected to drainage system	5 minutes
Well-developed districts with relatively flat slopes	10–15 minutes
Flat residential areas with widely spaced street inlets	20–30 minutes

duration equals the time of concentration. Once the intensity is obtained the design discharge can be found by a rational method:

$$Q_p = i \sum_{j=1}^{M} C_j A_j \qquad (5.53)$$

where i is the design rainfall intensity from the IDF curve, M the number of subareas above the stormwater pipe, A the drainage area of subarea j, C the runoff coefficient, and Q_p the design peak discharge. Finally, once the design discharge is determined, the Manning equation can be used to find the required pipe size. For circular pipes the formula is:

$$Dr = \left[\frac{nQ_p}{0.31\, k_n \sqrt{S_0}} \right]^{\frac{3}{8}} \qquad (5.54)$$

where Dr is the minimum diameter of pipe (actual size is next larger standard pipe size available), K_n is the conversion factor (1.0 m$^{1/3}$/s for SI units and 1.49 ft$^{1/3}$/s for US customary units), S_0 is the bottom slope of the sewer, and n the Manning roughness factor. The above formula is only valid under the assumption that the flow is full at the design discharge in the pipe. Usually, in addition there is a minimum velocity requirement for the flow in the pipe of 0.6–0.9 m/s to prevent the deposition of suspended materials and a maximum velocity of 3–4.5 m/s to prevent scouring.

Considering that the IDF curve in Figure 5.21 is updated for climate change, the fuzzy simulation procedure is suggested to address various uncertainties associated with such an update. The same procedure as represented by Equations (5.51)–(5.54) can be used with the fuzzified value of rainfall intensity, or the fuzzification process can start from the design discharge. In this example the latter has been implemented.

The input *flow* is partitioned into five simple partitions in the interval [0 m^3/s, 1 m^3/s], and the output *pipe size* is partitioned in the interval [−0.4888 m, 1 m] into five membership functions as shown in Figures 5.22 and 5.23 respectively.

The input variable *flow* corresponds to a fuzzy set that has five linguistic partitions describing a discharge flow, with the partitions labeled in Figure 5.22. The output variable *pipe size* corresponds to the pipe diameter and the fuzzy set partitions labeled in Figure 5.23. The triangular fuzzy set membership function shape

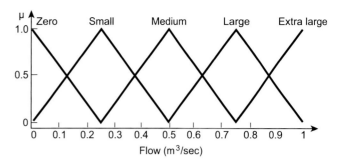

Figure 5.22 Five partitions for the input variable, flow (m^3/s).

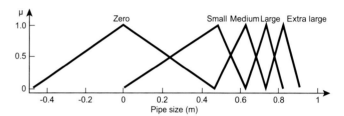

Figure 5.23 Five partitions for the output variable, pipe size (m).

has been assumed for illustrative simplicity. The *flow* input space and the *pipe size* output space parameters are subjectively chosen. The height of the triangle is defined by the full membership value and the base of the triangle is determined by the range of values holding some degree of membership in the fuzzy set.

In order to find the solution for the *pipe size* output a few input points are selected and the Mamdani graphical fuzzy inference method is employed. The centroid method is used for defuzzification.

Let us select eleven crisp singletons for inputs:

$$flow = \{0, 0.1, 0.2, 0.3, 0.4, 0.5, 0.6, 0.7, 0.8, 0.9, 1.0\}\ \text{m}^3/\text{s}$$

$$(5.55)$$

To illustrate the procedure, for flow input of 0.1 m^3/s, rules 1 and 2 are fired as shown in Figure 5.24. The resulting aggregated output after applying the union operator (disjunctive rules) is found and the fuzzy set is defuzzified using the centroid method yielding a result of 0.0847 m for pipe size as shown in Figure 5.24.

The results for each input, once aggregated and defuzzified, are summarized in Table 5.9.

5.4.2 Fuzzy linear programming for flood risk management under climate change

Flood risk management decisions under climate change are characterized by a set of decision alternatives (the decision space); a set of states of nature (the state space); a relation assigning to each pair of a decision and state a result; and, finally, an objective function that orders the results according to their desirability. When deciding under certainty, the flood manager knows which state to

Table 5.9 *Calculated pipe diameter (fuzzy simulation)*

Discharge (m³/s)	0.0	0.1	0.2	0.3	0.4	0.5	0.6	0.7	0.8	0.9	1.0
Pipe size (m)	0.000	0.085	0.233	0.391	0.425	0.620	0.654	0.693	0.751	0.781	0.823

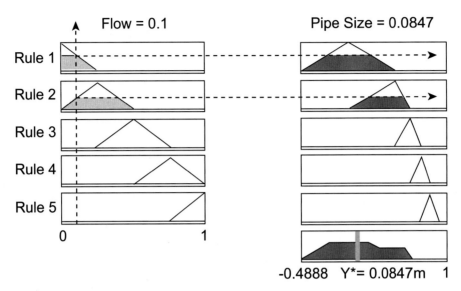

Figure 5.24 Graphical inference method: membership propagation and defuzzification.

expect and chooses the decision alternative with the highest value of the objective function (in the case of maximization, or the smallest value in the case of minimization), given the prevailing state of nature. When deciding under uncertainty, the decision-maker does not know exactly which state will occur. Only a probability function of the states may be known. Then decision-making becomes more difficult. The theory of probabilistic optimization is well developed and presented in Section 4.5, and applications in flood management decision-making can be found in the literature and practice.

The discussion in this chapter is focused on the expansion of the decision-making model to a fuzzy environment that considers a situation of decision-making under uncertainty, in which the objective function as well as the constraints are fuzzy. The fuzzy objective function and constraints are characterized by their fuzzy membership functions.

Since our aim is to optimize the objective function as well as the constraints, a decision in a fuzzy environment is defined by analogy with non-fuzzy environments as the selection of activities that simultaneously satisfy an objective function *and* constraints. According to the above definition and assuming that the constraints are "non-interactive" (not overlapping), the logical "*and*" corresponds to the intersection. The "decision" in a fuzzy environment is therefore viewed as the intersection of fuzzy constraints and a fuzzy objective function. The relationship between constraints and objective functions in a fuzzy environment is therefore

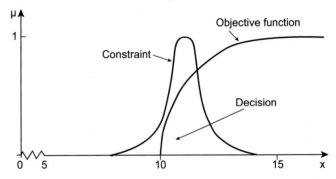

Figure 5.25 A fuzzy decision.

fully symmetrical: that is, there is no longer a difference between the former and the latter. Figure 5.25 illustrates the concept.

BASICS OF FUZZY LINEAR PROGRAMMING

The presentation in this section relies on Bellman and Zadeh (1970) and Zimmermann (1996). Linear programming models are considered as a special type of the general decision model discussed above: the decision space is defined by the constraints, the goal is defined by the objective function, and the type of decision is decision-making under certainty.

For transforming problems about LP decisions to be made in a fuzzy environment, quite a number of modifications are required to the deterministic form of LP (Simonovic, 2009). First of all the decision-making problem is transformed from actually maximizing or minimizing the objective function to reaching some aspiration levels established by the decision-maker. Second, the constraints are treated as vague. The $\leq$ sign is not treated in a strictly mathematical sense, and smaller violations are acceptable. In addition, the coefficients of the vectors b (right hand side of the constraints) or c (objective function coefficients) or of the matrix A (left hand side of the constraints) itself can have a fuzzy character, either because they are fuzzy in nature or because perception of them is fuzzy.

Finally, the role of the constraints is different from that in classical LP, where the violation of any single constraint by any amount renders the solution infeasible. Small violations of constraints are acceptable, and may carry different (crisp or fuzzy) degrees of importance. Fuzzy LP offers a number of ways to allow for all those types of vagueness, and I discuss some of them below.

In this text we accept Bellman and Zadeh's concept of a symmetrical decision model. We have to decide how a fuzzy "maximize" is to be interpreted. Our discussion will be limited to one approach for a fuzzy goal, and readers are directed to the literature for different interpretations. Finally, we have to decide where and how fuzziness enters the constraints. One way is to consider the coefficients of A, b, c as fuzzy numbers and the constraints as fuzzy functions. Another approach represents the goal and the constraints by fuzzy sets and then aggregates them in order to derive a maximizing decision. We use the latter. However, we still have to decide on the type of membership function characterizing the fuzzy sets representing the goal and constraints.

In classical LP the violation of any constraint makes the solution infeasible. Hence all constraints are considered to be of equal weight or importance. When departing from classical LP this is no longer true, and the relative weights attached to the constraints have to be considered.

Before we develop a specific model of LP for flood risk management under climate change in a fuzzy environment, it should be noted once more that by contrast to classical LP, *fuzzy LP is not* a uniquely defined type of model. Many variations are possible, depending on the assumptions or features of the real situation being modeled.

Let us now formulate the fuzzy LP problem. Let us assume that the decision-maker can establish an aspiration level, z_0, for the value of the objective function to be achieved, and that each of the constraints is modeled as a fuzzy set.

The fuzzy LP then finds x such that:

$$cx \precsim z_0$$
$$Ax \precsim b \qquad (5.56)$$
$$x \geq 0$$

where the symbol "$\precsim$" denotes a relaxed or fuzzy version of the ordinary inequality "$\leq$". The fuzzy inequalities represent the decision-maker's fuzzy goal and fuzzy constraints and mean that

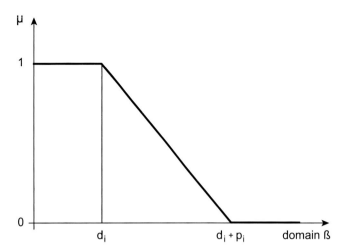

Figure 5.26 A linear membership function.

"the objective function cx should be essentially smaller than or equal to an aspiration level z_0 of the decision-maker" and "the constraints Ax should be essentially smaller than or equal to b," respectively. Furthermore, the fuzzy constraints and goal are viewed as equally important with respect to the fuzzy decision.

Zimmermann (1996) expressed the problem in simplified form for the fully symmetric objective and constraints:

$$Bx \precsim d$$
$$x \geq 0 \qquad (5.57)$$

where,

$$B = \begin{bmatrix} c \\ A \end{bmatrix}, \ d = \begin{bmatrix} z_0 \\ b \end{bmatrix} \qquad (5.58)$$

The following expression for the (monotonically decreasing) linear membership function illustrated in Figure 5.26 was proposed by Zimmermann for the ith fuzzy inequality $(Bx)_i \leq d_i$:

$$\mu_i((Bx)_i) = \left\{ \begin{array}{cc} 1; & (Bx)_i \leq d_i \\ 1 - \dfrac{(Bx)_i - d_i}{p_i}; & d_i \leq (Bx)_i \leq d_i + p_i \\ 0; & (Bx)_i \leq d_i + p_i \end{array} \right\}$$
$$(5.59)$$

where each d_i and p_i are the subjectively chosen constant values corresponding to the aspiration level and the violation tolerance of the ith inequality, respectively. If the constraints (including objective function) are well satisfied, the ith membership function value should be 1. If the constraint is violated beyond the limit of tolerance, p_i, then the value of μ will be 0 for all values greater than $(d_i + p_i)$. For values between d_i and $(d_i + p_i)$, the membership function value μ will change linearly between 1 and 0.

The membership function of the fuzzy set "decision" of the model in Equation (5.57) including the linear membership functions is shown below. The problem of finding the maximum decision is to choose x^* such that:

$$\mu_D(x^*) = \max_{x \geq 0} \min_{i=0,\ldots,m} \{\mu_i((Bx)_i)\} \qquad (5.60)$$

In other words, the problem is to find the $x^* \geq 0$ that maximizes the minimum membership function value. This value satisfies the fuzzy inequalities $(Bx)_i \precsim d_i$ with the degree of x^* (Sakawa, 1993).

Substituting the expression (5.59) for the linear membership function into Equation (5.60) yields:

$$\mu_D(x^*) = \max_{x \geq 0} \min_{i=0,\ldots,m} \left\{ 1 + \frac{d_i}{p_i} - \frac{(Bx)_i}{p_i} \right\} \quad (5.61)$$

The fuzzy set for decision can be transformed to an equivalent conventional LP problem by introducing the auxiliary variable λ:

$$
\begin{aligned}
&\textit{maximize} \quad \lambda \\
&\textit{subject to} \quad \lambda \leq 1 + \frac{d_i}{p_i} - \frac{(Bx)_i}{p_i}, \quad i = 0, \ldots, m \quad (5.62) \\
&\qquad\qquad\quad x \geq 0
\end{aligned}
$$

It should be emphasized that the above formulation (5.62) is for a minimization of the objective function and less than constraints, thus it should be modified appropriately for other conditions. So, the FLP problems can be easily solved using the LINPRO computer program. The formulation (5.62) is equivalent in form to a classical LP problem.

FUZZY OPTIMIZATION, CLIMATE CHANGE, AND FLOOD RISK MANAGEMENT

Very limited literature is available on the use of fuzzy optimization in climate change flood studies, similar to fuzzy simulation. The potential of the approach has not yet been fully utilized, creating opportunities to push the research and practical application boundaries ahead. The following are some related applications that illustrate the potential of fuzzy optimization.

Imprecision involved in the definition of reservoir loss functions is addressed using fuzzy set theory concepts by Teegavarapu and Simonovic (1999). A reservoir operation problem is solved using the concepts of fuzzy mathematical programming. Membership functions from fuzzy set theory are used to represent the decision-maker's preferences in the definition of shape of loss curves. These functions are assumed to be known and are used to model the uncertainties. Linear and non-linear optimization models are developed under a fuzzy environment. A new approach is presented that involves the development of compromise reservoir operating policies based on the rules from the traditional optimization models and their fuzzy equivalents while considering the preferences of the decision-maker. The imprecision associated with the definition of penalty and storage zones and uncertainty in the penalty coefficients are the main issues addressed through this study. The models developed are applied to the Green Reservoir, Kentucky. The main purpose of the reservoir is flood control. Simulations are performed to evaluate the operating rules generated by the models considering the uncertainties

in the loss functions. Results indicate that the reservoir operating policies are sensitive to change in the shapes of loss functions. This work does not address climate change but sets the stage for the following.

An FLP formulation is proposed for solving a reservoir operation problem and also to address the preferences attached to the direction and magnitude of climate change by the reservoir managers or decision-makers. No specific GCM-based scenario directly linked to the case study area is used but an overall reduction in streamflows is considered following the conclusions of a climate change study (Mulholland and Sale, 2002) conducted for the southeastern part of the USA. To demonstrate the utility of the FLP model in the current context, a short-term operation model is developed for Green Reservoir, Kentucky, USA. The primary objective of the reservoir is flood control in the Green River basin as well as in the downstream areas of the Ohio River. Secondary objectives include recreation and low flow augmentation. An optimization model formulation is developed and solved to fulfil these objectives.

Shrestha and Simonovic (2010) use a fuzzy set theory-based methodology for the analysis of uncertainty in the stage–discharge relationship. Individual components of stage and discharge measurement are considered as fuzzy numbers and the overall stage and discharge uncertainty is obtained through the aggregation of all uncertainties using fuzzy arithmetic. Building on the previous work – fuzzy discharge and stage measurements – we use fuzzy non-linear regression in this case study for the analysis of uncertainty in the stage–discharge relationship. The methodology is based on the fuzzy extension principle and considers input and output variables as well as the coefficients of the stage–discharge relationship as fuzzy numbers. The formulation leads to an optimization problem for evaluation of the coefficients. There are two different objective functions for the optimization: the least spread and the least absolute deviation. The results of the fuzzy regression analysis lead to a definition of lower and upper uncertainty bounds of the stage–discharge relationship and representation of discharge value as a fuzzy number. The methodology developed in this work is illustrated with a case study of Thompson River near Spences Bridge in British Columbia, Canada.

Teegavarapu (2010) provides an excellent review of modeling climate change uncertainties in water resources management models. He points out that the impact of climate change on the hydrologic design and management of water resources systems could be one of the important challenges faced by future practicing hydrologists and water resources managers. Many water resources managers currently rely on the historical hydrologic data and adaptive real-time operations without consideration of the impact of climate change on major inputs influencing the behavior of hydrologic systems and the operating rules. Issues such as risk, reliability, and robustness of water resources systems under different climate change scenarios were addressed in the past. However, water resources management with the decision-maker's preferences attached to climate change has never been dealt with. This review paper discusses issues related to impacts of climate change on water resources management and application of fuzzy set-based computational tools for climate-sensitive management of water resources systems.

Vucetic and Simonovic (2011) provide a comprehensive report on water resources decision-making under uncertainty. Uncertainty is in part about variability in relation to the physical characteristics of water resources systems. But uncertainty is also about ambiguity. Both variability and ambiguity are associated with a lack of clarity because of the behavior of all system components, a lack of data, a lack of detail, a lack of structure to consider water resources management problems, working and framing assumptions being used to consider the problems, known and unknown sources of bias, and ignorance about how much effort it is worth expending to clarify the management situation. Climate change, addressed in this report, is another important source of uncertainty that contributes to the variability in the input variables for water resources management. This report presents a set of examples that illustrate (i) probabilistic and (ii) fuzzy set approaches for solving various water resources management problems. The main goal of this report is to demonstrate how information provided to water resources decision-makers can be improved by using the tools that incorporate risk and uncertainty. The uncertainty associated with water resources decision-making problems is quantified using probabilistic and fuzzy set approaches. A set of selected examples is presented to illustrate the application of probabilistic and fuzzy simulation, optimization, and multi-objective analysis to water resources design, planning, and operations. Selected examples include flood levee design, sewer pipe design, optimal operations of a single purpose reservoir, and planning of a multi-purpose reservoir system. Demonstrated probabilistic and fuzzy tools can be easily adapted to many other water resources decision-making problems.

The following example illustrates the use of FLP and the computer package LINPRO (provided on the accompanying website) for its implementation.

Example 5

A city is upgrading flood control levees at four locations to accommodate climate change. It is also deciding on the choice of sources for riprap material being used. Four different sources (x_1 to x_4) are considered. The objective is to minimize cost, and the constraint is the need to supply all construction sites. That means a certain amount of material has to be delivered from different sources (quantity constraint) and a minimum material requirement per day has to be supplied from different sources (supply constraint). For other reasons, it is required that at least 6 units of material must come from the source x_1. The city management wants to use quantitative analysis and has agreed to the following

suggested LP model:

Minimize $41,400x_1 + 44,300x_2 + 48,100x_3 + 49,100x_4$

subject to constraints

$$0.84x_1 + 1.44x_2 + 2.16x_3 + 2.4x_4 \geq 170$$
$$16x_1 + 16x_2 + 16x_3 + 16x_4 \geq 1,300$$
$$x_1 \geq 6$$
$$x_2, x_3, x_4 \geq 0$$

$$(5.63)$$

Solution

The optimal solution to (5.63) is $x_1 = 6, x_2 = 16.29, x_3 = 0, x_4 = 58.96$. Minimum cost = 3,864,975. Use LINPRO to confirm the optimal solution. This part of Example 5 is on the website, in directory Fuzzy tools, sub-directory Optimization, sub-sub-directory LINPRO, sub-sub-sub-directory Examples, file Example1. linpro.

When the results are presented to the city management it turns out that they are considered acceptable but that the managers would rather have some flexibility in the constraints. They feel that because demand forecasts are used to formulate the constraints (and because forecasts are not always correct), there is a danger of not being able to meet higher demand by the construction sites and therefore having to pay penalties to the contractors.

The total budget of the city for the levees upgrade is 4.2 million, a figure that must not be exceeded. Since the city management feels it should use intervals instead of precise constraints, model (5.63) is selected to model the management's perceptions of the problem satisfactorily. The following parameters are estimated: the bounds of the tolerance interval for the objective function d_O and three constraints $d_i, i = 1, 2, 3$ are $d_O = 3,700,000, d_1 = 170, d_2 = 1,300$, and $d_3 = 6$; and the spreads of tolerance intervals are $p_O = 500,000, p_1 = 10, p_2 = 100$, and $p_3 = 6$. They define the simplest linear type of membership function.

Since the city problem is a minimization problem with greater than constraints, slight modification of the relationship (5.62), developed for a minimization problem with less than constraints, is required as follows:

$$\begin{aligned} maximize \quad & \lambda \\ subject\ to \quad & \lambda p_i + B_i x \leq p_i + d_i, \quad i = 0, \ldots, m+1 \\ & x \geq 0 \end{aligned}$$
$$(5.64)$$

If the optimal solution to (5.64) is the vector (λ, x_0), then x_0 is the maximizing solution (5.60) of model (5.56) assuming membership functions as specified in (5.59).

The new problem definition includes an objective function and four constraints. The first constraint is obtained from the modified formulation by replacing the given values into:

$$\lambda p_0 + B_0 x \leq p_0 + d_0 \qquad (5.65)$$

Table 5.10 *Levee upgrade problem solutions*

	Crisp	Fuzzy
Solution		
x_1	6	17.42
x_2	16.29	0
x_3	0	0
x_4	58.96	65.54
Z	3,864,975	3,939,202
Constraints		
1	170	171.93
2	1,300	1,327.36
3	6	17.42

where B_0 are the coefficients of the objective function. The remaining three constraints are simply obtained by replacing given values into:

$$B_i x - \lambda p_i \geq d_i, \quad i = 1, \ldots, m \qquad (5.66)$$

and using the membership functions with monotonically increasing linear shape (opposite from the one in Figure 5.26) because all constraints in the original formulation are greater than constraints. So our new problem is now:

maximize λ
subject to
$$\begin{aligned}
0.083x_1 + 0.089x_2 + 0.096x_3 + 0.098x_4 + \lambda &\leq 8.4 \\
0.084x_1 + 0.144x_2 + 0.216 + 0.24 - \lambda &\geq 17 \\
0.16x_1 + 0.16x_2 + 0.16x_3 + 0.16x_4 - \lambda &\geq 13 \\
0.167x_1 - \lambda &\geq 1 \\
\lambda, x_1, x_2, x_3, x_4 &\geq 0
\end{aligned}$$
$$(5.67)$$

Use LINPRO to find the optimal solution. This part of Example 5 is on the website, in directory Fuzzy tools, sub-directory Optimization, sub-sub-directory LINPRO, sub-sub-sub-directory Examples, file Example2.linpro.

The solution to the city's levee upgrade problem is shown in Table 5.10. As can be seen from the solution, flexibility has been provided with respect to all constraints at an additional cost of just below 2%.

The main advantage of the fuzzy formulation over the crisp problem formulation is that the decision-maker is not forced into precision for mathematical reasons. Linear membership functions (Figure 5.26) are obviously only a very rough approximation. Membership functions that monotonically increase or decrease in the interval of $(d_i, d_i + p_i)$ can also be handled quite easily.

FUZZY OPTIMIZATION OF A FLOOD CONTROL RESERVOIR

This section illustrates the use of fuzzy optimization for the analysis of a multi-purpose reservoir with flood control as the primary purpose. The Fanshawe Reservoir on the North Thames River (Ontario, Canada) is used as a case study. Most of the data and analyses are from Vucetic and Simonovic (2011) and Eum *et al.* (2009). The primary emphasis of the work by Eum *et al.* (2009) was linked to mitigation of climate change impacts on the operation of flood control reservoirs in the Thames River basin. However, in their optimization they did not use the fuzzy approach. The case study shown in this section is aimed at the introduction of the fuzzy optimization approach for reservoir optimization and will not directly address the climate change issues. Combination of the presented approach with the work of Eum *et al.* (2009) is suggested for future investigations.

Introduction Reservoir operation problems are challenging optimization problems. The reservoir operators must make long-term release decisions to accommodate incoming periods of floods and droughts so that the reservoir design goals are met. Generally, for reservoir operation optimization, inflow data must be known and discharge or release is considered as the optimization decision variable. The inflow data are usually from the historic records and are assumed to be an adequate representation of the future inflows. This assumption may be a source of critical error and the main source of uncertainty involved in making reservoir operation decisions. That justifies the use of an optimization approach able to deal with the reservoir operation problem under uncertainty. The optimization reservoir operation problems require a clear set of constraints that may not always be easy to formulate due to uncertainty in our knowledge. Use of fuzzy optimization may provide the solution for addressing the natural hydrologic variability and lack of knowledge related to complex reservoir operations decisions.

Mathematical formulation of the flood control reservoir optimization problem The flood control optimization problem is addressed here using a fuzzy optimization approach. This approach may be of significant value for climate change impact analyses when data may not be readily available and constraints are not formulated with a high level of precision.

Let us start with a traditional deterministic flood control reservoir optimization model formulation. The main objective is to determine the optimal release from the reservoir in various time intervals with the objective of minimizing flood damage due to excess storage of water in the reservoir. The flood damage is a function of storage and therefore minimization of reservoir active storage is an adequate representation for minimization of flood damage. In mathematical form, the flood control reservoir optimization problem can be stated as:

$$\begin{aligned}
\textit{Minimize}\quad & Z = S \\
\textit{subject to}\quad & S_{t-1} + i_t - R_t = S_t & t = 1, 2, \ldots, n \\
& 0 \leq R_t \leq R_{max} & t = 1, 2, \ldots, n \\
& S_{min} \leq S_t \leq C & t = 1, 2, \ldots, n \\
& S_0 \leq S_n
\end{aligned} \qquad (5.68)$$

where S is the active volume of water stored in the reservoir; S_t is the volume of water in the reservoir at time t; i_t is the inflow into the reservoir in the time interval $(t - 1, t)$; R_t is the amount of water released downstream in time interval $(t - 1, t)$; C is the reservoir capacity; S_{min} is the minimum operational reservoir storage level; S_n is storage at the end of the critical period; and S_0 is initial reservoir storage.

The first constraint represents the continuity equation. The remaining three constraints include: (i) deterministic constraint on the reservoir release; (ii) deterministic constraint that keeps the reservoir storage between the maximum reservoir capacity and minimum reservoir operational level; and (iii) deterministic constraint stating that the storage at the end of the critical period must be at least as great as the initial reservoir storage. All the constraints and objective function are linear in this formulation so that the optimal solution can be obtained using various software tools similar to LINPRO provided with this book.

The flood control reservoir operation optimization model formulation (5.68) is expanded to utilize the fuzzy linear optimization approach and address various uncertainties related to climate change, lack of knowledge, and insufficient data. The fuzzy reservoir optimization formulation does not require that all coefficients of the constraints are crisp numbers and that the objective function must be minimized or maximized (Zimmermann, 1996). Assuming the fuzzy objective function and fuzzy greater than constraints (as discussed in Section 5.4.2), the deterministic model (5.68) becomes:

Maximize λ

subject to $\dfrac{(Bx)_i}{p_i} + \lambda \le 1 + \dfrac{d_i}{p_i}, \quad i = 0, \dots, 360 \le R_t$

$\le R_{max}, t = 1, 2, \dots, n$

$\dfrac{(Bx)_i}{p_i} - \lambda \ge \dfrac{d_i}{p_i}, \quad i = 37, \dots, 60$

$x \ge 0$

$$(5.69)$$

Substituting for $(Bx)_i$ leads to:

Maximize λ
subject to objective function constraint

$\dfrac{\sum_{t=1}^{12} S_t}{p_i} + \lambda \le 1 + \dfrac{d_i}{p_i}, \quad i = 0$

and all other constraints

$\dfrac{(S_t + R_t - S_{t-1})}{p_i} + \lambda \le 1 + \dfrac{d_i}{p_i}, \quad i = 1, \dots, 12,$

$t = 1, \dots, 12,$

$\dfrac{S_t}{p_i} + \lambda \le 1 + \dfrac{d_i}{p_i}, \quad i = 13, \dots, 24, \; t = 1, \dots, 12,$

$\dfrac{R_t}{p_i} + \lambda \le 1 + \dfrac{d_i}{p_i}, \quad i = 25, \dots, 36, \; t = 1, \dots, 12,$

$\dfrac{S_t}{p_i} - \lambda \ge \dfrac{d_i}{p_i}, \quad i = 37, \dots, 48, \; t = 1, \dots, 12,$

$\dfrac{(S_t + R_t - S_{t-1})}{p_i} - \lambda \ge \dfrac{d_i}{p_i}, \quad i = 49, \dots, 60, \; t = 1, \dots, 12,$

$R_t \ge 0 \quad t = 1, 2, \dots, 12$

$$(5.70)$$

where the time step is one month and the optimization time horizon is one year.

The Fanshawe Reservoir case study The reservoir optimization case study is the Fanshawe Reservoir on the North Thames River located in Ontario, Canada (just outside the City of London). The reservoir data are provided by the UTRCA, which is responsible for the management of all flood control reservoirs in the basin. The Thames River basin, consisting of two major tributaries, has a drainage area of 3,482 km^2. The stream length is 273 km (from Tavistock to its mouth at Lake St. Clair). The North Branch flows southward through Mitchell, St. Marys, and eventually into London, and the East Branch flows through Woodstock, Ingersoll, and east London (Figure 5.27). The two branches join near the City of London and then the river flows westwards and exits the basin near Byron, eventually draining into Lake St. Clair. The basin receives about 1,000 mm of annual precipitation, 60% of which is lost through evaporation and/or evapotranspiration, stored in ponds and wetlands, or recharged as groundwater. The slope is about 1.9 m/km for most of its upper reaches, while its lower reaches are much flatter with a slope of less than 0.2 m/km. The Upper Thames River basin has three major flood management reservoirs: Wildwood, Pittock, and Fanshawe near St. Marys, Woodstock, and London, respectively (Figure 5.27). The primary goals of all reservoirs include flood control during the snowmelt period and summer storm season, low flow augmentation from May to October, and recreational uses during the summer season. Among these goals, the most important goal is flood control. Floods in the basin result from a combination of snowmelt and intensive precipitation during December to April. The basin has facilities such as extensive dyking systems and a diversion channel to prevent flood damage. Summer frontal storms are also known to produce severe flooding in and around the basin, but such storms are less frequent than spring floods. Drought conditions occur mostly from June to September, though they can occur at any time during the year.

The relevant data consist of physical constraints for the reservoir, such as the maximum and the minimum storage capacity and monthly historical inflow data (1953 to 2009).

The remaining data include: maximum reservoir capacity, $C = 0.22503 \times 10^8 \, m^3$; minimum reservoir storage, $S_{min} = 0.055 \times 10^8 \, m^3$; sill of dam elevation operator goal storage, $S_{GOAL} = 0.1235 \times 10^8 \, m^3$; initial reservoir storage, $S_0 = 0.1482 \times 10^8 \, m^3$; maximum possible release for non-flooding conditions, $R_{max} = 370 \, m^3/s$; and total maximum annual reservoir acceptable storage (to avoid inundation damage) of $1.6 \times 10^8 \, m^3$.

The release data are in Table 5.11. The reservoir inflow data (for year 2009) are in Table 5.12 and inflow statistics (mean μ and standard deviation σ) are in Table 5.13.

For the implementation of the fuzzy optimization approach we consider that the reservoir operators require some flexibility in the formulation of constraints to account for various uncertainties

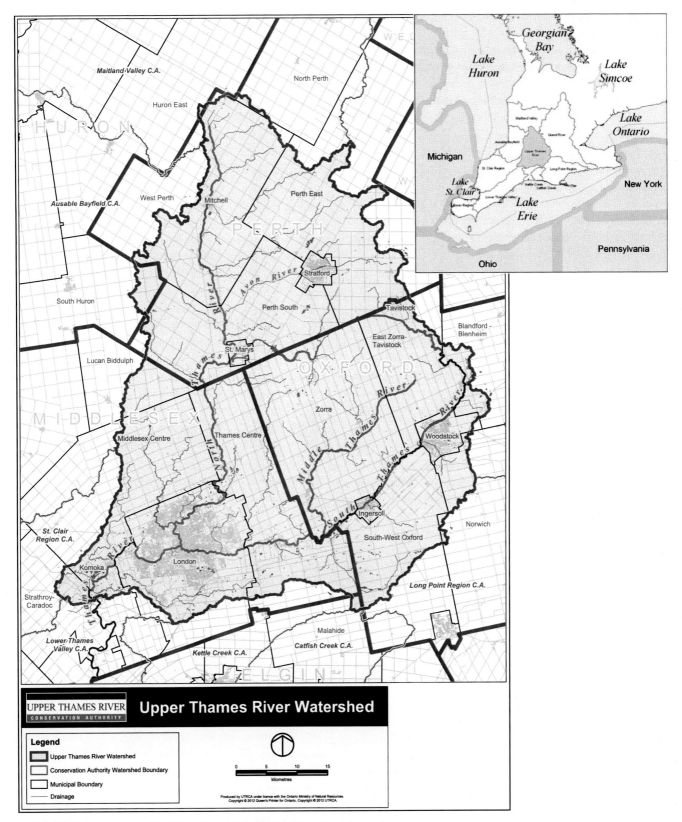

Figure 5.27 Schematic map of the Upper Thames River basin. See color plates section.

Table 5.11 *Maximum monthly Fanshawe Reservoir release (10^8 m^3)*

Month	1	2	3	4	5	6	7	8	9	10	11	12
R_{max}	9.91008	8.95104	9.91008	9.5904	9.91008	9.5904	9.91008	9.91008	9.5904	9.91008	9.5904	9.91008

Table 5.12 *Fanshawe Reservoir inflows (10^8 m^3)*

Month	1	2	3	4	5	6	7	8	9	10	11	12
Inflow 2009	0.34284	1.80472	1.21867	0.72058	0.54104	0.20062	0.12133	0.09508	0.07206	0.12294	0.10446	0.38033

Table 5.13 *Fanshawe reservoir inflow statistics (10^8 m^3)*

Month	1	2	3	4	5	6	7	8	9	10	11	12
μ	0.5036	0.5685	1.2499	0.9164	0.3722	0.1708	0.1329	0.1120	0.1797	0.2615	0.4689	0.6218
σ	0.3968	0.5231	0.5572	0.5551	0.2954	0.1395	0.1706	0.1157	0.2636	0.3161	0.4033	0.4592

associated with them. The lower bounds and the upper bounds of the tolerance interval, d_i, and spread of tolerance, p_i, were estimated by the reservoir operators (oral communication) and shown in Tables 5.14 and 5.15 respectively. The value of i in these tables corresponds to the constraint number from the formulation (5.70).

The parameters presented in Tables 5.14 and 5.15 form the linear membership functions to be used with the FLP formulation (5.70). For the continuity constraint a triangular membership function is used.

Results The Fanshawe Reservoir optimization is performed using the FLP formulation (5.70) and data presented in the previous section. The calculated optimal reservoir releases and storage are presented in Table 5.16. The optimal value of the objective function is 1.5867×10^8 m^3. The reservoir operation is successfully maintaining the goal storage with reservoir release staying within the given constraints. The flexibility obtained by the fuzzy formulation provides for effective reservoir operations.

Fuzzy linear formulation data for the Fanshawe case study are provided on the accompanying website, directory Fuzzy tools, sub-directory Optimization, sub-sub-directory LINPRO, sub-sub-sub-directory Examples, file Fanshawe.linpro.

5.4.3 Fuzzy multi-objective analysis for flood risk management under climate change

An earlier section presents limited work that has been done (and published) on stochastic multi-objective programming for flood risk management. Here we focus on the use of fuzzy sets in extending multi-objective analysis to uncertain conditions. Two conceptual tools are presented. The first is developed for the generation of the fuzzy non-dominated set (concept that extends the formulation (4.59) of the deterministic non-dominated set) that can be used by the decision-makers for the selection of the best compromise solution (Simonovic and Verma, 2008). The second presents an intuitive, and relatively interactive, decision tool for discrete flood management alternative selection, under various forms of uncertainty. This tool explores the application of fuzzy sets in conjunction with compromise programming. The adaptation of standard techniques to the fuzzy framework demands a different set of operators. We will explore how the fuzzy multi-objective compromise programming method (Bender and Simonovic, 2000) can be used in flood risk management that takes climate change into consideration.

Fuzzy decision-making techniques have addressed some uncertainties, such as the vagueness and conflict of preferences common in group decision-making. Their application, however, demands some level of intuitiveness for the decision-makers, and encourages interaction or experimentation. Fuzzy decision-making is not always intuitive to many people involved in practical decisions because the decision space may be some abstract measure of fuzziness, instead of a tangible measure of alternative performance. The alternatives to be evaluated are rarely fuzzy: it is their performance that is fuzzy. In other words, a fuzzy multi-objective decision-making environment may not be as generically relevant as a fuzzy evaluation of a decision-making problem. Most fuzzy multi-objective methods either concentrate on multi-objective LP techniques, or experiment with methods based on fuzzy relations (Simonovic, 2009).

METHOD FOR GENERATION OF FUZZY NON-DOMINATED SET OF SOLUTIONS
Methods for generating non-dominated (Pareto optimal) solutions remain very important in water resource decision-making when the decision-makers have to consider a wide range of options

Table 5.14 *Estimated Fanshawe reservoir lower bound parameters (10^8 m^3)*

i	d_i	p_i	Comments
0	1.1000	0.5000	Objective function
1	0.3428	0.0686	
2	1.8047	0.3609	
3	1.2187	0.2437	
4	0.7206	0.1441	
5	0.5410	0.1082	Continuity constraints
6	0.2006	0.0401	
7	0.1213	0.0243	
8	0.0951	0.0190	
9	0.0721	0.0144	
10	0.1229	0.0246	
11	0.1045	0.0209	
12	0.3803	0.0761	
13	0.2250	0.0001	
14	0.2250	0.0001	
15	0.2250	0.0001	
16	0.2250	0.0001	
17	0.2250	0.0001	Maximum reservoir capacity constraints
18	0.2250	0.0001	
19	0.2250	0.0001	
20	0.2250	0.0001	
21	0.2250	0.0001	
22	0.2250	0.0001	
23	0.2250	0.0001	
24	0.2250	0.0001	
25	9.9101	0.0001	
26	8.9510	0.0001	
27	9.9101	0.0001	
28	9.5904	0.0001	
29	9.9101	0.0001	
30	9.5904	0.0001	Maximum release constraints
31	9.9101	0.0001	
32	9.9101	0.0001	
33	9.5904	0.0001	
34	9.9101	0.0001	
35	9.5904	0.0001	
36	9.9101	0.0001	

Table 5.15 *Estimated Fanshawe reservoir upper bound parameters (10^8 m^3)*

i	d_i	p_i	Comments
37	0.1235	0.0900	
38	0.1235	0.0900	
39	0.1235	0.0900	
40	0.1235	0.0900	
41	0.1235	0.0900	
42	0.1235	0.0900	Minimum reservoir storage constraints
43	0.1235	0.0900	
44	0.1235	0.0900	
45	0.1235	0.0900	
46	0.1235	0.0900	
47	0.1235	0.0900	
48	0.1482	0.0000	
49	0.2743	0.0686	
50	1.4438	0.3609	
51	0.9749	0.2437	
52	0.5765	0.1441	
53	0.4328	0.1082	
54	0.1605	0.0401	Inflow constraints
55	0.0971	0.0243	
56	0.0761	0.0190	
57	0.0576	0.0144	
58	0.0984	0.0246	
59	0.0836	0.0209	
60	0.3043	0.0761	

multi-objective analysis has been focused on the development of fuzzy methods with prior articulation of preferences and fuzzy methods for progressive articulation of preferences.

In this method, we generate the fuzzy set of non-dominated optimal solutions for a multi-objective problem using positive and negative ideals (Lai *et al.*, 1994). The weights assigned to the objective functions are fuzzy and triangular in shape rather than crisp. Triangular fuzzy weights of different objective functions can be either in the same range of pessimistic and optimistic values or in a different range for different objective functions.

A general multi-objective decision-making problem has been expressed in mathematical form, Equation (4.58), as:

$$Max/Min[f_1(x),\ f_2(x), \ldots, f_k(x)]$$
$$subject\ to$$
$$x \in X = \{x \mid g_i(x)(\leq, =, \geq)0\}, \quad i = 2, \ldots, m \quad (5.71)$$

(alternatives) and criteria (objectives). Evaluation of many alternatives, as well as expression of preferences, is highly subjective in many cases. The use of fuzzy set theory (Zadeh, 1965) seems very appropriate here. However, most of the research in fuzzy

Table 5.16 *Fanshawe reservoir fuzzy optimization results (10^8 m^3)*

Month	Jan	Feb	Mar	Apr	May	Jun	Jul	Aug	Sep	Oct	Nov	Dec
Release	0.298	1.466	0.990	0.585	0.439	0.163	0.098	0.077	0.058	0.099	0.085	0.290
Storage	0.129	0.129	0.129	0.129	0.129	0.129	0.129	0.129	0.129	0.129	0.129	0.148

where x is an n-dimensional decision variable vector; $f_k(x)$ is a value of the objective k; and $g_i(x)$ is a value of the constraint i. The problem consists of n decision variables, m constraints, and k objectives.

For a multi-objective optimization problem like that represented by Equation (5.71), it is significant to note that multiple objectives are often non-commensurable and cannot be combined into a single one. The objectives usually conflict with each other and any improvement in one can be achieved only at the expense of another. There is no "optimal" solution for the multi-objective optimization as in the case of a single criterion problem. For the solution of problem (5.71) only a set of non-dominated, Pareto optimal solutions can be obtained. However, decisions represented by the Pareto set are not uniquely determined – different preferences associated with each of the objectives result in a different Pareto optimal solution. The final decision must include the selection from among the set of Pareto optimal solutions of one point that meets some additional objective, for example, the shortest distance to an ideal solution.

There are various techniques available to generate the non-dominated optimal solution set under certainty, amongst which the *weighting method* is the oldest, proposed by Zadeh in 1965. He demonstrated that if, within a multi-objective problem, a non-negative weight or parameter is associated to each criterion then the weighted criteria can be aggregated into a single objective function. The optimization of this objective function generates an extreme efficient point, one non-dominated optimal solution. The generating techniques can often leave many of the non-dominated optimal solutions unexplored. To avoid this, it is possible to reduce considerably the scale of the weights in the application of the weighting method. In any case, generating methods are devised simply to approximate the non-dominated optimal solution set.

In water resource decision-making practice the weights associated with the criteria are not always crisp. They are rather fuzzy in nature (Simonovic, 2009). With fuzzy weights the multi-objective problem given with Equation (5.71) becomes a fuzzy multi-objective decision-making problem. Weights assigned to the objectives are fuzzy weights in triangular form represented using a set of values (p, m, o), where p, m, and o are the pessimistic, modal, and optimistic values of the fuzzy weight membership function, as shown in Figure 5.28.

The multi-objective problem with fuzzy weights can be written as:

$$Max/Min(p, m, o)[f_1(x), f_2(x), \ldots, f_k(x)], \quad k = 1, 2, \ldots, K$$
subject to
$$x \in X = \{x \mid g_i(x)(\leq, =, \geq)0\}, \quad i = 1, 2, \ldots, m \quad (5.72)$$

Problem (5.72) can be further expressed as:

$$Max/Min[f_k^p(x), \ f_k^m(x), f_k^o(x)], \quad k = 1, 2, \ldots, K$$
subject to
$$x \in X = \{x \mid g_i(x)(\leq, =, \geq)0\}, \quad i = 2, \ldots, m \quad (5.73)$$

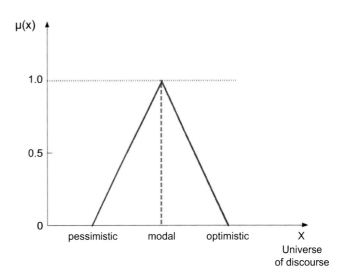

Figure 5.28 Fuzzy weight.

where $f_k^p(x)$, $f_k^m(x)$, $f_k^o(x)$ are the elements of the vector of objective associated with the fuzzy weights for each of K objectives.

To solve the fuzzy multi-objective decision-making problem, the fuzzy objective function with a triangular shape is used (for other shapes solutions can be obtained in a similar way as shown in Simonovic and Verma (2005)). This fuzzy objective is fully defined geometrically by three corner points $[f_k^p(x), 0]$, $[f_k^m(x), 1]$ and $[f_k^o(x), 0]$, $k = 1, 2, \ldots, K$ (see Figure 5.28).

In order to preserve the triangular membership shape of the solution space it is necessary to make a small change. Instead of maximizing/minimizing these three objectives simultaneously for all K, three optimization problems are solved:

$$Max/Min[f_k^m(x) - f_k^p(x)], \quad k = 1, 2, \ldots, K \quad (5.74)$$
$$Min/Max[f_k^m(x) - f_k^p(x)], \quad k = 1, 2, \ldots, K \quad (5.75)$$
$$Min/Max[f_k^o(x) - f_k^m(x)], \quad k = 1, 2, \ldots, K \quad (5.76)$$

The first and the last objectives measure the relative distance from $f_k^m(x)$ – the second objective. The solution of three new problems written for all K original objectives also supports the previous observation of shifting the triangular distribution to the left.

$$Max/Min \ F_k = [f_k^m(x) - f_k^p(x)], \quad k = 1, 2, \ldots, K$$
subject to
$$x \in X = \{x \mid g_i(x)(\leq, =, \geq)0\}, \quad i = 1, 2, \ldots, m \quad (5.77)$$

$$Min/Max \ F_k = f_k^m(x), \quad k = 1, 2, \ldots, K$$
subject to
$$x \in X = \{x \mid g_i(x)(\leq, =, \geq)0\}, \quad i = 1, 2, \ldots, m \quad (5.78)$$

$$Max/Min \ F_k = [f_k^o(x) - f_k^m(x)], \quad k = 1, 2, \ldots, K$$
subject to
$$x \in X = \{x \mid g_i(x)(\leq, =, \geq)0\}, \quad i = 1, 2, \ldots, m \quad (5.79)$$

So, the original fuzzy multi-objective problem (5.72) is replaced with the three new multi-objective problems (5.77), (5.78), and (5.79).

To solve the newly formulated fuzzy multi-objective problem the Zimmermann (1996) approach with positive and negative ideal solutions is used as follows. First, the positive ideal solution (PIS) is found for all the objectives. It is obtained by maximizing/minimizing the objective for the maximization/minimization problem.

$$F_k^{PIS} = Max/Min[f_k^m(x) - f_k^p(x)], \quad k = 1, 2, \ldots, K \quad (5.80)$$

$$F_k^{PIS} = Min/Max f_k^m(x), \qquad\qquad k = 1, 2, \ldots, K \quad (5.81)$$

$$F_k^{PIS} = Max/Min[f_k^o(x) - f_k^m(x)], \quad k = 1, 2, \ldots, K \quad (5.82)$$

Then, the negative ideal solution (NIS) is found for all the objectives. It is obtained by minimizing/maximizing the objective for the maximization/minimization problem.

$$F_k^{NIS} = Min/Max[f_k^m(x) - f_k^F(x)], \quad k = 1, 2, \ldots, K \quad (5.83)$$

$$F_k^{NIS} = Max/Min f_k^m, \qquad\qquad k = 1, 2, \ldots, K \quad (5.84)$$

$$F_k^{NIS} = Min/Max[f_k^o(x) - f_k^m(x)], \qquad k = 1, 2, \ldots, K \quad (5.85)$$

The best decisions are those that have the shortest distance from PIS and the largest distance from NIS, provide as-much-as-possible gain and avoid as-much-as-possible risk.

A multi-objective problem is reduced to a single-objective problem by the introduction of linear membership functions for each of the objectives:

$$\mu_{max} F_k(x)$$
$$= \begin{cases} 1 & if F_k(x) \geq F_k^{PIS} \\ 1 - \dfrac{F_k^{PIS}(x) - F_k(x)}{F_k^{PIS}(x) - F_k^{NIS}(x)} & if F_k^{PIS}(x) \leq F_k(x) \leq F_k^{NIS}(x) \\ 0 & if F_k(x) \leq F_k^{NIS}(x) \end{cases}$$
$$(5.86)$$

$$\mu_{min} F_k(x)$$
$$= \begin{cases} 1 & if F_k(x) \leq F_k^{PIS} \\ 1 - \dfrac{F_k(x) - F_k^{PIS}(x)}{F_k^{NIS}(x) - F_k^{PIS}(x)} & if F_k^{PIS}(x) \leq F_k(x) \leq F_k^{NIS}(x) \\ 0 & if F_k(x) \geq F_k^{NIS}(x) \end{cases}$$
$$(5.87)$$

Using Zimmermann's (1996) approach to solve the problem, the decision x^* is obtained by solving:

$$\mu_D(x^*) = max\{min \, \mu_k(x)\} \quad \forall_x \in X \quad (5.88)$$

Finally, let us introduce $\lambda = min \, \mu_k(x)$ to obtain the following equivalent model:

$$Max/Min \quad \lambda$$
$$subject \ to$$
$$\mu_k(x) \geq \lambda$$
$$x \in X = \{x | g_1(x)(\leq, =, \geq)0\}, \quad i = 1, 2, \ldots, m \quad (5.89)$$

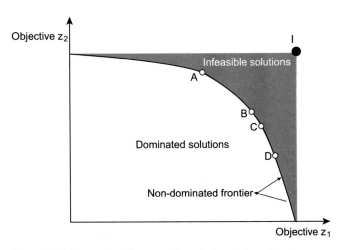

Figure 5.29 Illustration of compromise solutions in two-objective space.

where λ takes a value between 0 and 1 and represents the satisfaction level. Optimization problem (5.89) can be solved using software such as LINPRO. Multiple solutions of the problem provide a set of non-dominated solutions within the fuzzy interval. They are subject to the decision of the decision-maker in order to isolate one non-dominated solution that will be acceptable for the implementation. So, the subjective preferences of the decision-maker will play a role in the final choice of solution for the implementation.

FUZZY COMPROMISE PROGRAMMING

For compromise programming to address the vagueness in the flood risk decision-maker's value system and objective value uncertainty, a general fuzzy approach may be appropriate. Simply changing all inputs from crisp to fuzzy produces a definition for fuzzy compromise programming (FCP) analogous to the crisp original. Figure 5.29 shows a two-objective crisp problem. The solution for which both objectives (z_1, z_2) are maximized is point I (z_1^*, z_2^*) where z_i^* is the solution obtained by maximizing the objective i. It is clear that the solution I (named the ideal point) belongs to the set of infeasible solutions. Let us consider a discrete case with four solutions available as a non-dominated set: A, B, C, and D. The solutions identified as being closest to the ideal point (according to some measure of distance) are called *compromise solutions*, and constitute the *compromise set*. If we use a geometric distance, the set of compromise solutions may include a subset of the non-dominated set A and B. The fuzzy modification of multi-objective problem in Figure 5.29 can no longer be considered a single point for the ideal solution, and each alternative now occupies a small region to various degrees. Measurements of distances between the fuzzy ideal and the fuzzy performance of alternatives can no longer be given a single value, because many distances are at least somewhat valid. Choosing the shortest distance to the ideal is no longer a straightforward ordering of distance metrics, because of overlaps and varying degrees of possibility. The fuzzy multi-objective problem, however, contains a great amount of additional information about the consequences of a decision compared with the non-fuzzy counterpart.

A fuzzy distance metric possesses a valid range of values, each with a characteristic degree of possibility or membership, such that all possible values are a positive distance from the ideal solution (which also becomes fuzzy). Fuzzy inputs include the vagueness of criteria weights, $\tilde{\alpha}_i$, vagueness of both positive, $\tilde{z}_i^*$, and negative ideals, $\tilde{z}_i^{**}$, and vagueness in the appropriate distance metric exponent, $\tilde{p}$. Of course, if any of the inputs are known with certainty, then $\tilde{L}$ becomes less fuzzy:

$$\tilde{L}_p(x) = \left[\sum_{i=1}^{r} \tilde{\alpha}_i^{\tilde{p}} \left(\frac{\tilde{z}_i^* - \tilde{z}_i(x)}{\tilde{z}_i^* - \tilde{z}_i^{**}} \right)^{\tilde{p}} \right]^{1/\tilde{p}} \qquad (5.90)$$

The process of generating fuzzy sets for input is not trivial (Bender and Simonovic, 2000; Simonovic, 2009). Certainly, arbitrary assignment is simple and may cover the range of possibility, but it is possible to encode a lot of information and knowledge in a fuzzy set. The process of generating an appropriate fuzzy set, accommodating available data, heuristic knowledge, or conflicting opinions, should be capable of preserving and presenting information accurately both in detail and in general form. This topic is addressed in Section 5.2.5. Selection of an appropriate technique for fuzzy set generation is specific to the type of problem being addressed, the availability of different types of information, and the presence of different decision-makers.

In assuming fuzzy set membership functions for the various inputs to a distance metric calculation shown in Equation (5.90), a decision-maker must make a number of assumptions. Normal fuzzy sets are considered. They acknowledge that there is at least one completely valid value, analogous to the expected value case for probabilistic experiments. In circumstances where at least one modal point cannot be found, it is usually better to assign multiple modal points than to assign low membership values across the range of possible values (the universe of discourse), partly for the sake of interpreting evaluations. Multi-modal fuzzy sets may consist of multiple modal points or a continuous range of modes. The choice of boundaries for the universe of discourse also makes assumptions about available knowledge on the universe of discourse. Boundary and modal point selection, along with the shape of the fuzzy sets, define a degree of fuzziness that hopefully represents the characteristic fuzziness of real-world behavior.

In dealing with flood risk management problems under climate change, fuzzy sets describe a degree of possibility for valid values of a parameter. They do not possess properties such as conditional probabilities for stochastic applications, at least for simple applications. This is acceptable because typical sensitivity analyses explore all combinations of values anyway, and there is usually not enough information to form conditional properties. In an advanced fuzzy application, there is no reason not to provide conditional fuzzy sets.

It is possible, and may be desirable, to fuzzify all parameters in multi-objective problems formulated with a framework. Various shapes of input fuzzy sets can be used for criteria values, weights, positive ideals, and negative ideals for compromise programming. The FUZZYCOMPRO computer program provided on the website and discussed later in this chapter provides for the use of a wide range of linear and non-linear input fuzzy shapes.

Fuzzification of criteria/objective values is probably the most obvious use of fuzzy sets in flood risk decision-making problems. To capture the subtleties of relative performance of different alternatives from the perspective of a decision-maker, there may not be enough choices. Likewise, if a large number of choices are provided, the appreciation of subjectivity in linguistic terms disappears. Fuzzy sets are able to capture many qualities of relative differences in the perceived values of objectives among alternatives. Quantitative objectives present some slightly different properties from qualitative objectives. It can be assumed that quantitative objectives are measured in some way, either directly or through calculations based on some model. They may have stochastic properties that describe the probability of occurrence for values, based on future uncertainties for example. They also have some degree of imprecision in their measurement or modeling. In this way, quantitative objectives may have both stochastic and fuzzy properties. To prevent the complication of many flood risk decision-making problems, various uncertainties may be adequately represented with fuzzy sets. In general, the application of quantitative objectives within a fuzzy approach may assume that quantitative objectives are less fuzzy than the qualitative.

Weights associated with different objectives are an important aspect of the compromise programming method. Their assignment is completely subjective, usually with a rating on an interval scale. As a subjective value, weights may be more accurately represented by fuzzy sets. Generating these fuzzy sets is also a subjective process. It may be difficult to get honest opinions about the degree of fuzziness of a decision. It might actually be more straightforward to generate fuzzy sets for weights when multiple decision-makers are involved. Then, at least, voting methods and other techniques are available for producing a composite collective opinion. Regardless of this, more information can be provided about valid weights from fuzzy sets than from crisp weights.

The incorporation of vagueness into the ideal solution is an element that impacts the rankings of alternatives. When we incorporate fuzziness into the location of the ideal solution (both positive and negative), the valid area for the ideal point – in objective space – affects the measurement of distance to the alternatives. For example, if flood damage reduction is a criterion, then what is the ideal amount of damage reduction? Typically, (crisp) compromise programming applications use the largest objective value among the alternatives as the ideal value. This arbitrary placement is probably not valid, and also affects the relative distances to the overall ideal.

The distance metric exponent, p, is likely to be the most imprecise or vague element of the distance metric (5.90) calculation. There is no single acceptable value of p for almost any type of problem, and it can easily be misunderstood. Also, it is not related to problem information in any way except that it provides parametric control over the interpretation of distance. Fuzzification of the distance metric exponent, $\tilde{p}$, can take many forms. Using, for example, a triangular membership shape, the principal formulation of FCP provides a replacement for setting the range of values for parameter $1 \leq p \leq \infty$. The shape of the triangle and the values of the three parameters that define it allow for the solution of the FCP problem using only one solution of (5.90).

Traditional (non-fuzzy) compromise programming distance metrics measure the distance from an ideal point (Figure 5.29), where the ideal alternative would result in a distance metric, $L = 0$. In FCP, the distance is fuzzy, such that it represents all of the possible valid evaluations, indicated by the degree of possibility or membership value (Bender and Simonovic, 2000). Alternatives that tend to be closer to the ideal may be selected. This fuzzified distance metric is analogous to a sensitivity analysis for the non-fuzzy compromise programming case.

As an attempt to standardize a procedure for judging which $\tilde{L}$ is best among a set of alternatives, desirable properties can be defined. The most important properties to consider are: (i) possibility values tend to be close to the ideal, $x = 0$, distance; (ii) possibility values have a relatively small degree of fuzziness; (iii) modal values are close to the ideal; and (iv) possibility values tend to be far from poor solutions.

An experienced person may be able to visually distinguish the relative acceptability of alternatives, but in cases with many alternatives where each $\tilde{L}$ displays similar characteristics, it may be impractical or even undesirable to make a selection visually. A method for ranking alternatives, based on comparisons of $\tilde{L}$, will make summary ranking information more accessible, automating many of the visual interpretations and creating reproducible results. There is a theoretical presentation of various methods for comparing fuzzy sets in Section 5.2.4. We introduced the weighted center of gravity (Equation (5.22)) method introduced by Bender and Simonovic (2000) and the fuzzy acceptability measure for use with FCP.

FUZZY MULTI-OBJECTIVE ANALYSIS, CLIMATE CHANGE, AND FLOOD RISK MANAGEMENT

Very limited literature is available on the use of fuzzy multi-objective analysis in climate change flood studies. The potential of the approach has not yet been fully investigated, creating plenty of opportunities for future work in this area. The following are some applications that can be found in related research areas.

Bender and Simonovic (2000) developed the FCP technique. The demonstration of the utility of the new fuzzy multi-objective technique was done using the Tisza River basin in Hungary. The example is taken to select the most desirable water management

system alternative, either as a best compromise or as a robust choice. The example redefines the problem in fuzzy terms to demonstrate the added value of adopting a fuzzy compromise approach.

Raju and Kumar (2001) used FCP for the selection of the best compromise alternative plan in irrigation development. Their case study was the Sri Ram Sagar Project in Andra Pradesh, India. The study deals with three conflicting objectives, namely, net benefits, crop production, and labor employment. Their analysis combines multi-objective optimization, cluster analysis, and multi-criterion decision-making. Compromise programming and multi-criterion Q analysis are employed to select the best compromise alternative.

Prodanovic and Simonovic (2002) used FCP to evaluate discrete alternatives in the context of water resources decision-making. All uncertain variables (subjective and objective) are modeled by way of fuzzy sets. Fuzzy set ranking methods are employed to compare, rank, and/or sort the fuzzy output produced by FCP. Literature suggests that many ranking methods are available; however, not all may be appropriate for water resources decision-making. The objective of this paper is to compare fuzzy set ranking methods that can be implemented with FCP. Nine such ranking methods are considered in this research, of which two are fully tested using case studies from the literature (Bender and Simonovic, 2000). It was found that for all case studies the ranking of alternatives was not very sensitive to changes in the expert's degree of risk acceptance, or to the changes in the ranking methods themselves.

Prodanovic and Simonovic (2003) used a multi-criterion FCP in combination with a methodology known as "group decision-making under fuzziness" to come up with a new technique that supports decision-making with multiple objectives and multiple participants (or experts). All criteria (qualitative and quantitative) are modeled by way of fuzzy sets, utilizing the fact that criteria values in most water resources problems are vague, imprecise and/or ill defined. The involvement of multiple experts in the decision process is achieved by incorporating each participant's perception of objective weights, best and worst objective values, relative degrees of risk acceptance, as well as other parameters into the problem. The proposed methodology is illustrated with a case study taken from the literature (Bender and Simonovic, 2000), combined with the input of four expert individuals with diverse backgrounds. After processing the input from the experts, a group compromise decision is formulated.

Akter and Simonovic (2005) dealt with various uncertainties inherent in modeling a reservoir operation problem. Two of these are addressed in this study: uncertainty involved in the expression of reservoir penalty functions, and uncertainty in determining the target release value. Fuzzy set theory is used to model these uncertainties where the preferences of the decision-maker for the fuzzified parameters are expressed as membership functions. Non-linear penalty functions are used to determine the penalties due to deviations from targets. The optimization is performed using a GA with the objectives of minimizing the total penalty and maximizing the level of satisfaction of the decision-maker with fuzzified input parameters. The proposed formulation has been applied to the problem of finding the optimal release and storage values, taking Green Reservoir in Kentucky as a case study. The approach offers more flexibility to reservoir decision-making by demonstrating an efficient way to represent subjective uncertainties, and to deal with non-commensurate objectives under a fuzzy multi-objective environment.

Akter *et al.* (2004) researched the Red River basin in Canada, which faces periodic flooding. The flood management decision-making problems in the basin are often associated with multiple objectives and multiple stakeholders. To enable a more effective and acceptable decision outcome, it is required that more participation is ensured in the decision-making process. The challenge remains of how to obtain and use the diversified opinions of a large number of stakeholders where uncertainty plays a major role. In response to this challenge, a methodology has been proposed to capture and aggregate the views of multiple stakeholders using fuzzy set theory and fuzzy logic. Three possible different response types: scale (crisp), linguistic (fuzzy), and conditional (fuzzy) are analyzed to obtain the aggregated input by using the fuzzy expected value. The methodology has been tested for flood management in the Red River basin using a generic case study. While the results show successful application of the methodology, they also show significant differences in preferences of the stakeholders as a function of their location in the basin. Thus, the paper provides alternative ways for collecting and aggregating the input of multiple stakeholders to assist the fuzzy multi-objective flood management decision-making process.

Simonovic and Nirupama (2005) looked into flood risk decision-making as a spatial problem. Topographical features of the region, location of flood risk management infrastructure, interaction between the water resources system and other social and ecological systems, and impact of different water resources regulation measures are all variables with considerable spatial variability. In this paper a new technique named spatial fuzzy compromise programming (SFCP) is developed to enhance our ability to address different uncertainties in spatial water resources decision-making. A general multi-objective FCP technique, when made spatially distributed, proved to be a powerful and flexible addition to the list of techniques available for flood risk decision-making where multiple objectives are used to evaluate multiple alternatives. All uncertain variables (subjective and objective) are modeled by way of fuzzy sets. Through a case study of the Red River floodplain near the City of St. Adolphe in Manitoba, Canada, it has been illustrated that the new technique provides measurable improvement in management of floods.

Fu (2008) presented a fuzzy optimization method based on the concept of ideal and anti-ideal points to solve multi-objective decision-making problems under fuzzy environments. The quantitative objectives for each alternative are represented by triangular fuzzy numbers, and their qualitative counterparts and the weight of each objective are described using linguistic terms, which are also be expressed as triangular fuzzy numbers in the proposed method. With the definition of fuzzy ideal and anti-ideal weight distances, an objective function is constructed to derive the optimal evaluation for each alternative denoted by a fuzzy membership degree. The ranking of alternatives and the best one can be determined directly on the basis of the fuzzy membership degrees without the need to compare fuzzy numbers. The evaluation process is simple and easy to use in practice. A case study of reservoir flood control operation is given to demonstrate the proposed method's effectiveness.

Schumann and Nijssen (2011) investigated the effectiveness of technical flood control measures that strongly depend on multiple characteristics of floods. Copulas can be applied for multivariate statistical descriptions of flood scenarios. However, the parameterization of these multi-variate statistical models involves many uncertainties that are handled as imprecise probabilities. These imprecise probabilities are described by fuzzy numbers and integrated in a multi-objective decision-making framework. This work demonstrates the application of the fuzzy multi-objective framework for flood retention planning in a river basin. A fuzzified version of the well-known analytic hierarchy process method (AHP) is used. Original AHP is based on two groups of subjective evaluations: comparison of the objectives according to their relative importance, and pairwise comparison of the outcomes of alternatives for each objective. The uncertainties in these comparisons are addressed using a fuzzy scale instead of the AHP rating scale.

The following example (modified after Bender and Simonovic, 2000) illustrates the application of FCP.

Example 6

Compare five major flood risk management alternatives for the Tisza River basin according to 12 criteria/objectives. Table 5.17 lists the crisp form of the available input information. The last eight criteria in the table are subjective, and have linguistic evaluations assigned to them. The criteria for water quality, recreation, flood protection, manpower impact, environmental architecture, and development possibility are all considered on a scale with five linguistic options {excellent, very good, good, fair, bad}. The last two criteria are judged by different linguistic scales. First, international cooperation has a subjective scale {very easy, easy, fairly difficult, difficult}. Finally, the sensitivity criterion also uses a subjective scale with four categories (although one of them is not chosen) {very sensitive, sensitive, fairly sensitive, not sensitive}.

The data as presented do not address uncertainty issues involved with the decision-making problem. The weighting of relative importance is also an issue of uncertainty. Criteria weights are provided from the set of {1, 2}. All criteria were weighted as 2 except land and forest use, manpower impact, development possibility, and sensitivity, which were given a weight of 1.

Table 5.17 *Crisp data for comparison of flood risk management alternatives*

Criteria	Alternatives				
	I	II	III	IV	V
1. Total annual cost	99.6	85.7	101.1	95.1	101.8
2. Probability of water shortage	4	19	50	50	50
3. Energy (reuse factor)	0.7	0.5	0.01	0.1	0.01
4. Land and forest use (1000 ha)	90	80	80	60	70
5. Water quality	very good	good	bad	very good	fair
6. Recreation	very good	good	fair	bad	bad
7. Flood protection %	good	excellent	fair	excellent	bad
8. Manpower impact	very good	very good	good	fair	fair
9. Environmental architecture	very good	good	bad	good	fair
10. Development possibility	very good	good	fair	bad	fair
11. International cooperation	very easy	easy	fairly difficult	difficult	fairly difficult
12. Sensitivity	not sensitive	not sensitive	very sensitive	sensitive	very sensitive

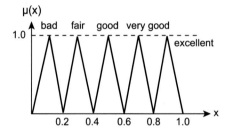

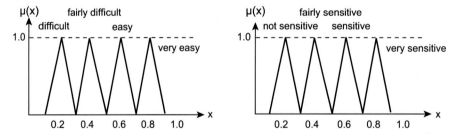

Figure 5.30 Fuzzy subjective criteria for the Tisza River flood risk management problem.

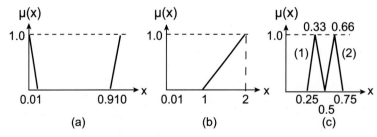

Figure 5.31 Fuzzy input for the Tisza River flood risk management problem.

Bender and Simonovic (2000) introduced the fuzzy definitions shown in Figure 5.30 for linguistic terms used in assessing subjective criteria. Quantitative criteria are also fuzzified, but generally are less fuzzy. Other fuzzy inputs include the expected ranges of criterion values (Figure 5.31a) and the form of distance metric or degree of compensation, $\tilde{p}$, among criteria for different alternatives (Figure 5.31b). Criterion weights, $\tilde{\alpha}_i$, are fuzzified on a range of [0,1] (Figure 5.31c).

Table 5.18 *Rankings of flood risk management alternatives*

Rank	Alternative	WCoG ($q = 1$)
2	I	1.045
1	II	0.985
4	IV	2.231
3	III	1.841
5	V	2.241

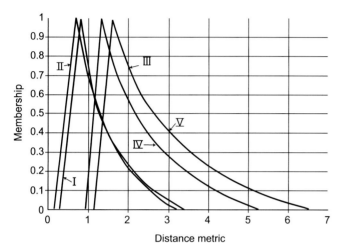

Figure 5.32 Distance metrics for the flood risk management problem.

Solution

Assuming the fuzzy definition for the distance metric exponent ($\tilde{p}$), and knowing the form of criterion values and weights to be triangular, the fuzzy distance metrics ($\tilde{L}_i$) are calculated using Equation (5.90). The final results are shown in Figure 5.32. As can be observed, the fuzzy distance metrics have near linearity below the mode, and a somewhat quadratic polynomial curvature above the mode. Although the degree of fuzziness (range of valid distances from the ideal solution) is similar for all five alternatives, some of the alternatives are clearly inferior.

Ranking these alternatives is reasonably straightforward because of the simplicity of the shapes, and similarity in degree of fuzziness. We use the weighted centroid measure (Equation (5.22) in Section 5.2.4). The results are shown in Table 5.18. Rankings are insensitive to changes in levels of risk aversion, as is expected from visual inspection. The resulting ranks confirm that alternatives I and II dominate III, IV, and V.

You can use the FUZZYCOMPRO program provided on the website to confirm the solution in Table 5.18. Example 5 data are in the folder Fuzzy tools, sub-folder Multi-objective analysis, sub-sub-folder FUZZYCOMPRO, sub-sub-sub folder Examples, file Example1.fcompro.

FUZZYCOMPRO COMPUTER PROGRAM

The accompanying website includes the FUZZYCOMPRO software. The folder Fuzzy tools, sub-folder Multi-objective analysis, sub-sub-folder FUZZYCOMPRO contains two sub-sub-sub-folders, FUZZYCOMPRO, and Examples. The readme file contains instructions for installation of the FuzzyCompro software, and a detailed tutorial for its use is a part of the FUZZYCOMPRO Help menu.

FUZZYCOMPRO facilitates the multi-objective analysis of discrete problems using the fuzzy compromise programming technique. The approach allows various sources of uncertainty, and is intended to provide a flexible form of group decision support. Fuzzy compromise programming allows a family of possible conditions to be reviewed, and supports group decisions through fuzzy sets designed to reflect collective opinions and conflicting judgments. The transformation of a distance metric to a fuzzy set is accomplished by changing all inputs from crisp to fuzzy and applying the fuzzy extension principle. The measure of distance between an ideal solution and the perceived performance of an alternative is used for comparison among alternatives. An operational definition (5.90) is used for computation of the distance metrics. Choosing the shortest distance to the ideal is no longer a straightforward ordering of distance metrics, because of overlaps and varying degrees of possibility. Ranking of alternatives is accomplished with fuzzy ranking measures designed to illustrate the effect of risk tolerance differences among decision-makers. FUZZYCOMPRO uses a centroid ranking measure (Equation (5.22) in Section 5.2.4) to rank alternatives according to the fuzzy distance metric value. It can be used to identify the best compromise alternative under various uncertainties.

FUZZY MULTI-OBJECTIVE ANALYSIS OF FLOOD MANAGEMENT ALTERNATIVES IN THE RED RIVER BASIN, MANITOBA, CANADA

This example is based on the collaborative research performed by the University of Western Ontario and the communities in the Red River basin. The requirement of the study was set to develop a methodology for a multi-objective, multi-participant decision analysis, which would be able to: (i) evaluate potential flood management alternatives based on multiple objectives under uncertainty; (ii) accommodate the high diversity and uncertainty inherent in human preferences; and (iii) handle information collected from a large number of stakeholders (IJC, 2000; Akter *et al.*, 2004; Simonovic, 2004, 2009, 2011; Akter and Simonovic, 2005). One of the major problems at the planning stage in the Red River basin is the complex, large-scale problem of ranking potential flood risk management alternatives. Decision-making uncertainty addressed through the development of the presented methodology includes climate change. For illustration in this text I will focus only on that part of the overall work dealing with the application of fuzzy multi-objective compromise programming in ranking flood risk management alternatives.

Study area Situated in the geographic center of North America, the Red River originates in Minnesota and flows north (one of eight rivers in North America that flow north). The Red River basin covers 116,500 km^2 (exclusive of the Assiniboine River and its tributary, the Souris) of which nearly 103,600 km^2 are in the USA (Figure 5.33). The basin is remarkably flat. The elevation at Wahpeton, North Dakota, is 287 m above sea level. At Lake Winnipeg, the elevation is 218 m. The basin is about 100 km across at its widest. The Red River floodplain has natural levees at points both on the main stem and on some tributaries. Because of the flat terrain, when the river overflows these levees, the water can spread out over enormous distances without stopping or pooling, exacerbating flood conditions. During major floods, the entire valley becomes a floodplain. The type of soil, largely clay with characteristic low absorptive capacity, also contributes to flooding. Water tends to sit on the surface for extended periods of time. In general, the climate of southeastern Manitoba is classified as sub-humid to humid continental with resultant extreme temperature variations. Annually, most of the precipitation received is in the summer rather than the winter. Approximately three-quarters of the 50 cm of annual precipitation occur from April to September. Consequently, in most years spring melt is well managed by the capacities of the Red River and its tributaries. However, periodically weather conditions exist that instead promote widespread flooding through the valley. The most troublesome conditions are as follows: (i) heavy precipitation in the autumn, (ii) hard and deep frost prior to snowfall, (iii) substantial snowfall, (iv) late and sudden spring thaw, and (v) wet snow/rain during spring breakup of ice. Changing climatic conditions are another source of uncertainty that is expected to affect the frequency and magnitude of flooding in the basin.

In Manitoba, almost 90% of the residents of the Red River/Assiniboine basin live in urban centers. Metropolitan Winnipeg contains approximately 700,000 people, and another 50,000 live along the Red River north and south of the city. The Red River valley is a highly productive agricultural area serving local, regional, and international food needs. There has been an extensive and expanding drainage system instituted in the basin to help agricultural production by increasing arable land.

The basin floods regularly. Early records show several major floods in the 1800s, the most notable being those of 1826, 1852, and 1861. In the last century, major floods occurred in 1950, 1966, 1979, 1996, and 1997. The Red River basin has 25 sub-basins, which have different topography, soils, and drainage that result in different responses during flood conditions. One common characteristic is overland flow during times of heavy runoff. Water overflows small streams and spreads overland, returning to those streams or other watercourses downstream. Existing monitoring and forecasting systems do not track these flows well, leading to unanticipated flooding. The earliest recorded flood in the basin was in 1826, although anecdotal evidence refers to larger floods in the late 1700s. The flood of 1826 is the largest flood on record.

A pivotal event in the Red River flood history was the 1950 flood, which was classified a great Canadian natural disaster based on the number of people evacuated and affected by the flood. A very cold winter and heavy snowpack in the USA, combined with heavy rain during runoff, were the primary causes. All towns within the flooded area in the upper valley had to evacuate. More than 10,000 homes were flooded in Winnipeg and 100,000 people evacuated. A plan to evacuate all 350,000 people in Winnipeg was prepared, although luckily it did not have to be used.

Most of the flood management planning in Manitoba was initiated after the 1950 flood. This flood was the turning point in the history of flooding and flood control in Manitoba's portion of the Red River basin. Construction of elevated boulevards (dykes) within the City of Winnipeg and associated pumping stations was initiated in 1950. The current flood control works for the Red River valley consist of the Red River Floodway, the Portage diversion and Shellmouth Dam on the Assiniboine River, the primary dyking system within the City of Winnipeg, and community dyking in the Red River valley (Simonovic, 2004).

The large 1979 flood was primarily the result of a rapid thaw and wet spring. Half of the upper valley was evacuated. Homes just south of the flood control system were very hard hit yet again. Winnipeg was largely spared. The 1997 flood was a true test of the flood control system throughout the valley. Extreme snowpack (98th percentile), extreme cold north and south of the border, high topsoil moisture, unfavorable time of runoff, and an April blizzard combined to cause the inundation. The peak discharge at Emerson, Manitoba (at the border), was 3,740 m^3/s; in the 1950 flood it was 2,670 m^3/s. At the Floodway Inlet (just south of Winnipeg) the peak was 4,587 m^3/s compared to 3,958 m^3/s in 1950. Floodwaters at the inlet actually crested 0.45 to 0.60 m higher than the forecast range pronounced; unexpected overland flooding was a major contributor to the error in forecasting, and ultimately increased damage. The 1997 flood was the highest recorded in the twentieth century. An estimated 1840 km^2 of land was flooded as the Red River rose 12 m above winter levels. Structural measures such as the dyking systems and the Red River Floodway are known to have prevented enormous losses, as did emergency dyking. Estimates of those prevented damages run as high as C$6 billion. Eight valley towns with ring dykes remained dry; however, one town, one urban fringe community, and numerous farm properties were flooded with subsequent damage.

Flood risk management problem The problem addressed in this section deals with large-scale ranking of potential flood risk management alternatives. During the evaluation of alternatives, it is proposed to consider multiple objectives/criteria that may be

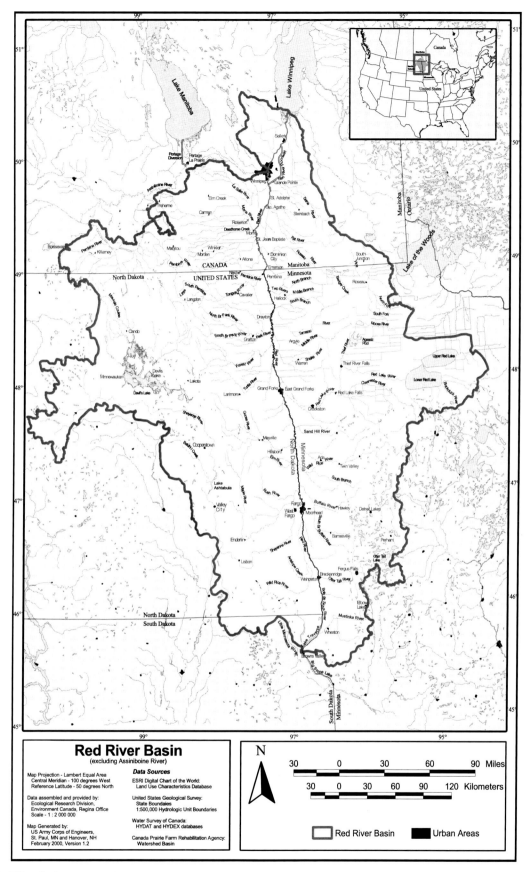

Figure 5.33 Red River basin.

both quantitative and qualitative, various uncertainties associated with the flood risk decision-making, and the views of numerous stakeholders.

At present the Government of Manitoba, Canada, is responsible for making decisions about flood management measures (Simonovic, 2004). The decision-making process involves consulting different organizations for their technical input. The concerns of the general public about the alternatives are gathered through public hearings and workshops. Economic analysis plays an important role in formulating plans for reducing flood damage and making operational decisions during the emergency. One of the main limitations of the current flood risk management is its high emphasis on the economic criterion. Very minor attention is given to the various uncertainties that characterize the problem.

The general public have shown increasing concern about decisions on flood risk management measures. During the 1997 flood some stakeholders in the basin, particularly the floodplain residents, felt they did not have adequate involvement in flood risk management decision-making. They expressed particular dissatisfaction about evacuation decisions during the emergency management, and about compensation decisions during post-flood recovery (IJC, 2000).

Flooding and its impact on communities are inevitable in the Red River basin. Dealing with such aspects as communicating with multiple stakeholders to decide what flood risk management measure needs to be adopted and where, presents a level of complexity almost ignored in traditional decision-making techniques. Any flood risk management measure implies varied effects on the different communities along the Red River. A common complaint against the government is its lack of consideration towards the rural communities around Winnipeg. Rural community residents claim that the government's misuse of the flood risk management structures has intensified the backwater effect, leaving their communities to deal with the increased water levels outside the Winnipeg perimeter – an attitude they refer to as "perimeter thinking." So it is clear that different stakeholders have different objectives to be satisfied by flood risk management activities. Implementing a multi-objective, multi-stakeholder decision-making tool definitely provides a means to incorporate the views and concerns of all communities who are affected by future developments in flood risk management and uncertainty arising from climate change.

Uncertain risk perception expressed through the preferences of decision-makers is highly affected by the potential impacts that climate change can bring to the basin by changing the frequency and magnitude of future floods. Fuzzy context is selected in order to collect, integrate and use in multi-objective decision-making various inputs that contain subjective and objective uncertainties.

Description of the method and data Three generic alternatives for improved flood risk management are considered. A flood risk management pay-off (decision) matrix with relevant criteria/objectives and generic alternatives is developed. Each of the criteria categories includes quantitative assessment of potential climate change impacts and subjective perception of climate change impacts on flood risk management.

The three generic options considered are structural alternatives, non-structural alternatives, and a combination of both. The selection of criteria against which the alternatives are ranked is one of the most difficult but important tasks of any multi-objective decision analysis. Here the selection is based mainly on prior studies of the Red River flooding (IJC, 2000) and climate change-related issues raised by the residents of the basin after the large flood of 1997. In addition, studies of the Red River flood of 1997 and numerous more recent interviews with stakeholders made it clear that it was of prime importance to consider climate change impacts if a flood risk management policy in the Red River basin was to be implemented successfully. A special emphasis is placed on the climate-related social impacts.

A detailed survey was conducted in the basin to collect information on these social criteria (Akter *et al.*, 2004). It used a survey questionnaire to allow stakeholders to express their views in an easy way. Therefore, the remainder of this case study focuses only on the application of the fuzzy multi-objective analysis using the three generic alternatives and real data on the social objectives. In addition to social objectives, the comprehensive multi-objective analysis includes two more types of criteria: economic (cost, damage, benefits, etc.) and environmental (chemical contamination, inter-basin transfer of alien invasive species, and protection and enhancement of the floodplain environment). They are straightforward to quantify and are not presented here.

The detailed survey used to define the social objectives used a set of questions rather than a single preference value:

(1) What is the level of opportunity provided by each alternative during the planning stage of flood protection?

(2) What is the level of opportunity provided by each alternative to address the main impacts of climate change during the time of flooding?

(3) To what degree does each alternative induce a sense of complacency to rely heavily on the existing predictions of flooding impacts (no consideration of climate change)?

(4) What is the level of climate change knowledge that you would be able to provide for each alternative?

(5) How much training is required for each alternative to be actively involved in flood management activities under changing climatic conditions?

(6) What is your level of willingness to participate in such activity for your personal estate?

(7) What is your level of willingness to participate in such activity for your local community?

(8) What is your level of willingness to participate in such activity for the city of Winnipeg?

(9) What is the role of leadership in the successful execution and implementation of each alternative?

Table 5.19 *Interview responses for community awareness objective and structural alternative*

Participant No.	Region	Flood experience	Community awareness: structural									
			Question									
			1	2	3	4	5	6	7	8	9	10
1	Winnipeg	Technical	1	4	4	4	5	4	4	3	5	4
2	Winnipeg	Technical	1	2	5	5	5	5	5	5	5	3
3	Winnipeg	Technical	2	1	4	1	5	4	4	4	5	5
4	Winnipeg	Technical	4	5	2	5	4	4	4	5	5	5
5	Winnipeg	Technical	3	2	4	4	4	5	5	5	4	4
6	Winnipeg	Technical	5	2	4	4	5	5	5	5	5	4
7	Winnipeg	Technical	5	3	4	2	2	4	4	5	5	5
8	Winnipeg	Non-Tech	3	4	4	1	5	5	5	5	5	5
9	Winnipeg	Technical	3	2	4	4	4	3	4	5	5	4
10	Winnipeg	Non-Tech	4	3	4	1	4	5	4	4	4	4
11	MB Conv	Technical	3	3	4	4	4	5	5	5	4	4
12	MB Conv	Technical	3	2	3	4	4	4	4	4	4	4
13	MB Conv	Technical	3	2	4	3	4	5	5	5	4	3
14	MB Conv	Technical	5	5	5	5	5	5	5	5	5	5
15	MB Conv	Technical	2	2	4	4	4	5	4	3	3	3
16	MB Conv	Technical	5	2	5	5	1	5	1	1	5	5
17	MB Conv	Technical	2	1	4	4	4	5	2	5	4	4
18	MB Conv	Technical	5	3	4	4	4	4	4	4	4	3
19	MB Conv	Non-Tech	5	5	5	2	5	5	5	5	5	2
20	MB Conv	Technical	4	4	3	4	4	4	3	3	4	4
21	St. Adol	Non-Tech	3	4	4	3	3	5	5	3	5	5
22	St. Adol	Non-Tech	4	2	5	3	5	5	5	1	5	2
23	St. Adol	Non-Tech	4	2	4	2	4	5	5	5	5	4
24	St. Adol	Non-Tech	5	3	2	4	5	5	3	1	5	5
25	St. Adol	Non-Tech	5	5	3	4	3	5	5	3	5	4
26	St. Adol	Non-Tech	–	–	–	3	1	5	5	4	5	4
27	St. Adol	Non-Tech	4	5	3	3	4	5	4	2	4	4
28	St. Adol	Non-Tech	4	5	2	4	2	5	5	4	5	5
29	St. Adol	Non-Tech	2	1	4	3	4	4	4	1	5	3
30	Morris	Non-Tech	2	3	2	3	3	4	3	1	5	4
31	Morris	Non-Tech	4	4	2	4	3	4	4	4	5	4
32	Selkirk	Non-Tech	4	3	3	4	2	2	4	3	4	3
33	Selkirk	Non-Tech	5	1	1	1	1	5	5	5	5	5
34	Selkirk	Non-Tech	3	4	2	2	3	5	5	5	5	3
35	Selkirk	Non-Tech	4	4	2	2	3	4	4	4	5	4

MB Conv, Manitoba Conservation; St. Adol, St. Adolphe

(10) Rate the alternatives according to the degree to which they promote local leadership and community awareness in addressing climate change impacts.

(11) What is the severity of flood loss under climate change (land, homestead, and business) at a personal level for each alternative?

(12) Rate the degree of impact on personal health each alternative would expose the public to during a flood under changing climatic conditions.

(13) Rate the level of stress induced in the daily lives of the public by each alternative during the planning and preparation for climate change-affected flooding.

(14) Rate the level of stress induced in the daily lives of the public by each alternative during a flood.

(15) What is the level of personal safety provided by each alternative if the conditions change?

(16) Rate the level of control an individual has over the flood protection measures to be implemented.

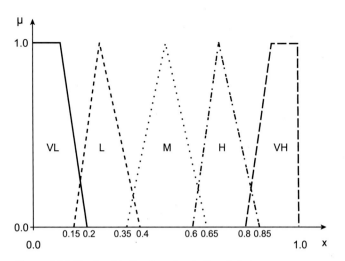

Figure 5.34 Membership functions for the linguistic terms.

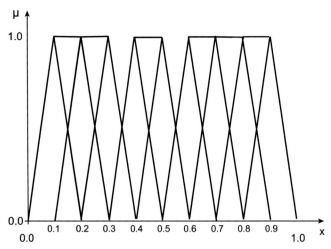

Figure 5.35 Membership functions for fuzzy percentages.

These 16 questions were used to develop two evaluation objectives/criteria. The first ten questions formed the community climate change awareness criterion and the remaining six questions, a personal loss criterion.

Thirty-five respondents were interviewed. As part of the survey, each participant was asked to answer each question using linguistic answers that are processed using fuzzy sets. This response type complicates data processing but allows participants to respond with familiar preferences in the form of words. Linguistic fuzzy sets allow stakeholder preferences to vary in degree of membership, better modeling human views and opinions. The participants could use one of the following five linguistic terms, very low, low, medium, high, and very high, to answer each question (see Table 5.19).

Table 5.19 shows a sample of collected data for the community awareness objective and structural alternative. Note that the following numbering is used for linguistic responses: 1, very low; 2, low; 3, medium; 4, high; and 5, very high. These linguistic terms are converted to a set of membership functions. Figure 5.34 depicts the set of membership functions as used in this study. The construction of membership functions is a very important step in the implementation of the fuzzy approach, and should be based on proper investigation when applied for a real problem.

The x axis of Figure 5.34 represents the universe of discourse from 0 to 1, which is the range of the values assigned to the linguistic terms. For example, the term very low (VL) contains the range from 0 to 0.2. The y axis represents the membership values, which indicate the degree of membership of the range of values assigned to the term very low. It is shown that the range 0–0.1 holds the full membership ($\mu = 1$), and for the range 0.1–0.2, the membership value is linearly decreasing from 1 to 0. For the term low (L), the range is 0.15–0.4, where the membership value increases linearly to 1 for the range 0.15–0.2, and then decreases linearly to 0 for the range 0.2–0.4. Similarly, other

linguistic terms (medium, high, and very high) are also given membership functions as shown in the figure.

The membership functions for fuzzy percentages are also developed for processing linguistic type inputs. Figure 5.35 shows the membership functions for percentage values. All inputs obtained from all the stakeholders are processed (aggregated) using the fuzzy expected value method (Akter *et al.*, 2004; Simonovic, 2009). Table 5.20 summarizes the final results of the fuzzy expected value aggregation.

The three objectives are ranked according to two objectives, based on the aggregated data, using fuzzy compromise programming (Equation (5.90)) and set of weights in Table 5.21. Since the stakeholders were located in different regions of the basin, the weights are obtained in such a way that fuzzy compromise programming can be applied for multiple rankings of three alternatives.

Analyses and results The fuzzy expected values in Table 5.20 were used to rank the three generic alternatives. All questions were considered to carry the same weight. A set of ranking experiments was conducted to evaluate the impact of different stakeholder groups on the final rank of alternatives: (i) experiment 1 – all stakeholders interviewed; (ii) experiment 2 – stakeholders from the city of Winnipeg; (iii) experiment 3 – stakeholders from the Morris area (south of Winnipeg); and (iv) experiment 4 – stakeholders from the Selkirk area (north of Winnipeg).

The fuzzy inputs for Equation (5.90) are: the best, the worst, and the actual values assigned to each objective. The membership functions for the best and the worst values are obtained by using very high as the best and very low as the worst. With all these inputs and using the equation of fuzzy compromise programming (Equation (5.90)), we obtain the fuzzy distance matrices for all three alternatives. The distance matrices obtained for all

Table 5.20 *Aggregated fuzzy expected values*

	Alternative question	Structural FEV	Non-structural FEV	Combined FEV
Objective 1: Community awareness	1	0.650	0.650	0.625
	2	0.517	0.517	0.570
	3	0.700	0.625	0.625
	4	0.650	0.650	0.650
	5	0.700	0.650	0.650
	6	0.825	0.770	0.825
	7	0.770	0.717	0.770
	8	0.700	0.650	0.700
	9	0.825	0.850	0.825
	10	0.717	0.650	0.700
Objective 2: Personal loss	11	0.770	0.700	0.717
	12	0.570	0.650	0.625
	13	0.570	0.625	0.570
	14	0.717	0.717	0.717
	15	0.770	0.650	0.717
	16	0.570	0.570	0.570

Table 5.21 *Collective objective weights from the stakeholders*

	Winnipeg	Morris	Selkirk	All
Obj. 1	0.70	0.90	0.90	0.700
Obj. 2	0.70	0.80	0.80	0.771

four groups (all, Winnipeg, Morris, and Selkirk) are shown in Figure 5.36.

The final results of four ranking experiments with three generic alternatives and two criteria are shown in Table 5.22 (the defuzzified distance metric value with the rank in brackets). Table 5.22 shows the ranking results for the three alternatives using the weighted center of gravity method (Equation (5.22) from Section 5.2.4). The values in the table are the weighted centroid measures of input distance matrices for structural, non-structural, and combination alternatives for four groups: all, Morris, Winnipeg, and Selkirk. The numbers within the parentheses indicate the rank of the alternative based on obtained values, where the smallest distance value is given the highest rank. It is obvious that the final rank varies with the experiment, thereby confirming that the preferences of different stakeholders are captured by the developed methodology.

Conclusions This case study documents an innovative methodology that provides alternative ways to extract and aggregate the inputs from a large number of stakeholders for flood risk management decision-making under climate change. Fuzzy expected value is used as a method to aggregate those inputs and generate the elements of the multi-criterion decision matrix for further analysis. A linguistic type of response for flood risk management is used. The fuzzy compromise programming technique is used to analyze the alternative flood risk management options.

The analyses of flood risk management options in the Red River basin show the applicability of the methodology for a real flood risk management decision-making problem. The stakeholders can now express their concerns regarding climate change-caused flood hazard in an informal way, and that can be incorporated into the multi-objective decision-making model. This methodology helps solve the problem of including a large number of stakeholders in the flood risk decision-making process.

5.5 CONCLUSIONS

As most of the climate change-related decision-making processes take place in situations where the goals, the constraints, and the consequences are not known precisely, it is necessary to include these types of uncertainty in the decision-making methodology. Fuzzy set and fuzzy logic techniques have been used successfully to represent the imprecise and vague information in many fields, and so are considered an effective way to represent uncertainties.

Three sets of tools that use fuzzy sets are presented: fuzzy simulation, fuzzy optimization, and fuzzy multi-objective analysis. Demonstration of their utility is presented through the examples of (i) fuzzy simulation for generic design of municipal water infrastructure; (ii) FLP of flood control reservoir in the Thames River

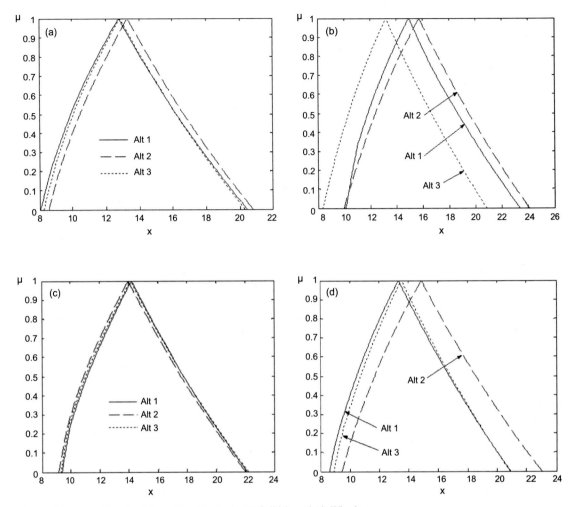

Figure 5.36 Fuzzy distance matrices for (a) all data, (b) Morris, (c) Selkirk, and (d) Winnipeg.

basin (Ontario, Canada); and (iii) fuzzy multi-objective compromise programming in the Red River basin (Manitoba, Canada).

Although flood risk management plans can be designed and achieved without stakeholders' participation, they cannot be implemented without it. The presented tools and techniques provide support for active participation of stakeholders.

As in the case of the probabilistic tools, there is ample opportunity for wider use of the presented techniques in flood risk management under climate change.

5.6 EXERCISES

5.1. What are the main differences between the probabilistic and the fuzzy set approaches for addressing flood risk management uncertainties?

5.2. Model the following expressions as fuzzy sets:

    a. Large flow.

    b. Minimum flow.

    c. Reservoir release between 10 and 20 m³/s.

Table 5.22 *Final rank of flood risk management measures*

Participants	Structural	Non-structural	Combination
All	13.46 (1)	13.94 (3)	13.52 (2)
Morris	15.73 (2)	16.30 (3)	13.87 (1)
Selkirk	14.87 (3)	14.68 (1)	14.83 (2)
Winnipeg	13.98 (1)	15.48 (3)	14.14 (2)

5.3. Assume that $X = \{1, 1.5, 2, 2.5, 3, 3.5, 4, 4.5, 5\}$ is the set of possible Simon River stages in meters. Using a fuzzy set describe "river stage" according to the following assumptions:

a. The river stage membership function is unimodal and convex.
b. The river stage membership function reaches a maximum value of 1 for a stage value of 3 m.
c. The river stage membership function reaches a minimum value of 0 for a stage value of 1 m, as well as for a stage value of 5 m.
d. Show the river stage membership function in a graphical form.

5.4. What is the support for a river stage fuzzy set?

5.5. What is the $\alpha = 0.5$ level set of a river stage fuzzy set?

5.6. Assume that the small municipality of Trident protects itself from the Simon River flooding by the levee along the river. Let the fuzzy set = {(2.5, 0),(6, 1)} represent the Trident levee height. Show the Trident levee height membership function in graphical form.

5.7. Find the intersection of the Simon River available flow and the Trident city domestic water demand. What is the linguistic interpretation of the obtained fuzzy set?

5.8. Find the union of the Simon River stage and the Trident city levee. What is the linguistic interpretation of the fuzzy set obtained?

5.9. Find the complement of the Simon River stage and the Trident city levee. What is the linguistic interpretation of the fuzzy set obtained?

5.10. Let $\tilde{A}$ = {(1,0.3), (2,1), (3,0.4)} and $\tilde{B}$ ={(2,0.7), (3,1), (4,0.2)}:

a. Compute $\tilde{A}$ (+) $\tilde{B}$.
b. Compute $\tilde{A}$ (−) $\tilde{B}$.
c. Compute $\tilde{A}$ (×) $\tilde{B}$.
d. Compute $\tilde{A}$ (/) $\tilde{B}$.

5.11. Assess the compatibility of the two fuzzy membership functions from Exercise 5.10.

a. Find the possibility $Poss(\tilde{A}, \tilde{B})$.
b. Find the necessity $Nec(\tilde{A}, \tilde{B})$.

5.12. Using as a guide the case study from Section 5.2.5 (Example 2):

a. Develop a concept hierarchy (similar to that in Figure 5.6) of "flood evacuation" using a maximum of three hierarchical levels and three issues at each level.

b. Discuss your hierarchy with an expert in flood evacuation and document the discussion.

5.13. Using your own words, define a partial failure and give an example in flood risk management.

5.14. When constructing a flood wall, there is a direct correlation between the block widths (W), length (L), and wall strength (S). The following two rules apply:

Rule 1: IF W is small, and L is small, THEN S is small.

Rule 2: IF W is large and L is small, THEN S is medium.

Use symmetric triangles to construct the membership functions. For the width W, use a triangle centered on the interval [0, 20] cm for Small, and use a triangle centered on the interval [10, 25] cm for Large. For the length L, use a triangle centered on the interval [0, 40] cm for Small. For the strength S, use a triangle centered on the interval [0, 28] N/mm^2 for Small, and use a triangle centered on the interval [6, 35] N/mm^2 for Medium. Conduct a simulation for the inputs W = 15 cm and L = 25 cm.

5.15. Using data from Example 3 in Section 5.4.1, change the shape of the membership functions for all three input variables (flow, flow*, and safety). Simulate the levee height.

5.16. Consider the levee upgrade optimization problem from Example 5 in Section 5.4.2. Assume the following parameters for the fuzzy formulation: $d_0 = 3{,}500{,}000$; $d_1 = 170$; $d_2 = 1{,}300$; $d_3 = 6$; and $p_0 = 1{,}000{,}000$; $p_1 = 120$; $p_2 = 150$; $p_3 = 4$.

a. Reformulate the fuzzy optimization problem.
b. Solve the fuzzy optimization problem using FUZZYLINPRO on the website.
c. Compare your solutions with those listed in Table 5.10. Discuss the difference. Comment on the impact of uncertainty on this decision-making problem.

5.17. For the Tisza River problem presented in Example 6, Section 5.4.3:

a. Change the fuzzy definitions for linguistic terms and quantitative criteria. Keep the rest of the inputs as given in Section 5.4.3.
b. Explain the reasoning for the proposed change.
c. Use FUZZYCOMPRO to solve the problem.
d. Compare your solution to the one in Table 5.18. Discuss the difference.

5.18. For the Tisza River problem presented in Example 6, Section 5.4.3:

a. Develop a new fuzzy definition of degree of compensation, $\tilde{p}$, which is different from the one presented in Figure 5.31b. Keep the rest of the inputs as given in Section 5.4.3.

b. Explain the reasoning for the proposed change.

c. Use FUZZYCOMPRO to solve the problem.

d. Compare your solution to the one in Table 5.18. Discuss the difference.

5.19. For the Tisza River problem presented in Example 6, Section 5.4.3:

a. Develop a new fuzzy definition of weights, $\tilde{\alpha}_i$, that is different from the one presented in Figure 5.31c. Keep the rest of the inputs as given in Section 5.4.3.

b. Explain the reasoning for the proposed change.

c. Use FUZZYCOMPRO to solve the problem.

d. Compare your solution to the one in Table 5.18. Discuss the difference.

Part IV
Future perspectives

6   Future perspectives

Flood risk management systems include people, infrastructure, and environment. These elements are interconnected in complicated networks across broad geographical regions. Each element is vulnerable to change in the climatic conditions. However, the discussion of flood risk management under climate change is very much dependent on the broader context of the climate change debate. The effectiveness of future decisions that may address the increasing risk of flooding is directly related to our success in communicating the message of climate change urgency.

6.1   UNDERSTANDING CLIMATE CHANGE AND FLOOD RISK MANAGEMENT

Chapter 2 of the book explains the link between climate change and risk of flooding, providing the best scientific knowledge that links them. Policies to manage complex systems that include people, infrastructure, and environment must be based on the best available scientific knowledge. The IPCC provides carefully vetted information on climate change to policy-makers. In spite of the strong scientific consensus on the causes and risks of climate change there is still widespread confusion and complacency among the general public (Leiserowitz, 2007). There is a gap between the scientific consensus and public perception of the importance of climate change. Why is that, and why is it important? In democratic societies, the beliefs of the public, not just those of experts, affect government decision-making.

Effective climate change-caused flood risk communication is based on deep understanding of the mental models of policy-makers and the general public (Sterman, 2008). A mental model is, in the context of this discussion, an explanation of someone's thought process about how something works in the real world. It is a representation of the surrounding world, the relationships between its various parts and a person's intuitive perception about his or her own acts and their consequences. Our mental models help shape our behavior and define our approach to solving problems and carrying out tasks. Studies show an apparent contradiction in the principal mental models shaping people's beliefs about climate change and its consequences (including the risk

of flooding). Leiserowitz (2007) points out that the majority in many countries have heard of climate change and say they support action to address it. Yet climate change ranks far behind the economy, war, and terrorism among the greatest concerns. Large majorities oppose policies that would cut greenhouse gas emissions by raising fuel prices. Many surveys in the USA, Canada, Russia, China, and India advocate a "wait-and-see" or "go-slow" approach to emissions reductions. For most people, uncertainty about the risks of climate change means delaying costly actions to mitigate the risks. If climate change begins to harm the economy, risk mitigation policies can then be implemented. However, it is clear that due to the long delays in the climate's response to anthropogenic forcing such reasoning is flawed. The wait-and-see approach assumes that climate is approximately a first-order linear system with short delays. However, in reality, climate is a complex dynamic system with long delays, multiple positive feedbacks (that are destabilizing the system), and non-linearities that may cause abrupt, costly, and irreversible changes.

There is an even deeper problem of poor understanding of stocks and flows and the concept of accumulation (Sterman, 2008; Simonovic, 2009, Chapter 8). Accumulation is omnipresent in everyday life: (i) water is accumulated in different reservoirs through the inflow less the outflow; (ii) a bank account accumulates money through deposits less withdrawals; (iii) a person's weight is managed through the inflows and outflows of calories coming from diet and exercise; etc. Research shows that most people have difficulty relating the flows into and out of a stock to the level of the stock. Instead, people access system dynamics using pattern matching, assuming that the output of a system should be correlated with its inputs.

Weak understanding of accumulations leads to serious errors in mental models of climate change. Training in science does not prevent these errors. Building public support for action on climate change requires overcoming the obstacles discussed above. Science is needed to help us recognize: (i) how greenhouse gas emissions cause warming of the atmosphere; (ii) how warming of the atmosphere increases its capacity to hold water and accelerates many processes involved in the redistribution of moisture; and (iii) how increases in frequency and magnitude of flooding can harm future generations. However, there is no purely technical solution to climate change flood risk. For climate change flood risk management to be grounded in the results of science, it is now time to turn our attention to the dynamics of social and political change.

In summary:

(1) *There are complex linkages among emissions, concentrations, climate change, and flood impacts.* Projecting future climate change requires understanding numerous linkages among human activities, greenhouse gas (GHG) emissions, changes in atmospheric composition, the response of the climate system, and impacts on human and natural systems. The basic links in this chain are well understood, but some elements (in particular, projecting specific impacts at specific times and places) are much less so. As a result, the outcomes of actions to reduce emissions or to reduce the vulnerabilities of human and natural systems must often be presented in probabilistic or qualitative terms, rather than as certain predictions.

(2) *There are significant time lags in the climate system.* It takes a very long time (decades to millennia) for some aspects of the climate system to respond fully to changes in atmospheric greenhouse gas concentrations. This is because the world's oceans can store a large amount of heat and because impacts such as sea level rise and the melting of ice sheets can take several centuries or even millennia to be fully expressed.

(3) *There are also significant time lags in human response systems.* Flood-related climate change impacts are to a large extent built into societal infrastructure (e.g., buildings, power plants, settlement and transportation patterns) and into human habits and organizational routines, few of which change quickly. Preparing to adapt to the flooding effects of climate change will require transformative changes, for instance, in how the country produces and uses energy, builds buildings and transportation infrastructure, and manages water and other natural resources. Overcoming the inertia of the status quo in advancing these sorts of transformations will pose challenges for government and individual citizens.

(4) *Risks, judgments about risk, and adaptation needs are highly variable across different contexts.* Different regions, economic and resource sectors, and populations will experience different impacts from climate change-caused flood risk increase. Their ability to tolerate and adapt to such impacts will hence generate differences in their judgments about the potential risks posed by climate change.

(5) *Many factors complicate and delay public understanding of flood risk due to climate change.* Public understanding of flood risk is important because public opinion affects policy and because the public – as consumers, employers, and community members – can initiate, implement, and support

actions to reduce flood damage and encourage adaptation. Fully understanding climate change-caused flood risk increase is a difficult task even for scientific experts. For instance, personal experience powerfully influences people's understanding of their environment. But this can be misleading in the context of climate change-caused flooding because long-term change is difficult to detect against natural variability, and because judgments of varying phenomena are strongly influenced by memorable and recent extreme events.

6.2 ADAPTIVE FLOOD RISK MANAGEMENT UNDER CLIMATE CHANGE

Chapter 3 provides the main contribution of this book by presenting the flood risk management process as adaptation to climate change. Flood risk management builds on the expanded definition of integrated disaster management provided by Simonovic (2011): *Integrated disaster management is an iterative process of decision-making regarding prevention of, response to, and recovery from a disaster.* A six-step procedure is offered (see Section 3.1.1) for flood risk management that provides a practical guide for the decision-makers. Iterative risk management – which emphasizes taking action now, but in doing so, being ready to learn from experience and adjust these efforts later on – offers the most useful approach for guiding flood risk management under climate change. Iterative risk management is a system for assessing risks, identifying options that are robust across a range of possible futures, and assessing and revising those choices as new information emerges. In cases where uncertainties are substantial or risks cannot be reliably quantified, one can pursue multiple, complementary actions known as a "hedging strategy." Ideally, this approach includes mechanisms for integrating scientific and technical analysis with active participation of the stakeholders most affected by any given decision.

Many high level advisory groups worldwide, including, for example, the IPCC, the United Nations Development Programme, the World Bank, the Australian Greenhouse Office, and the UK Climate Impacts Programme, suggest that for a problem as complex as climate change, risk management should be implemented through a process of "adaptive governance" that involves assuring adequate coordination among the institutions and actors involved in responding to climate change, sharing information with decision-makers across different levels and sectors, ensuring that decisions are regularly reviewed and adjusted in light of new information, and designing policies that can adapt and last over time (National Academy of Sciences, 2011). These concepts may be of value for the future of flood risk management under climate change.

Even if substantial progress in combating climate change is achieved, human societies will need to adapt to some degree of change. Choosing to do nothing now and simply adapting to climate change flood impacts as they occur, for example by accepting losses, would be an unwise choice for a number of reasons. Even a moderate increase in flood risk due to climate change will be associated with a wide range of impacts on both human and natural systems, and the possibility of severe impacts with a host of adverse outcomes cannot be completely eliminated. Unforeseen events may pose serious adaptation challenges for regions, sectors, and groups that may not now seem particularly vulnerable. *A proactive approach to mobilize the capacity to adapt to future flooding caused by climate change is required.*

Many types of flood risk management decisions would benefit from improved insight into adaptation needs and options, but most notable are decisions with long time horizons. These include, for example, decisions about selecting appropriate flood protection infrastructure, land use, siting major facilities, conserving natural areas, managing water resources, and developing coastal zones. In these activities, *adaptation is less about doing different things than about doing familiar things differently.*

In many cases, *adaptation options are available now that both help to manage longer term risks of climate change flood impacts and also offer near-term benefits for other development and environmental management goals.* Experience to date indicates that if adaptation planning is pursued collaboratively among local governments, the private sector, non-governmental organizations, and community groups, it often feeds into broader thinking about alternative futures, which is itself a considerable benefit.

A sound strategy for managing climate-related flood risks requires *sustained investments in developing new knowledge and new technological capabilities as well as investments in the institutions and processes* that help ensure such advances are actually used. There is, in particular, an important role for support of basic climate-related research and pre-commercial technology development that will not be able to attract the support of the private sector and individuals.

Effective flood risk management under climate change requires *information support* to help decision-makers develop effective responses, including the following: (i) information on climate change, vulnerability, and impacts in different regions and sectors; (ii) the timely production and delivery of useful flood data, information, and knowledge to decision-makers; (iii) information for greenhouse gas management and accounting; (iv) information on energy efficiency and emissions; (v) public communication that rests on high quality information and clearly conveys climate science and climate choices; and (vi) support for international information systems and assessments because many climate-related flood management choices occur in an international context.

The success of an iterative adaptive flood risk management approach is based on the establishment of *processes that bring together scientific/technical experts and government officials with key stakeholders* in the private sector and the general public. This is because these other stakeholders make important contributions to flood risk management efforts through their daily choices; because they are an important source of information and perspectives in assessing policy options and in setting priorities for research and development; and because they determine the direction and viability of most governmental policies over the long term.

6.3   RISK COMMUNICATION

I offer in this book a six-step process of flood decision-making under uncertainty (see Section 3.1.1 and Figure 3.2) and a set of analytical tools to support that process. The process, as well as the analytical tools, will have limited value without proper risk communication. The early work of Slovic (1993) and more recent work of Leiss (2001, 2010) highly motivate the discussion here.

Risk communication can be broadly seen as a process that includes actions, words, and other interactions that incorporate and respect the perceptions of the information recipients, intended to help people make more informed decisions about threats to their health and safety. Risk communication will be more effective if it is performed as a dialogue, not instruction. The main goal of risk communication should be to encourage certain behaviors, not simply assume that the provided information will lead to expected action. A very important role in risk communication is played by people's perceptions of risk. The perception of risk is a dual process of fact and feeling. We use the information we have and a set of instincts that help us gage how frightening something feels. Instinctive factors that influence public concern about the risk of flooding include life threat, economic loss, unknowability, trust, tradeoff between risks and benefits, choice, and control.

Climate change and flooding as its consequence refer to events with probability and consequences that are outside the bounds of normal expectations. Normal expectations are usually defined by prior experience or by accepted theories. More specifically, these events belong to the category of events that are *more likely* to occur than is normally expected, and if they occur they will have *catastrophic dimensions.* Climate change risk management has become one of the most politicized and contentious issues of our times. Polarized views, controversy, and open conflict have become omnipresent. Lack of proper risk communication plays a central role in this perspective. Slovic (1993) stipulates that the conflicts and controversies surrounding risk management are not due to public ignorance or irrationality but, instead, are seen as a side effect of our remarkable form of participatory democracy, amplified by powerful technological and social changes that systematically destroy trust. Giving proper attention to the importance of trust and understanding the *dynamics of the system* that destroys trust will shape risk management in the future.

The most likely solution for efficient flood risk management due to climate change and its communication may require a degree of openness and involvement with the public that goes far beyond current practice of public relations and two-way communication. *Future flood risk management (including communication) will require high levels of power sharing and public participation in decision-making* that have rarely been seen up to now. Considering the polarization of positions regarding climate change, we have to accept that trust and conflict resolution will take time. So, the question is, how can the tools and techniques presented in this book be of value in this process?

Some of the answers are in the basic characteristics of the approach behind all the tools presented in the book – the systems approach. A systems approach to problems focuses on interactions among the elements of a system and on the effects of these interactions. Systems theory recognizes multiple and interrelated causal factors, emphasizes the dynamic character of processes involved, and is particularly interested in a system change with time – be it a flood, floodplain, or a disaster-affected community. The traditional view is typically linear and assumes only one, linear, cause-and-effect relationship at a particular time. A systems approach allows a wider variety of factors and interactions to be taken into account. Using a systems view, Simonovic (2011) states that flood disaster losses are the result of interaction among three systems and their many subsystems: (i) the Earth's physical systems (the atmosphere, biosphere, cryosphere, hydrosphere, and lithosphere); (ii) human systems (e.g., population, culture, technology, social class, economics, and politics); and (iii) the constructed systems (e.g., buildings, roads, bridges, public infrastructure, housing).

Section 1.5 of this book presents the seven key management systems principles that provide for effective climate change-caused flood risk management and communication: (i) To achieve sustainable management of flood risk, interactions among the four subsystems – individual, organization, society, and environment – must be appropriately integrated. (ii) Two flows, resource flows and information flows, link the individual, organization, society, and environment subsystems. Value systems are the means through which different values are attached to information and resource flows. (iii) The ongoing need of subsystems for resources from one another sets the limits of their exploitation of one another and of the environment, and is a determinant of behavior within the system. (iv) Information is used by subsystems to make decisions intended to ensure fit with the needs of other subsystems and the environment. (v) Values provide meaning to information flows that are then used to determine resource use by subsystems. (vi) The most effective management strategies for sustainable management of flood risk are those that condition access to resources. (vii) More intensive focus on the systems view of flood risk management will accelerate understanding of which management strategies work, and particularly why they might work.

In flood risk management, the probabilistic approach is used on historical data. In spite of the fact that climate change probabilities are subjective, based on the degree of belief that change will occur, the probabilistic approach still has its place in the consideration of climate change. It is bounded by the knowledge about likelihoods and knowledge about outcomes. Probabilistic simulation can quantify uncertainties of climate change flood impacts. Better understanding of uncertainties provides for more effective communication of resulting risks. Probabilistic optimization on the other hand can arrive at the optimal solutions associated with a particular level of risk. Communication of optimal flood protection infrastructure design values or operational strategies that carry the information about the associated level of risk offers a new dimension for power sharing and public participation in the process of flood risk management. Stochastic multi-objective analysis is the tool well suited for participatory decision-making and conflict resolution. Direct involvement of decision-makers in the process of implementing this set of tools provides for establishment of close links between the best understanding of risks and their consequences on the jointly made decisions.

The fuzzy set approach is probably the most innovative concept presented in this book. It is well suited for addressing the subjective uncertainties that underpin the issue of climate change and its flood consequences. The approach is slowly getting the recognition it deserves and there is hope that this book will provide some contribution to its broader application. Fuzzy simulation is an appropriate approach to include various inherent uncertainties of flood risk management systems in the simulation process. There is a great potential for fuzzy simulation in the design of fuzzy socio-economic-physical scenarios that can be investigated in a collaborative context. Fuzzy optimization is the process of decision-making in which the objective function as well as the constraints are fuzzy. The fuzzy objective function and constraints are characterized by their fuzzy membership functions. The "decision" in a fuzzy environment is therefore viewed as the intersection of fuzzy constraints and a fuzzy objective function. This mathematical context describes the risk communication strategy that is the basis of the whole optimization process. Fuzzy multi-objective decision-making techniques have addressed some uncertainties, such as the vagueness and conflict of preferences common in group decision-making. Their application, however, demands some level of intuitiveness from the decision-makers, and encourages interaction or experimentation. Fuzzy multi-objective decision-making is not always intuitive to many people involved in practical decisions because the decision space may be some abstract measure of fuzziness, instead of a tangible measure of alternative performance. The alternatives to be evaluated are rarely fuzzy: it is their performance that is fuzzy.

In other words, a fuzzy multi-objective decision-making environment may not be as generically relevant as a fuzzy evaluation of a decision-making problem.

6.4 CONCLUSIONS

A flood myth is a mythical or religious story of a "great flood" sent by a deity or deities to destroy civilization as an act of divine retribution. It is a theme widespread among many cultures, though it is perhaps best known in modern times through the biblical account of Noah's Ark, the foundational myths of the Maya peoples, the Deucalion in Greek mythology, the Utnapishtim in the Epic of Gilgamesh, or the Hindu story of Manu, which has some very strong parallels with the story of Noah. Parallels are often drawn between the floodwaters of these myths and the primeval waters found in some creation myths, since the floodwaters are seen to cleanse humanity in preparation for rebirth. Most flood myths also contain a cultural hero who strives to ensure this rebirth.

This myth links humanity to water and its power over humans. In relation to climate change, Garrett (2011) states:

> Effectively, civilization is in a double-bind. If civilization does not collapse quickly this century, then CO_2 will likely end up exceeding 1,000 ppm; but, if CO_2 levels rise by this much, then the danger is that civilization will gradually tend towards collapse.

This discussion of the coupled evolution of the economy and the atmosphere ends with the gloomy conclusions:

> One route for constraining CO_2 growth is to reduce the growth rate of wealth. This can be done by slowing the technological advancements that would enable society to grow into new energy reservoirs. Alternatively, society could increase its exposure to environmental predation. Unfortunately, both of these options necessitate inflationary pressures, so it is hard to see how democratically elected policy makers would willingly prescribe either of these things. Otherwise, civilization must rapidly de-couple its growth from CO_2-emitting sources of energy. There is an important caveat however, which is that such decarbonization does not slow CO_2 accumulation by as much as might be anticipated. Decarbonizing civilization promotes civilization wealth by alleviating the rise in dangerous atmospheric CO_2 levels. But if the growth of wealth is supported, then energy consumption accelerates, and this acts to accelerate CO_2 emissions themselves. Thus, civilization appears to be in a double-bind with no obvious way out. Only a combination of extremely rapid decarbonization and civilization collapse will enable CO_2 concentrations to be maintained below 500 ppm within this century.

This book is written in the hope that we will be able to mobilize the creative powers of humankind in order to respond to the challenge of climate change. Facts and figures are not on our side and mitigation actions will need to be combined with adaptation. The methods and tools of this book are aimed at using flood risk management as an adaptation to climate change. I suggest that the material from the book be used with various forms of education and training – short courses, workshops, formal training, development of case studies, and similar.

The final words, as always, are for my two children and their children – I hope that some of this work will provide you with the "Noah's Ark" for surviving the "great floods" of the future.

References

Ahmad, S., and S. P. Simonovic (2007). A methodology for spatial fuzzy reliability analysis, *Applied GIS Journal*, **3**(1): 1–42.

Ahmad, S., and S. P. Simonovic (2011). A three dimensional fuzzy methodology for flood risk analysis, *Journal of Flood Risk Management*, **4**: 53–47.

Akter, T., and S. P. Simonovic (2005). Aggregation of fuzzy views of a large number of stakeholders for multi-objective flood management decision making, *Journal of Environmental Management*, **77**(2): 133–143.

Akter, T., S. P. Simonovic, and J. Salonga (2004). Aggregation of inputs from a large number of stakeholders for flood management decision making in the Red River basin, *Canadian Water Resources Journal*, **29**(4): 251–266.

Alexander, L. V., X. Zhang, T. C. Peterson, *et al.* (2006). Global observed changes in daily climate extremes of temperature and precipitation, *Journal of Geophysical Research: Atmospheres*, **111**, D05109.

Altay, N., and W. G. Green III (2006). OR/MS research in disaster operations management, *European Journal of Operational Research*, **175**: 475–493.

Ang, H.-S., and H. Tang (1984). *Probability Concepts in Engineering Planning and Design*. Hoboken, NJ: John Wiley.

Arrow, K. (1963). *Social Choice and Individual Values*. New Haven, CT: Yale University Press.

Bardossy, A., L. Duckstein, and I. Bogardi (1995). Fuzzy rule-based classification of atmospheric circulation patterns, *International Journal of Climatology*, **15**: 1087–1097.

Bardossy, A., I. Bogardi, and I. Matyasovszky (2005). Fuzzy rule-based downscaling of precipitation, *Theoretical and Applied Climatology*, **82**: 119–129.

Bates, G. D. (1992). Learning how to ask questions, *Journal of Management in Engineering*, **8**(1): editorial.

Bellman, R., and L. A. Zadeh (1970). Decision-making in a fuzzy environment, *Management Science*, **17**(B): 141–164.

Bender, M. J., and S. P. Simonovic (2000). A fuzzy compromise approach to water resources planning under uncertainty, *Fuzzy Sets and Systems*, **115**(1): 35–44.

Beven, K., and S. Blazkova (2004). Flood frequency estimation by continuous simulation of subcatchment rainfalls and discharges with the aim of improving dam safety assessment in a large basin in the Czech Republic, *Journal of Hydrology*, **292**(1–40): 153–172.

Bhutiyani, M. R., V. S. Kale, and N. J. Pawar (2008). Changing streamflow patterns in the rivers of northwestern Himalaya: Implications of global warming in the 20th century, *Current Science*, **95**(5): 618–626.

Birge, J. R., and F. Louveaux (1997). *Introduction to Stochastic Programming*. New York: Springer.

Booij, M. J. (2005). Impact of climate change on river flooding assessed with different spatial model resolutions, *Journal of Hydrology*, **303**: 176–198.

Borgman, L. E. (1963). Risk criteria, *Journal of Waterways and Harbors Division, ASCE*, **89**(WW3): 1–35.

Caesar, J., L. Alexander, and R. Vose (2006). Large-scale changes in observed daily maximum and minimum temperatures: Creation and analysis of a new gridded data set, *Journal of Geophysical Research*, **111**, D05101.

Cameron, D., K. Beven, and P. Naden (2000). Flood frequency estimation by continuous simulation under climate change (with uncertainty), *Hydrology and Earth System Sciences*, **4**(3): 393–405.

Canadian Standards Association (1997). *Risk Management: Guidelines for Decision-Makers*, CAN/CSA-Q850–97, reaffirmed in 2009.

Canadian Standards Association (2010). *Development, Interpretation and Use of Rainfall Intensity-Duration-Frequency (IDF) Information: Guideline for Canadian Water Resources Practitioners*, Draft Standard Plus 4013 Technical Guide.

Chacko, G. (1991). *Decision-Making Under Uncertainty: An Applied Statistics Approach*. New York: Praeger.

Chang, F. J., and L. Chen (1998). Real-coded genetic algorithm for rule-based flood control reservoir management, *Water Resources Management*, **12**: 185–198.

Chang, M.-S., Y.-L. Tseng, and J.-W. Chen (2007). A scenario planning approach for the flood emergency logistics preparation problem under uncertainty, *Transportation Research Part E*, **43**: 737–754.

Charnes, A., and W. W. Cooper (1961). *Management Models and Industrial Applications of Linear Programming*, Vols. I and II. New York: John Wiley & Sons.

Choi, G., D. Collins, G. Ren, *et al.* (2009). Changes in means and extreme events of temperature and precipitation in the Asia-Pacific Network region, 1955–2007, *International Journal of Climatology*, **29**(13): 1956–1975.

Chongfu, H. (1996). Fuzzy risk assessment of urban natural hazards, *Fuzzy Sets and Systems*, **83**: 271–282.

The Copenhagen Diagnosis (2009). *Updating the World on the Latest Climate Science*. I. Allison, N. L. Bindoff, R. A. Bindschadler, *et al.*, The University of New South Wales Climate Change Research Centre (CCRC), Sydney, Australia.

CRED (Centre for Research on the Epidemiology of Disasters) (2009). www. emdat.be (last accessed November 29, 2011).

Cunderlik, J. M., and T. B. M. J. Ouarda (2009). Trends in the timing and magnitude of floods in Canada, *Journal of Hydrology*, **375**(3–4): 471–480.

Cunderlik, J., and S. P. Simonovic (2005). Hydrologic extremes in southwestern Ontario under future climate projections, *Journal of Hydrologic Sciences*, **50**(4): 631–654.

Cunderlik, J., and S. P. Simonovic (2007). Inverse flood risk modeling under changing climatic conditions, *Hydrological Processes Journal*, **21**(5): 563–577.

Dankers, R., and L. Feyen (2008). Climate change impact on flood hazard in Europe: an assessment based on high resolution climate simulations, *Journal of Geophysical Research*, **113**, D19105, doi:10.1029/2007JD009719.

Dankers, R., and L. Feyen (2009). Flood hazard in Europe in an ensemble of regional climate scenarios, *Journal of Geophysical Research*, **114**, D16108, doi:10.1029/2008JD011523.

Dartmouth Flood Data Observatory (2010). http://floodobservatory.colorado.edu/ (last accessed November 29, 2011).

Delgado, J. M., H. Apel, and B. Merz (2009). Flood trends and variability in the Mekong river, *Hydrology and Earth System Sciences*, **6**: 6691–6719.

Despic, O., and S. P. Simonovic (2000). Aggregation operators for soft decision making in water resources, *Fuzzy Sets and Systems*, **115**(1): 11–33.

Dessai, S., and M. Hulme (2003). *Does Climate Policy Need Probabilities?* Tyndall Centre for Climate Change Research, University of East Anglia, Norwich, UK. Working Paper 34.

El-Baroudy, I., and S. P. Simonovic (2004). Fuzzy criteria for the evaluation of water resources systems performance, *Water Resources Research*, **40**(10): W10503.

Engineers Canada (2008). *Adapting to Climate Change: Canada's First National Engineering Vulnerability Assessment of Public Infrastructure*. Ottawa, www.pievc.ca/e/Adapting_to_climate_Change_Report_Final.pdf (last accessed November 10, 2010).

Eum, H.-I., and S. P. Simonovic (2010). *City of London: Vulnerability of Infrastructure to Climate Change, Background Report 1 – Climate and Hydrologic Modelling*, Water Resources Report no. 068, Department of Civil and Environmental Engineering, London, Ontario, Canada.

Eum, H.-I., V. Arunachalam, and S. P. Simonovic (2009). *Integrated Reservoir Management System for Adaptation to Climate Change Impacts in the Upper Thames River Basin*. Water Resources Report no. 062, Department of Civil and Environmental Engineering, London, Ontario, Canada. Available online at www.eng.uwo.ca/research/iclr/fids/publications/products/62.pdf (last accessed September 2011).

Eum, H.-I., D. Sredojevic, and S. P. Simonovic (2011). Engineering input for the assessment of flood risk due to the climate change in the Upper Thames River Basin, *ASCE Journal of Hydrologic Engineering*, **16**(7): 608–612.

Fischhoff, B., and G. Morgan (2009). The science and practice of risk ranking, *Horizons*, **10**(3): 40–47.

Foued, B. A., and M. Sameh (2001). Application of goal programming in a multi-objective reservoir operation model in Tunisia, *European Journal of Operational Research*, **133**: 352–361.

Fu, G. (2008). A fuzzy optimization method for multicriteria decision making: An application to reservoir flood control operation. *Expert Systems with Applications*, **34**: 145–149.

Garrett, T. J. (2011). No way out? The double-bind in seeking global prosperity along with mitigated climate change, *Earth System Dynamics*, **2**(1): 315–354.

Global Disaster Information Network (1997). *Harnessing Information and Technology for Disaster Management*. Disaster Information Task Force, Washington, USA.

Goicoechea, A., D. R. Hansen, and L. Duckstein (1982). *Multiobjective Decision Analysis with Engineering and Business Applications*. New York: John Wiley & Sons.

GTZ (2002). *Disaster Risk Management: Working Concept*. German Ministry for Economic Cooperation and Development, Eschborn, Germany.

GTZ (2004). *Guidelines: Risk Analysis – A Basis for Disaster Risk Management*. German Ministry for Economic Cooperation and Development, Eschborn, Germany.

Gumbricht, T., P. Wolski, P. Frost, and T. S. McCarthy (2004). Forecasting the spatial extent of the annual flood in the Okavango Delta. *Journal of Hydrology*, **290**: 178–191.

Hall, J., G. Fu, and J. Lawry (2007). Imprecise probabilities of climate change: aggregation of fuzzy scenarios and model uncertainties, *Climatic Change*, **81**: 265–281.

Hannaford, J., and T. J. Marsh (2008). High-flow and flood trends in a network of undisturbed catchments in the UK, *International Journal of Climatology*, **28**(10): 1325–1338.

Hanson, S., R. J. Nicholls, P. Balson, *et al.* (2010). Capturing coastal geomorphological change within regional integrated assessment: An outcome-driven fuzzy logic approach, *Journal of Coastal Research*, **26**(5): 831–842.

Hewlett, J. D. (1982). Forests and floods in the light of recent investigations. In *Proceedings of the Canadian Hydrological Symposium*, 14–15 June 1982, Fredericton, New Brunswick. Ottawa: National Research Council, pp. 543–560.

Hirabayashi, Y., and S. Kanae (2009). First estimate of the future global population at risk of flooding, *Hydrologic Research Letters*, **3**: 6–9.

Hirabayashi, Y., S. Kanae, S. Emori, T. Oki, and M. Kimoto (2008). Global projections of changing risks of floods and droughts in a changing climate, *Hydrological Sciences Journal*, **53**(4): 754–773.

Homer-Dixon, T. (2006). *The Upside of Down: Catastrophe, Creativity, and the Renewal of Civilization*. Toronto: Alfred A. Knopf Canada.

Horton, R. E. (1933). The role of infiltration in the hydrologic cycle, *Transactions, American Geophysical Union*, **14**: 446–460.

IJC (International Joint Commission) (2000). *Living with the Red*. Ottawa, Washington. Available online at www.ijc.org/php/publications/html/living.html (last accessed September 2011).

IOM (International Organization for Migration) (2008). *Migration and Climate Change*, IOM Report 37, Geneva.

IPCC (2007). Summary for Policymakers. In *Climate Change 2007: The Physical Science Basis. Contribution of Working Group I to the Fourth Assessment Report of the Intergovernmental Panel on Climate Change (IPCC AR4)*. S. Solomon *et al.*, eds. Cambridge, UK: Cambridge University Press.

Jiang, T., Z. W. Kundzewicz, and B. Su (2008). Changes in monthly precipitation and flood hazard in the Yangtze River Basin, China, *International Journal of Climatology*, **28**(11): 1471–1481.

Kaplan, S., and B. J. Garrick (1981). On the quantitative definition of risk, *Risk Analysis*, **1**(1): 165–188.

Karamouz, M., O. Abesi, A. Moridi, and A. Ahmadi (2009). Development of optimization schemes for floodplain management: A case study, *Water Resources Management*, **23**: 1743–1761.

Karimi, I., and E. Hullermeier (2007). Risk assessment system of natural hazards: A new approach based on fuzzy probability, *Fuzzy Sets and Systems*, **158**: 987–999.

Kaufmann, A., and M. M. Gupta (1985). *Introduction to Fuzzy Arithmetic: Theory and Applications*. New York: Van Nostrand Reinhold Company.

Kelman, I. (2003). Defining risk, *FloodRiskNet Newsletter*, Issue **2**, Winter, 6–8.

Kim, K., and K. Park (1990). Ranking fuzzy numbers with index of optimism, *Fuzzy Sets and Systems*, **35**: 143–150.

Kumar, D. N., and M. J. Reddy (2006). Ant colony optimization for multi-purpose reservoir operation, *Water Resources Management*, **20**: 879–898.

Kundzewicz, Z. W., Y. Hirabayashi, and S. Kanae (2010). River floods in the changing climate: Observations and projections, *Water Resources Management*, **24**: 2633–2646.

Lai, Y.-J., T.-Y. Liu, and C.-L. Hwang (1994). TOPSIS for MODM, *European Journal of Operational Research*, **76**: 486–500.

Leiserowitz, A. (2007). *Public Perception, Opinion and Understanding of Climate Change: Current Patterns, Trends and Limitations*, UNDP Human Development Report 2007/2008, Occasional Paper, **31**.

Leiss, W. (2001). *In the Chamber of Risks: Understanding Risk Controversies*. Montreal and Kingston, Canada: McGill-Queens University Press.

Leiss, W. (2010). *The Doom Loop in the Financial Sector: And Other Black Holes of Risk*. Ottawa, Canada: University of Ottawa Press.

Levy, J. K., and J. Hall (2005). Advances in flood risk management under uncertainty, *Stochastic Environmental Research Risk Assessment Journal*, **19**: 375–377.

Lund, J. R. (2002). Floodplain planning with risk-based optimization, *ASCE Journal of Water Resources Planning and Management*, **128**(3): 202–207.

Mahabir, C., F. E. Hicks, and A. Robinson Fayek (2007). Transferability of a neuro-fuzzy river ice jam flood forecasting model, *Cold Regions Science and Technology*, **48**: 188–201.

Makkeasorn, A., A. B. Chang, and X. Zhou (2008). Short-term streamflow forecasting with global climate change implications: A comparative study between genetic programming and neural network models, *Journal of Hydrology*, **352**: 336–354.

Mamdani, E. H., and S. Assilian (1975). Advances in linguistic synthesis of fuzzy controllers, *International Journal of Man-Machine Studies*, **7**: 1–13.

Marengo, J. A., M. Rusticucci, O. Penalba, and M. Renom (2009). An intercomparison of observed and simulated extreme rainfall and temperature events during the last half of the twentieth century. Part 2: Historical trends, *Climatic Change*, doi:10.1007/s10584-009-9743-7.

MATLAB (2011). *User's Guide*. MathWorks, Massachusetts, USA. Available online at www.mathworks.com/help/techdoc/matlab_product_page.html (last accessed June 2011).

McCarthy, J., T. Gumbricht, T. S. McCarthy, *et al.* (2003). Flooding patterns of the Okavango wetland in Botswana between 1972 and 2000, *Ambio*, **32**(7): 453–457.

Meehl, G. A., C. Tebaldi, G. Walton, D. Easterling, and L. McDaniel (2009). The relative increase of record high maximum temperatures compared to record low minimum temperatures in the U.S., *Geophysical Research Letters*, **36**, L23701.

Mileti, D. S. (1999). *Disasters by Design*. Washington, DC: Joseph Henry Press.

Milly, P. C. D., J. Betancourt, M. Falkenmark, *et al.* (2008). Stationarity is dead: whither water management? *Science*, **319**(5863): 573–574, doi:10.1126/science.1151915.

Moberg, A., P. D. Jones, D. Lister, *et al.* (2006). Indices for daily temperature and precipitation extremes in Europe analyzed for the period 1901–2000, *Journal of Geophysical Research: Atmospheres*, **111**, D22106.

Mujumdar, P., and N. Kumar (2012). *Floods in a Changing Climate: Hydrologic Modeling*. Cambridge, UK: Cambridge University Press.

Mulholland, P. J., and M. J. Sale (2002). Impacts of climate change on water resources: Findings of the IPCC regional assessment of vulnerability for North America. In Fredrick, K. D., ed., *Water Resources and Climate Change*. Cheltenham, UK: Edward Elgar, pp. 10–14.

Munich Re (2011). www.munichre.com/en/reinsurance/business/non-life/georisks/natcatservice/default.aspx (last accessed November 29, 2011).

Muzik, I. (2002). A first-order analysis of the climate change effect on flood frequencies in a subalpine watershed by means of a hydrological rainfall–runoff model, *Journal of Hydrology*, **267**: 65–73.

National Academy of Sciences (2011). *America's Climate Choices*. Washington, DC: NAS.

National Research Council (NRC) (2008). *Canadian Code Center*, http://irc.nrc-cnrc.gc.ca/codes/index_e.html (last accessed August 25, 2010).

NOAA (2010). *Trends in Atmospheric Carbon Dioxide*. Earth System Research Laboratory, global monitoring division. Available online at www.esrl.noaa.gov/gmd/ccgg/trends/ (last accessed October 20, 2011).

Noble, D., J. Bruce, and M. Egener (2005). *An Overview of the Risk Management Approach to Adaptation to Climate Change in Canada*. Global Change Strategies International, Ottawa, Canada. Report prepared for Natural Resources Canada – Climate Change Impacts and Adaptation Directorate.

Park, J. (2008). Assessing climate change under uncertainty: A Monte Carlo approach, Masters project submitted in partial fulfillment of the requirements for the Master of Environmental Management degree in the Nicholas School of the Environment and Earth Sciences of Duke University, USA.

Parker, D. J. (editor) (2000). *Floods*, Volumes I and II. London: Routledge.

Peck, A., E. Bowering, and S. P. Simonovic (2010). Assessment of climate change risk to municipal infrastructure: A City of London case study, *11th International Environmental Specialty Conference*, 2010 CSCE Annual General Meeting and Congress: Engineering a Sustainable World, CD-ROM Proceedings: EN-11-1–EN-11-9.

Peck, A., E. Bowering, and S. P. Simonovic (2011). *Impact of Climate Change to Municipal Infrastructure: A City of London Case Study*. Water Resources Report no. 74, Department of Civil and Environmental Engineering, London, Ontario, Canada.

Pedrycz, W. and F. Gomide (1998). *An Introduction to Fuzzy Sets*. Cambridge, MA: MIT Press.

Peterson, T. C., X. Zhang, M. Brunet-India, and J. L. Vazquez-Aguirre (2008). Changes in North American extremes derived from daily weather data, *Journal of Geophysical Research*, **113**, DO7113.

Petrow, T., and B. Merz (2009). Trends in flood magnitude, frequency and seasonality in Germany in the period 1951–2002, *Journal of Hydrology*, **371**(1–4): 129–141.

Pinter, N., A. A. Jemberie, J. W. F. Remo, R. A. Heine, and B. S. Ickes (2008). Flood trends and river engineering on the Mississippi River system, *Geophysical Research Letters*, **35**, doi:10.1029/2008GL035987.

Plate, E. J. (2002). Risk management for hydraulic systems under hydrological loads. In *Risk, Reliability, Uncertainty, and Robustness of Water Resources Systems*, J. J. Bogardi and Z. W. Kundzewicz, eds., Cambridge, UK: Cambridge University Press.

Population Division of the Department of Economic and Social Affairs of the United Nations Secretariat (2009). *World Population Prospects: The 2008 Revision. Highlights*. New York: United Nations.

Prodanovic, P., and S. P. Simonovic (2002). Comparison of fuzzy set ranking methods for implementation in water resources decision-making, *Journal of Civil Engineering, CSCE*, **29**: 692–701.

Prodanovic, P., and S. P. Simonovic (2003). Fuzzy compromise programming for group decision making, *IEEE Transactions on Systems, Man, and Cybernetics. Part A: Systems and Humans*, **33**(3): 358–365.

Prodanovic, P., and S. P. Simonovic (2008). Intensity duration frequency analysis under changing climatic conditions, CD Proceedings, *4th International Symposium on Flood Defence: Managing Flood Risk, Reliability and Vulnerability*, S. P. Simonovic, and P. Bourget, eds., Paper 142.

Prudhomme, C., J. Dorte, and C. Svensson (2003). Uncertainty and climate change impact on the flood regime of small UK catchments, *Journal of Hydrology*, **277**: 1–23.

Rahmstorf, S., A. Cazenave, J. A. Church, *et al.* (2007). Recent climate observations compared to projections, *Science*, **316**: 709.

Raju, S. K., and D. N. Kumar (1998). Application of multiobjective fuzzy and stochastic linear programming to Sri Ram Sagar irrigation planning project of Andhra Pradesh. In *Proceedings, National Systems Conference-98*, December 11–13, Calicut, Kerala, pp. 423–428.

Raju, K. S., and D. N. Kumar (2001). Multicriterion Q-analysis and compromise programming for irrigation planning, *Journal of Institution of Engineers (India), Civil Engineering Division*, **82**(CV-1): 57–62.

Ross, T. J. (2010). *Fuzzy Logic with Engineering Applications*, third edition. Chichester, UK: John Wiley & Sons.

Sakawa, M. (1993). *Fuzzy Sets and Interactive Multiobjective Optimization*. New York: Plenum Press.

Samuels, P. (2005). *Language of Risk: Project Definitions, FLOODs*. Report: T32–04-01, HR Wallingford Ltd., UK.

Sayers, P. B., J. W. Hall, and I. C. Meadowcroft (2002). Towards risk-based flood hazard management in the UK, *Civil Engineering, Proceedings of ICE*, **150**: 36–42, Paper 12803.

Scheraga, J. D., and A. E. Grambsch (1998). Risks, opportunities, and adaptation to climate change, *Climate Research*, **10**: 85–95.

Schipper, L., and I. Burton (editors) (2008). *The Earthscan Reader on Adaptation to Climate Change*. London: Earthscan.

Schumann, A. H., and D. Nijssen (2011). Application of scenarios and multi-criteria decision making tools in flood polder planning. In *Flood Risk Assessment and Management*, A. H. Schumann, ed., Dordrecht: Springer, pp. 249–275.

Shafiei, M., and O. B. Haddad (2005). Optimization of levee's setback: A new GA approach. In *Proceedings 6th WSEAS International Conference on Evolutionary Computing*, Lisbon, Portugal, June 16–18, pp. 400–406.

Sharif, M., and D. H. Burn (2006). Improved K-nearest neighbor weather generating model, *ASCE Journal of Hydrologic Engineering*, **12**(1): 42–51.

Shaw, S. B., and S. J. Riha (2011). Assessing possible changes in flood frequency due to climate change in mid-sized watersheds in New York State, USA, *Hydrological Processes*. Published online in Wiley Online Library (wileyonlinelibrary.com) doi: 10.1002/hyp.8027.

Shiklomanov, A. I., R. B. Lammers, M. A. Rawlins, L. C. Smith, and T. M. Pavelsky (2007). Temporal and spatial variations in maximum river discharge from a new Russian data set, *Journal of Geophysical Research, Biogeosciences*, **112**, G04S53.

Shrestha, B., and L. Duckstein (1997). A fuzzy reliability measure for engineering applications. In *Uncertainty Modelling and Analysis in Civil Engineering*, B. M. Ayyub, ed., Boca Raton, FL: CRC Press, pp. 120–135.

Shrestha, R. R., and S. P. Simonovic (2010). Fuzzy nonlinear regression approach to stage–discharge analyses, *ASCE Journal of Hydrologic Engineering*, **15**(1): 49–56.

Simonovic, S. P. (2004). *Canada: Flood Management in the Red River, Manitoba*. Integrated Flood Management Case Study, World Meteorological Organization: The Associated Programme on Flood Management, Geneva. Available online at www.apfm.info/case_studies.htm#namerica (last accessed September 15, 2011).

Simonovic, S. P. (2008). Engineering literature review: Water resources – infrastructure impacts, vulnerabilities and design considerations for future climate change, in *Adapting to Climate Change, Canada's First National Engineering Vulnerability Assessment of Public Infrastructure*, Appendix C Literature Reviews (www.pievc.ca/e/Appendix_C_Literature_Reviews.pdf), Engineers Canada (last accessed November 9, 2010).

Simonovic, S. P. (2009). *Managing Water Resources: Methods and Tools for a Systems Approach*. Paris, France: UNESCO, and London, UK: Earthscan James & James.

Simonovic, S. P. (2010). A new methodology for the assessment of climate change impacts on the watershed scale, *Current Science*, **98**(8): 1047–1055.

Simonovic, S. P. (2011). *Systems Approach to Management of Disasters: Methods and Applications*. Hoboken, NJ: John Wiley & Sons Inc.

Simonovic, S. P., and S. Ahmad (2007). A new method for spatial fuzzy reliability analysis of risk in water resources engineering management, *Open Civil Engineering Journal*, **1**: 1–12. Available online at www.bentham.org/open/tociej/openaccess2.htm (last accessed November 5, 2010).

Simonovic, S. P., and Nirupama (2005). A spatial multi-objective decision making under uncertainty for water resources management, *Journal of Hydroinformatics*, **7**(2): 117–133.

Simonovic, S. P., and A. Peck (2009). *Updated Rainfall Intensity Duration Frequency Curves for the City of London under the Changing Climate*. Water Resources Research Report no. 065, Department of Civil and Environmental Engineering, London, Ontario, Canada. Available online at www.eng.uwo.ca/research/iclr/fids/publications/products/65.pdf (last accessed August 30, 2011).

Simonovic, S. P., and R. Verma (2005). A new method for generating fuzzy Pareto optimal set for multicriteria decision making in water resources. In *Innovative Perspectives of Integrated Water Resources Management Under a Changing World*, Proceedings of special seminar, XXXI IAHR Congress, Seoul, Korea, September 11–16, pp. 69–78.

Simonovic, S. P., and R. Verma (2008). A new methodology for water resources multi-objective decision making under uncertainty, *Physics and Chemistry of the Earth*, **33**: 322–329.

Singh, V. P., S. K. Jain, and A. Tyagi (2007). *Risk and Reliability Analysis: A Handbook for Civil and Environmental Engineers*, Reston, VA: ASCE Press.

Skipper, H. D., and W. Jean Kwon (2007). *Risk Management and Insurance: Perspectives in a Global Economy*. Oxford, UK: Blackwell Publishing.

Slovic, P. (1993). Perceived risk, trust and democracy, *Risk Analysis*, **13**(6): 675–682.

Slovic, P. (2000). *The Perception of Risk*. London, UK: Earthscan.

Smit, B., B. Burton, R. J. T. Klein, and J. Wandel (2000). An anatomy of adaptation to climate change and variability, *Climatic Change*, **45**: 223–251.

Smith, K., and R. Ward (1998). *Floods: Physical Processes and Human Impacts*. New York: John Wiley & Sons.

Sredojevic, D., and S. P. Simonovic (2010). *City of London: Vulnerability of Infrastructure to Climate Change, Background Report 2: Hydraulic Modelling and Floodplain Mapping*, Water Resources Report no. 069, Department of Civil and Environmental Engineering, London, Ontario, Canada.

Stallings, R. A. (editor) (2002). *Methods of Disaster Research*. Philadelphia, PA: Xlibris Corporation Publishing.

Statistics Canada (2011). www12.statcan.ca/english/census01/Products/Reference/dict/geo021.htm (last accessed November 29, 2011).

Sterman, J. D. (2008). Risk communication on climate: Mental models and mass balance, *Science*, **322**: 532–533.

Stern, N. (2007). *Stern Review on the Economics of Climate Change. Part III: The Economics of Stabilisation*. HM Treasury, London. Available online at http://hm-treasury.gov.uk/sternreview_index.htm (last accessed October 20, 2010).

Stirling, A. (1998). Risk at turning point? *Journal of Risk Research*, **1**(2): 97–109.

Suchen (2006). Flood risk assessment and management for the Thames Estuary. In *Sustainable Chinese Entrepreneurship*, essays, available online at www.suchenglobal.org/pdf/flood_essay.pdf (last accessed, June 2011).

Taha, H. A. (1976). *Operations Research: An Introduction*. New York: Macmillan Publishing.

Tannert, C., H. D. Elvers, and B. Jandrig (2007). The ethics of uncertainty. In the light of possible dangers, research becomes a moral duty, *EMBO Reports*, **8**(10): 892–896.

Teegavarapu, R. S. V. (2010). Modeling climate change uncertainties in water resources management models, *Environmental Modelling & Software*, **25**: 1261–1265.

Teegavarapu, R. S. V. (2012). *Floods in a Changing Climate: Extreme Precipitation*. Cambridge, UK: Cambridge University Press.

Teegavarapu, R. S. V., and S. P. Simonovic (1999). Modeling uncertainty in reservoir loss functions using fuzzy sets, *Water Resources Research*, **35**(9): 2815–2823.

USACE (2005). *HEC-GeoRAS, GIS Tools for Support of HEC-RAS Using ArcGIS, User's Manual, Version 4*. United States Army Corps of Engineers, Hydrologic Engineering Center, Davis, CA.

USACE (2006). *HEC-RAS, River Analysis System, User's Manual, Version 4.0*. United States Army Corps of Engineers, Hydrologic Engineering Center, Davis, CA.

USACE (2008). *Hydrologic Modeling System, HEC-HMS User's Manual*. United States Army Corps of Engineers, Hydrologic Engineering Center, Davis, CA.

Vick, S. G. (2002). *Degrees of Belief: Subjective Probability and Engineering Judgment*. Reston, VA: ASCE Press.

Vrijling, J. K., V. W. Van Hengel, and R. J. Houben (1995). A framework for risk evaluation, *Journal of Hazardous Materials*, **43**: 245–261.

Vucetic, D., and S. P. Simonovic (2011). *Water Resources Decision Making Under Uncertainty*. Water Resources Research Report no. 073, Department of Civil and Environmental Engineering, London, Ontario, Canada. Available online at www.eng.uwo.ca/research/iclr/fids/publications/products/73.pdf (last accessed September 15, 2011).

Ward, R. C. (1978). *Floods: A Geographical Perspective*. London, UK: Macmillan.

Wardlaw, R., and M. Sharif (1999). Evaluation of genetic algorithms for optimal reservoir system operation, *ASCE Journal of Water Resources Planning and Management*, **125**(1): 25–33.

WCED (World Commission on Environment and Development) (1987). *Our Common Future*. Oxford: Oxford University Press.

Whitfield, P. H. (2012). River flooding under changing climate, *International Journal of Flood Risk Management*, under review.

WMO (2009). *Integrated Flood Management: Concept Paper*, WMO-No. 1047, Geneva.

Wolski, P., T. Gumbricht, and T. S. McCarthy (2002). Assessing future change in the Okavango Delta: the use of regression model of the maximum annual flood in a Monte Carlo simulation. In *Proceedings of the Conference on Environmental Monitoring of Tropical and Subtropical Wetlands*, 4–8 December, Maun, Botswana.

Yates, D., S. Gangopadhyay, B. Rajagopalan, and K. Strzepek (2003). A technique for generating regional climate scenarios using a nearest-neighbor algorithm, *Water Resources Research*, **39**(7), SWC 7-1–SWC 7-14.

Zadeh, L. A. (1965). Fuzzy sets, *Information Control*, **8**: 338–353.

Zimmermann, H. J. (1996). *Fuzzy Set Theory and Its Application*, second revised edition. Boston, MA: Kluwer Academic Publishers.

Zwiers, F. W., X. Zhang, and Y. Feng (2011). Anthropogenic influence on long return period daily temperature extremes at regional scales, *Journal of Climate*, **24**: 881–892.

Index

Printed in the United States
By Bookmasters